SCIENTIFIC
MASCULINITIES

EDITED BY

Erika Lorraine Milam
and Robert A. Nye

OSIRIS | 30

A Research Journal Devoted to the
History of Science and Its Cultural Influences

Osiris

Series editor, 2013–2023

ANDREA RUSNOCK, *University of Rhode Island*

Volumes 28 to 32 in this series are designed to connect the history of science to broader cultural developments, and to place scientific ideas, institutions, practices, and practitioners within international and global contexts. Some volumes address new themes in the history of science and explore new categories of analysis, while others assess the "state of the field" in various established and emerging areas of the history of science.

28   ALEXANDRA HUI, JULIA KURSELL, & MYLES W. JACKSON, EDS., *Music, Sound, and the Laboratory from 1750 to 1980*
29   MATTHEW DANIEL EDDY, SEYMOUR H. MAUSKOPF, WILLIAM R. NEWMAN, EDS., *Chemical Knowledge in the Early Modern World*

Series editor, 2002–2012

KATHRYN OLESKO, *Georgetown University*

17   LYNN K. NYHART & THOMAS H. BROMAN, EDS., *Science and Civil Society*
18   SVEN DIERIG, JENS LACHMUND, & J. ANDREW MENDELSOHN, EDS., *Science and the City*
19   GREGG MITMAN, MICHELLE MURPHY, & CHRISTOPHER SELLERS, EDS., *Landscapes of Exposure: Knowledge and Illness in Modern Environments*
20   CAROLA SACHSE & MARK WALKER, EDS., *Politics and Science in Wartime: Comparative International Perspectives on the Kaiser Wilhelm Institute*
21   JOHN KRIGE & KAI-HENRIK BARTH, EDS., *Global Power Knowledge: Science and Technology in International Affairs*
22   GREG EGHIGIAN, ANDREAS KILLEN, & CHRISTINE LEUENBERGER, EDS., *The Self as Project: Politics and the Human Sciences*
23   MICHAEL D. GORDIN, KARL HALL, & ALEXEI KOJEVNIKOV, EDS., *Intelligentsia Science: The Russian Century, 1860–1960*
24   CAROL E. HARRISON & ANN JOHNSON, EDS., *National Identity: The Role of Science and Technology*
25   ERIC H. ASH, ED., *Expertise and the Early Modern State*
26   JAMES RODGER FLEMING & VLADIMIR JANKOVIC, EDS., *Klima*
27   ROBERT E. KOHLER & KATHRYN M. OLESKO, EDS., *Clio Meets Science: The Challenges of History*

INTRODUCTION

PROFESSIONALIZATION AND GENDER NORMS

SCIENTIFIC LABOR AND GENDERED SPACES

MEASURES AND METAPHORS OF GENDER

NOTES ON CONTRIBUTORS

INDEX

Acknowledgments

This project has come to fruition with the help of many people. We wish to thank the Philadelphia Area Center for History of Science for hosting our workshop in the summer of 2012, Babak Ashrafi and Simon Joseph for their incredible logistical support throughout that process, and Jason Oakes and Peter Collopy for keeping diligent notes so that we had a written record of our conversations while revising our drafts. We were delighted by the enthusiastic and vigorous suggestions and critiques offered by all the participants and workshop observers. Our introduction has benefited in addition from a variety of friends, colleagues, and collaborators, including Michael Gordin, Nathan Ha, Eugenia Lean, Mary Terrall, and Zeb Tortorici.

An Introduction to
Scientific Masculinities

by Erika Lorraine Milam* and Robert A. Nye†

ABSTRACT

This volume seeks to integrate gender analysis into the global history of science and medicine from the late Middle Ages to the present by focusing on masculinity, the part of the gender equation that has received the least attention from scholars. The premise of the volume is that social constructions of masculinity function simultaneously as foils for femininity and as methods of differentiating between "kinds" of men. In exploring scientific masculinities without taking the dominance of men and masculinity in the sciences for granted, we ask, What is masculinity and how does it operate in science? Our answers remind us that gender is at once an analytical category and a historical object. The essays are divided into three sections that in turn emphasize the importance of gender to the professionalization of scientific, technological, and medical practices, the spaces in which such labor is performed, and the ways that sex, gender, and sexual orientation are measured and serve as metaphors in society and culture.

The substantial literature on gender studies of science and women in science provides a wide array of tools for understanding complex gender dynamics and amply demonstrates the gendered nature of all human activity, including the importance of gender in the lived experiences of scientists, engineers, and medical practitioners. Historians who developed gender as a category of analysis sought to move beyond dualisms like male/female or masculine/feminine to create sliding scales of sex, gender, and sexuality. In this they succeeded marvelously, yet scholarly attention within the history of science has largely continued to engage with elite men—the largest category of historical actors about whom we write—as a foil for theorizing the gendered, classed, or racialized experiences of others, but otherwise as belonging to an unmarked social category. In light of the ubiquitous presence of men as scientists, engineers, and doctors throughout history, then, what are the consequences of changing the kinds of questions we ask about the scientific enterprise from "why did scientists think X" to "why did *male* scientists think X"? Or, more exactly, what does it add to our understanding of the sciences if we factor in masculine social and cultural perspectives of time and place?

Recently, historians have begun to explore the masculine cultures of the "field"

*Department of History, 136 Dickinson Hall, Princeton University, Princeton, NJ 08544-1017; emilam@princeton.edu.
†Department of History, Oregon State University, Corvallis, OR 97331; nyer@onid.orst.edu.

sciences, engineering, and technology, as well as mathematics, the physical sciences, computer science, medicine and its specialties, genetics, and other fields encompassing many historical periods and cultures.[1] Gender, as these histories illustrate, is not just a topic we should address episodically in our teaching and research; it is a powerful analytical tool with which to rethink science as a fundamentally gendered activity, whether or not women are present. The predominantly male groups that shaped the work of science, technology, and medicine also authorized the construction of gendered and sexed bodies. Entirely male enterprises were haunted by the specter of femininity because distinctions between bodies, behavior, social origin, and race fell along a readily feminized spectrum of "more or less" masculine. Similarly, scientific and medical norms took their dominant form from men's concerns about feminine or effeminizing "pathologies" or departures from ideals of male beauty or health.[2]

This volume holds in tension the masculine cultures of doing science and the study of sexed bodies within the sciences and medicine. Our challenge was to bring to light the ways that these scientific masculinities have operated over time and within different cultures without reenacting history by excluding women or femininity from the story.

The first efforts to open up the issue of gender in the history of science in the 1960s were feminist analyses of the causes and reasons for the exclusion of women from professional science, technology, and medicine. Several myths needed to be exploded: science was an objective enterprise unsuited to women; women had contributed historically little to scientific developments; when they did contribute, it was in fields appropriate to their natures and skills. The job of unpacking these myths fell to philosophers, sociologists, and historians of science. Philosophers have worked to discredit the notion of the gender neutrality of valid science, sociologists have illuminated the

---

[1] David Noble, *A World without Women: The Christian Clerical Culture of Western Science* (New York, 1992); Sharon Traweek, *Beamtimes and Lifetimes: The World of High Energy Physics in Japan* (Cambridge, Mass., 1988); Elizabeth Lunbeck, *The Psychiatric Persuasion: Knowledge, Gender and Power in Modern America* (Princeton, N.J., 1994); Bruce Hevly, "The Heroic Science of Glacier Motion," *Osiris* 11 (1996): 66–86; Naomi Oreskes, "Objectivity or Heroism? On the Invisibility of Women in Science," *Osiris* 11 (1996): 87–116; Robert A. Nye, "Medicine and Science as Masculine Fields of Honor," *Osiris* 12 (1997): 60–79; Ruth Oldenziel, *Making Technology Masculine: Men, Women, and Modern Machines in America, 1870–1945* (Amsterdam, 1999); Roger Horowitz, *Boys and Their Toys? Masculinity, Technology, and Class in America* (New York, 2001); Jane Margolis and Allan Fisher, *Unlocking the Clubhouse: Women in Computing* (Cambridge, Mass., 2003); Andrew Warwick, *Masters of Theory: Cambridge and the Rise of Mathematical Physics* (Chicago, 2003); Kristen Haring, *Ham Radio's Technical Culture* (Cambridge, Mass., 2007); Grace Yen Shen, "Taking to the Field: Geological Fieldwork and National Identity in Republican China," *Osiris* 24 (2009): 231–52; Ellen S. More, Elizabeth Fee, and Manon Parry, eds., *Women Physicians and the Cultures of Medicine* (Baltimore, 2009); Henrika Kuklick, "Personal Equations: Reflections on the History of Fieldwork, with Special Reference to Sociocultural Anthropology," *Isis* 102 (2011): 1–33; Ian Nicholson, "'Shocking' Masculinity: Stanley Milgram, 'Obedience to Authority,' and the 'Crisis of Manhood' in Cold War America," *Isis* 102 (2011): 238–68; Rob Boddice, "The Manly Mind? Revisiting the Victorian 'Sex in Brain' Debate," *Gend. & Hist.* 23 (2011): 321–40.

[2] From the plentiful literature on these issues, see Ann Laura Stoler, *Race and the Education of Desire: Foucault's History of Sexuality and the Colonial Order of Things* (Durham, N.C., 1995); Joanna Bourke, *Dismembering the Male: Men's Bodies, Britain and the Great War* (London, 1996); George L. Mosse, *The Image of Man: The Creation of Modern Masculinity* (New York, 1996); Charlotte Furth, *A Flourishing Yin: Gender in China's Medical History, 960–1665* (Berkeley and Los Angeles, 1999); Alexandra Shepard, *Meanings of Manhood in Early Modern England* (Oxford, 2003), 47–69; and Christina S. Jarvis, *The Male Body at War: American Masculinity during World War II* (DeKalb, Ill., 2004).

practical obstacles and constricted pathways leading to careers for women in science, and historians have studied the diversity of historical, cultural, and disciplinary situations in which previously "invisible" women contributed to scientific and medical research and teaching.[3]

Though much of the early work on science and gender was about men, patriarchy, and a masculinist domination of nature, the need to identify how and why women had been marginalized necessarily took precedence over using the tools of gender analysis to understand the particular characteristics of masculine scientific and professional cultures.[4] Because women collectively and as individuals had been invisible, historians left largely unexamined the discriminatory hierarchies within all-white male cultures, which advanced the careers of some men while excluding or marginalizing other men (and, by extension, women) on the basis of class, race, religion, or sexual orientation.[5] Gender theory teaches us that masculine/feminine binaries are conceptually conjoined; to define the masculine as not feminine or vice versa evokes the other as an inevitable component of identity. This has meant in practice that imputations of effeminacy have been mixed together with other justifications for excluding qualified men from scientific cultures who did not seem to be the "right" kind of man. We must take these different perspectives seriously: social constructions of masculinity function simultaneously as foils for femininity and as methods of differentiating between kinds of men.

In the last decades, "gender" has thus evolved from merely a grammatical category

---

[3] Some important philosophical interventions in this area are Evelyn Fox Keller, *Reflections on Gender and Science* (New Haven, Conn., 1985); Helen Longino, *Science as Social Knowledge* (Princeton, N.J., 1990); Sandra Harding, *The Science Question in Feminism* (Ithaca, N.Y., 1986); see also Muriel Lederman and Ingrid Bartsch, eds., *The Gender and Science Reader* (New York, 2001). Key sociological works include Henry Etkowitz, Carol Kemelgor, and Brian Uzzi, eds., *Athena Unbound: The Advancement of Women in Science and Technology* (Cambridge, 2000); Sue V. Rosser, *The Science Glass Ceiling* (New York, 2004); Margaret A. Eisenhart and Elizabeth Finkel, *Women's Science: Learning and Succeeding from the Margins* (Chicago, 1998). Pioneers in history have been Carolyn Merchant, *The Death of Nature: Women, Ecology, and the Scientific Revolution* (San Francisco, 1980); Londa Schiebinger, *Has Feminism Changed Science?* (Cambridge, Mass., 1999); Margaret Rossiter, *Women Scientists in America: Struggles and Strategies to 1940* (Baltimore, 1982); Rossiter, *Women Scientists in America: Before Affirmative Action, 1940–1972* (Baltimore, 1995); Rossiter, *Women Scientists in America: Forging a New World since 1972* (Baltimore, 2012); Ellen S. More, *Restoring the Balance: Women Physicians and the Profession of Medicine, 1850–1995* (Cambridge, Mass., 1999); in this same vein, see also Sally Gregory Kohlstedt and Helen E. Longino, eds. *Women, Gender, and Science: New Directions*, vol. 12 of *Osiris* (1997). The most comprehensive source for appreciating the depth and global scope of recent historical studies of gender is Teresa A. Meade and Merry E. Wiesner-Hanks, *A Companion to Gender History* (Oxford, 2004).

[4] See, e.g., the retrospective article in the thirty-five-year-old journal *Sex Roles*: Joan C. Chrisler, "In Honor of *Sex Roles*: Reflections on the History and Development of the Journal," *Sex Roles* 63 (2010): 299–310.

[5] Michael S. Kimmel, *The History of Men: Essays in the History of British and American Masculinities* (Albany, N.Y., 2005); R. W. Connell, *Masculinities* (London, 1995). Historians have studied race as both a historical object and a marker of social discrimination and privilege; see, e.g., Nancy Leys Stepan, *The Hour of Eugenics: Race, Gender, and Nation in Latin America* (Ithaca, N.Y., 1991); Keith Wailoo, *Dying in the City of the Blues: Sickle Cell Anemia and the Politics of Race and Health* (Chapel Hill, N.C., 2001); and Alondra Nelson, *Body and Soul: The Black Panther Party and the Fight against Medical Discrimination* (Minneapolis, 2011). Historians of colonialism and eugenics have also interrogated the construction of whiteness as race in much the same way that we explore masculinity as gender: Warwick Anderson, *The Cultivation of Whiteness: Science, Health, and Racial Destiny in Australia* (Carlton South, Victoria, 2002); Alexandra Minna Stern, *Eugenic Nation: Faults and Frontiers of Better Breeding in Modern America* (Berkeley and Los Angeles, 2005); Julian B. Carter, *The Heart of Whiteness: Normal Sexuality and Race in America, 1880–1940* (Durham, N.C., 2007).

to a crucial aspect of modern selfhood.[6] The history of this evolution illuminates the many ways scientific language constructs our perceptions of the world and is itself constructed by material and cultural change. Psychologist John Money and his colleagues at Johns Hopkins first used gender as a nongrammatical term in the 1950s to create a protocol for rearing intersex children whose genitals had been surgically "improved" to resemble typical male or female organs.[7] In this protocol, Money and his associates invoked "gender" to denote the assigned sex of rearing, as they assumed the power of the environment would prove stronger than the children's ambiguous sex. Psychoanalyst Robert Stoller later coined "gender identity" to describe the end point of a successful transition for intersex and transsexual individuals. Feminist theorists later appropriated "gender" to emphasize the means by which culturally enforced norms created and maintained different (and fundamentally unequal) expectations for how men and women were supposed to behave.[8] Ironically, in the last two decades, psychiatrists, doctors, scientists, and other scholars have begun to use gender as a valid substitute for sex, transforming gender into a foundational marker of personal identity. That is another, and very complicated, story, one in which physicians and scientists appear to recapitulate in language an essentializing difference between masculinity and femininity.[9]

Feminist scholars have also deployed gender as an analytical tool for interpreting the past. A recent forum in the *American Historical Review* on Joan Scott's epochal 1986 article, "Gender: A Useful Category of Historical Analysis," revealed historians' prodigious use of gender analysis in their work in the intervening years.[10] In her own remarks, Scott foregrounded the linguistic turn as a powerful lesson for her and her generation, teaching her "to understand that differences of sex were not set by nature but were established through language, and to analyze language as a volatile, mutable system whose meanings could never finally be secured." She continued by inviting her readers "to think critically about how the meanings of sexed bodies are produced,

---

[6] Of course, many languages don't have gender and when they do, gender doesn't always track with sex; Greville G. Corbett, *Gender* (New York, 1991).

[7] For an account of the work of Money and his colleagues, see Elizabeth Reis, *Bodies in Doubt: An American History of Intersex* (Baltimore, 2009), 115–52.

[8] Robert Stoller, *Sex and Gender* (New York, 1968); Joanne Meyerowitz tells part of this story in "A History of 'Gender,'" *Amer. Hist. Rev.* 113 (2008): 1346–56; on the dilemma of being stuck with the older term when "gender" would be preferable, see Irene Hanson Frieze and Joan C. Chrisler, "Editorial Policy in the Use of the Two Terms 'Sex' and 'Gender,'" *Sex Roles* 64 (2011): 789–90; the medical history may be found in Anne Fausto-Sterling, *Sexing the Body: Gender Politics and the Construction of Sexuality* (New York, 2000), 45–78. For the clinical process of sex determination, see Sandra Eder, "The Volatility of Sex: Intersexuality, Gender and Clinical Practice in the 1950s," *Gend. & Hist.* 22 (2010): 692–707.

[9] See the article by the evolutionary biologist David Haig, "The Inexorable Rise of Gender and the Decline of Sex: Social Change in Academic Titles," *Arch. Sexual Behav.* 33 (2004): 87–90; see also Robert A. Nye, "The Biosexual Foundations of Our Modern Concept of Gender," in *Sexualized Brains: Scientific Modeling of Emotional Intelligence from a Cultural Perspective*, ed. Nicole C. Karafyllis and Gotlind Ulshöfer (Cambridge, Mass., 2008), 69–80. The most up-to-date scientific research on the sex/gender distinction is reviewed in Anne Fausto-Sterling, *Sex/Gender: Biology in a Social World* (New York, 2012). See also Rebecca Jordan-Young's lucid elaboration of sex, gender, and sexuality as inextricably intertwined three-ply yarn in *Brain Storm: Flaws in the Science of Sex Differences* (Cambridge, Mass., 2010).

[10] Scott's article was incorporated into her book, *Gender and the Politics of History* (New York, 1988); see "AHR Forum: Revisiting 'Gender: A Useful Category of Historical Analysis,'" *Amer. Hist. Rev.* 113 (2008): 1344–1429.

deployed, and changed."[11] We keep her invitation in mind when exploring how and why scientists constructed gender in particular social, material, and cultural contexts.

Masculine cultures of science influenced scientists' construction of normative male (and female) anatomy, sexual orientation, and behavior and provided frameworks from which to draw upon modern technology to define masculinity and declare technology a masculine subject. These cultures also supplied masculine perspectives on popular writing about exploration, science, and technology, even where many of the writers or readers of these works were women. Depending on the sociocultural context, male scientists chose from among a variety of masculine roles, including laboratory-based scientist-heroes,[12] outdoor, self-reliant men,[13] sensitive and sympathetic readers of nature,[14] and family men.[15] When women did enter the scientific workplace—in domestic settings, laboratories, field sites, and classrooms—gendered models from the broader culture often determined how these integrated spaces were organized, perhaps along patriarchal or familial lines or, more ordinarily, following the ambient norms of gender segregation.[16] Although popular conceptions of science mirrored the gendered realities of contemporary scientific culture, they also preserved and popularized certain images of masculinity at the expense of others, particularly scientists who were also men of action—just think of Richard Feynman, L. S. B. Leakey, or, as an extreme example, Steve Irwin. Further, through film, advertising, journalistic attention, and mere happenstance, these visions of science and scientists help generate cultural values and popular interests.[17]

Most historians of gender, however, eschew a radical nominalism that considers bodily sex to be a mere "effect" of language.[18] Though some scholars have embraced cultural explanations of naturalized bodies loosely drawn from Judith Butler's concept of gender performativity, others prefer the French sociologist Pierre Bourdieu's notion of habitus, according to which femininity and masculinity are embedded in

---

[11] Joan Scott, "Unanswered Questions," *Amer. Hist. Rev.* 113 (2008): 1422–9, on 1423.

[12] Natasha Myers, "Pedagogy and Performativity: Rendering Laboratory Lives in the Documentary *Naturally Obsessed: The Making of a Scientist*," *Isis* 101 (2010): 817–28.

[13] Donna Haraway, "Teddy Bear Patriarchy: Taxidermy in the Garden of Eden, New York City, 1908–1936," in *Primate Visions: Gender, Race, and Nature in the World of Modern Science* (New York, 1989), 26–58; Gregg Mitman, "Cinematic Nature: Hollywood Technology, Popular Culture, and the American Museum of Natural History," *Isis* 84 (1993): 637–61.

[14] Janet Browne, "I Could Have Retched All Night: Charles Darwin and His Body," in *Science Incarnate: Historical Embodiments of Natural Knowledge*, ed. Christopher Lawrence and Steven Shapin (Chicago, 1998), 240–87; Jim Endersby, "Sympathetic Science: Charles Darwin, Joseph Hooker, and the Passions of Victorian Naturalists," *Victorian Stud.* 51 (2009): 299–320; Rob Boddice, "Vivisecting Major: A Victorian Gentleman Defends Animal Experimentation, 1876–1885," *Isis* 104 (2011): 215–37.

[15] John Tosh, *A Man's Place: Masculinity and the Middle-Class Home in Victorian England* (New Haven, Conn., 1999).

[16] Marsha L. Richmond, "The Domestication of Heredity: The Familial Organization of Geneticists at Cambridge University, 1895–1910," *J. Hist. Biol.* 39 (2006): 565–605; Michael Hoskin, *Discoverers of the Universe: William and Caroline Herschel* (Princeton, N.J., 2011); Steven J. Peitzman, *A New and Untried Course: Women's Medical College and Medical College of Pennsylvania* (New Brunswick, N.J., 2000).

[17] Bernard Lightman, *Victorian Popularizers of Science: Designing Nature for New Audiences* (Chicago, 2007); Salim Al-Gailani, "Magic, Science and Masculinity: Marketing Toy Chemistry Sets," *Stud. Hist. Phil. Sci. Pt. A* 40 (2009): 372–81; David A. Kirby, *Lab Coats in Hollywood: Science, Scientists, and Cinema* (Cambridge, Mass., 2011).

[18] See, e.g., three recent texts on gender history: Kathleen Canning, *Gender History and Practice* (Ithaca, N.Y., 2006), esp. 168–92; Laura Lee Downs, *Writing Gender History*, 2nd ed. (New York, 2010); Sonya O. Rose, *What Is Gender History?* (London, 2010).

bodies and structures in ways that perpetually reconfigure gender difference.[19] In Bourdieu's sociological account, the capacity of language to summon ontological categories (like gender) into being "does not lie . . . in the language itself, but in the group that authorizes and recognizes it and, with it, authorizes and recognizes itself"—dubbed the "officialization effect."[20] This process implants particular qualities in bodies, which are felt and perceived by others to be natural, endowing gender with a lived corporeality.

A similar process worked to pathologize same-sex desire, even before discursive distinctions between heterosexual and homosexual sexual orientation emerged in late nineteenth-century medicine and science. Heterosexual desire in this discourse could be identified by a number of somatic and psychic "signs," which were regularly aligned with the prevailing norms of male bodies and masculine comportment, while the bodies of homosexuals were scrutinized for morphological defects and evidence of effeminacy.[21] The prestige of Western medicine was such that this model was replicated by many of the medical professions of modernizing nations elsewhere in the world.[22]

As Nelly Oudshoorn has argued, early feminist critiques of the masculine construction of female bodies perpetuated the notion that male bodies were a stable, unmarked category, "untouched by time and place."[23] Their research revealed how, in comparison to men's bodies, scientists often defined women's bodies as the immature, "default" pattern from which maleness differentiated.[24] In the modern era, a growing scientific and medical understanding of the building blocks of sex has roughly coincided with a new cultural self-consciousness about the constructedness and mutability of bodies. Resistance to this new understanding remains intact, not least due to some of the research that helped establish the myriad pathways of sex development. However, the possibility of alternate developmental outcomes, together with cultural

---

[19] Judith Butler, *Gender Trouble: Feminism and the Subversion of Identity* (New York, 1990); Butler, "Performativity's Social Magic," in *Bourdieu: A Critical Reader*, ed. Richard Shusterman (Oxford, 1999), 113–28. See also Brooke Holmes, *Gender: Antiquity and Its Legacy* (Oxford, 2012), on the ways in which Butler and other theorists worked with stories from ancient Greece and Rome to problematize strict boundaries between cultural "gender" and biological "sex."

[20] Pierre Bourdieu, *The Logic of Practice*, trans. Richard Nice (Stanford, Calif., 1990), 109–10; see also Bourdieu, *La domination masculine* (Paris, 1998).

[21] See on this history Vernon A. Rosario, *Science and Homosexualities* (London, 1997); Alice Dreger, *Hermaphrodites and the Medical Invention of Sex* (Cambridge, Mass., 1998); Jennifer Terry, *An American Obsession: Science, Medicine, and Homosexuality in Modern Society* (Chicago, 1999); Henry L. Minton, *Departing from Deviance: A History of Homosexual Rights and Emancipatory Science in America* (Chicago, 2002). See also Luis Campos, "Mutant Sexuality: The Private Lives of Plants," in *Making Mutations: Objects, Practices, Contexts*, ed. Luis Campos and Alexander von Schwerin, Max Planck Institute for the History of Science, Preprint 393 (Berlin, 2010), 49–70.

[22] Sabine Frühstück, *Colonizing Sex: Sexology and Social Control in Modern Japan* (Berkeley and Los Angeles, 2003); Frank Dikötter has also argued that Western models of homosexuality, spermatorrhea, and masturbation were adopted by Chinese medical professionals in the Republican period to reinforce procreative sexuality: Dikötter, *Sex, Culture and Modernity in China: Medical Science and the Construction of Sexual Identities in Early Republican China* (Honolulu, 1994); and Dikötter, *Imperfect Conceptions: Medical Knowledge, Birth Defects and Eugenics in China* (New York, 1998); cf. Susan L. Mann, *Gender and Sexuality in Modern Chinese History* (New York, 2011).

[23] Nelly Oudshoorn, "On Bodies, Technologies, and Feminisms," in *Feminism in Twentieth-Century Science, Technology, and Medicine*, ed. Angela N. H. Creager, Elizabeth Lunbeck, and Londa Schiebinger (Chicago, 2001), 199–213, on 206.

[24] For example, see Cynthia Eagle Russett, *Sexual Science: The Victorian Construction of Womanhood* (Cambridge, Mass., 1989).

acceptance of the idea that we can refashion our bodies, has turned scholars' discriminating gazes to the construction of male bodies.[25]

Scientific understandings of male bodies have never been entirely fixed. Medievalists and early modernists, for example, were inspired by Greek medicine, which encouraged many contemporaries to think of bodies as governed by the influences of internal humors and the environment and therefore subject to change.[26] Neither modern science nor modernity itself succeeded in eliminating this mutability; instead they have further complicated the problem and raised the stakes. As historian Christopher Forth has written, the conditions of modernity "at once reinforce and destabilize the representation of masculinity as an unproblematic quality of male anatomy." In his analysis, "the double logic of modern civilization" became a process promoting and supporting male interests while simultaneously "threatening to undermine those interests by eroding the corporeal foundations of male privilege."[27]

Constitutionally, the masculinity of bodies defined by scientific research functioned at overlapping metaphorical levels, including the physical body (or individual health), the social body (or integrity of the community), and the national body (and its status relative to that of other countries).[28] In this way, the sciences of masculinity have existed in both familiar contexts, where the idea of masculine bodies is naturalized according to our contemporary perspectives, and in the less familiar ones, where these very categories can be problematic and thus provide another means by which to deconstruct the powerful normative claims of science.

As historians we must attend to both the gendered linguistic conventions of scientific cultures and the cultural history of scientific and medical research on male bodies. This volume seeks to develop a set of tools for understanding the pervasive role of masculinity in shaping normative scientific research. In exploring scientific masculinities without taking the dominance of men and masculinity in the sciences for granted, we ask, What is masculinity and how does it operate in science? Our answers remind us that gender is at once an analytical category and a historical object. Joan Scott is right—the language of gender is volatile and highly mutable—yet masculinity persists as a marker, maker, and companion of privilege. The deep irony revealed by Forth's analysis is that despite this close association, the privilege of masculinity seems perpetually under siege.

In assembling the contributions to this volume, we have attempted to reach a global focus by including colonial Latin America, modern China, and the Soviet Union.

[25] On this point, see Terrance MacMullan, "Introduction: What Is Male Embodiment?" in *Revealing Male Bodies*, ed. Nancy Tuana, William Cowling, Maurice Hamington, Greg Johnson, and Terrance MacMullan (Bloomington, Ind., 2002), 1–16, on 2. Of particular importance to this new visibility was the discovery of the hormones; Nelly Oudshoorn, *Beyond the Natural Body: An Archaeology of Sex Hormones* (London, 1994); David Serlin, *Replaceable You: Engineering the Body in Postwar America* (Chicago, 2004); and Chandak Sengoopta, *The Most Secret Quintessence of Life: Sex, Glands, and Hormones, 1850–1950* (Chicago, 2006). On the operationalization of "masculine" hormones, see John Hoberman, *Testosterone Dreams: Rejuvenation, Aphrodisia, Doping* (Berkeley and Los Angeles, 2005).

[26] On late antique and medieval bodies, see Peter Brown, *The Body and Society: Men, Women, and Sexual Renunciation in Early Christianity* (1988; New York, 2008); and Joan Cadden, *Meanings of Sex Difference in the Middle Ages* (Cambridge, 1993).

[27] Christopher Forth, *Masculinity in the Modern West: Gender, Civilization and the Body* (New York, 2008), 5. See also Christopher Forth and Ivan Crozier, eds., *Body Parts: Critical Explorations in Corporeality* (Lanham, Md., 2005).

[28] Mary Douglas, *Natural Symbols: Explorations in Cosmology* (New York, 1970).

Similarly, we sought to span a wide array of chronologies by incorporating perspectives from the medieval and early modern periods in addition to those from the nineteenth and twentieth centuries. We are not trying to provide a single or unified perspective on scientific masculinities but instead to illustrate that scientific masculinities vary with respect to historical and cultural context. We thus see the broadly selective nature of the contributions to this volume (in both space and time) as a distinctive strength.

The easy dichotomies with which scholars have interrogated the gendered world—such as domestic/public, feminine/masculine, and popular/professional—break down under close scrutiny.[29] In place of these simple binaries, this volume delves into the nuanced privileges of scientific masculinities that were created by, and which sometimes resisted, overlapping discourses of morality, family life, education, class, disciplinary affiliations, and cultural identity. Owing to the mutability of masculinities, disparate generations and cultures reconstitute gendered privilege in novel terms. So-called crises of masculinity exist in each scientific culture we have investigated and cannot be reduced to specific moments of social upheaval or anxiety.[30] Despite the myriad historical incarnations of scientific cultures, masculinity tends to retain its hegemonic place.

When read collectively, these essays explore the mobilization of gendered discourses in the sciences, and the functions they served for professionals, amateurs, and popularizers who deployed, maintained, and implemented them within masculine cultures. For reasons of conceptual clarity, we have chosen to examine the gendering of scientific activities in three important domains: the professionalization of science, the practices of physical and intellectual scientific labor, and the acts of measuring and theorizing sexed bodies.[31] As the reader can see from the table of contents, we have accordingly grouped essays in three sections, but in each, we reference essays that overlap significantly with the central subject in order to illustrate the volume's unity of theme and analysis.

PROFESSIONALIZATION AND GENDER NORMS

Gender, whether conceived as an identity or normative description, consists of negotiable constructions that scientists mobilize for particular tasks.[32] These identities invariably overlap in and between persons and communities, but despite the often haphazard ascriptions of gender in the material world, there has been a persistent effort to

---

[29] See, e.g., James Secord, "Knowledge in Transit," *Isis* 95 (2004): 654–72; Katherine Pandora, "Popular Science in National and Transnational Perspective: Suggestions from the American Context," *Isis* 100 (2009): 346–58.

[30] Robert A. Nye, "Locating Masculinity: Some Recent Work on Men," *Signs* 30 (2005): 1937–62.

[31] Until sometime after the publication of Anne Witz's *Professions and Patriarchy* (London, 1992), which pioneered gender analysis in the field, historians and sociologists of the professions—when they considered women at all—treated them as a "given" composed of their sex and its characteristic emotional and intellectual characteristics (2–3). These include important books such as Eliot Freidson, *Profession of Medicine: A Study of the Sociology of Applied Knowledge* (New York, 1970); Samuel Haber, *The Quest for Authority and Honor in the American Professions, 1750–1900* (Chicago, 1991); William M. Sullivan, *Work and Integrity: The Crisis and Promise of Professionalism in America* (New York, 1995); Kees Gispen, *New Profession, Old Order: Engineers and German Society* (Cambridge, 1989).

[32] On the "paradoxes" of the continuous historical creation and maintenance of difference see Judith Lorber, *Paradoxes of Gender* (New Haven, Conn., 1994), 5–6.

maintain clear gender boundaries in social and cultural life. In other words, although nature loves variety, societies hate it. Medical and scientific specialists have played important roles in this process by establishing and defending natural laws of male and female difference, normal and pathological sexual orientation, and fixed gender norms using the tools of their professional expertise. They have then applied these laws to their own professional domains as criteria for admission and advancement and exported them to others. There is a clear dialectical relationship between the establishment of gender norms and professionalization processes, as several of these essays illustrate, including, as we have indicated, how some men benefited professionally at the expense of others.[33] Professional success required not only scientific acumen but in addition the ability to negotiate the gendered cultural norms of scientific society.

Several essays survey historical actors and cultures that refused to acknowledge the fluidity of sex/gender in natural or human law and sought to demarcate "natural" boundaries between male and female bodies and sexual behavior. In her investigation of hermaphroditism in the Middle Ages, Leah DeVun scrutinizes medieval surgical manuals for insights into how and why doctors "corrected" the errors of nature in the genitals of hermaphrodites. Nathan Ha considers the Cold War–era research of the Czech-born Canadian sexologist Kurt Freund, who used penile plethysmography to measure male sexual desire and orientation, a procedure with important implications for defining and (at the time) correcting gender identity.

The fluidity of gender we so readily discover in analyzing masculine scientific cultures is often the result of new intellectual challenges. The emergence of novel technologies or professional opportunities may require adaptations in gendered spaces or in the identities of the scientists and technicians who must perform new tasks. Nathan Ensmenger discusses the rugged individualism favored by male computer programmers in a context of professionalization and gender competition in America from the 1950s to the 1970s. Ensmenger's male computer experts fashioned for themselves a unique masculinity that permitted a form of professional self-presentation allowing them to seem both heroic and (for women) impossible to emulate. Eugenia Lean examines periodicals aimed at women readers in early Republican China to identify the importance of feminine domesticity as a cultural trope that would entice elite men to engage in scientific and technological work in industrializing China. Lean's male propagandists for household industry thus chose to represent themselves in the feminine voices of the domestic sphere. Michael Robinson writes about the way that late nineteenth- and early twentieth-century arctic explorers embarked on their expeditions with the clear intention of later popularizing their exploits as adventures in masculine self-fashioning. Robinson's essay reveals that these travelers identified new sources of patronage for their work by recruiting female journalists to relate their essentially masculine narratives to a broader culture.[34]

---

[33] See, e.g., Sheila Jasanoff, ed., *State of Knowledge: The Co-Production of Science and Social Order* (New York, 2004); Mary Terrall, *The Man Who Flattened the Earth: Maupertuis and the Sciences in the Enlightenment* (Chicago, 2006); Lynn Nyhart, *Modern Nature: The Rise of the Biological Perspective in Germany* (Chicago, 2009); Eric Ash, ed., *Expertise: Knowledge and the Early Modern State*, vol. 25 of *Osiris* (2010).

[34] For some interesting recent work on the gender implications of popularizing science to mixed audiences, see Rebekah Higgitt and Charles W. J. Withers, "Science and Sociability: Women as Audience at the British Association for the Advancement of Science, 1831–1901," *Isis* 99 (2008): 1–27; Veronica della Dora, "Making Mobile Knowledges: The Educational Cruises of the *Revue Générale des Sciences Pures et Appliquées*, 1897–1914," *Isis* 101 (2010): 467–500.

Perhaps more typically, however, masculine gender ideals have been deployed as part of a rear-guard defense of established monopolies in professional life. In this vein, Alexandra Rutherford revisits mid-twentieth-century American experimental psychologists' commitment to the popular and professional assumption that the minds of men were uniquely suited to scientific inquiry. The Harvard psychologist Edwin Boring, the principal subject of Rutherford's essay, defended his besieged model of experimental psychology against the perceived feminization of his discipline. Erika Lorraine Milam focuses on the research and institution-building work of the social anthropologists Robin Fox and Lionel Tiger to understand the parallels between pop-anthropological theories of male bonding (and aggression) and the predominantly male professional identity some anthropologists feared to lose. Milam's nexus of social anthropologists, who were threatened by feminists in contemporary life and by emerging scientific fields beyond the scope of their expertise, created an evolutionary picture of males and females that reinforced traditional gender arrangements. In each of these cases, scientists countered a perilously fluid professional situation by rearticulating gendered norms of behavior.

When women began to penetrate the boundaries of these scientific domains at the dawn of the twentieth century, the shock to the men who monopolized most intellectual fields must have been stupendous.[35] As historian Paul Lucier has noted, before "scientists" came to describe those who did scientific work, Americans typically used the phrase "men of science." In the debate over the professionalization of science and what to call the men who worked in its precincts, women were essentially excluded from public conversations about scientific expertise.[36] Yet evidence also suggests that many all-male groups in the twentieth century forged self-sufficient friendships.[37] Within these groups, some men might not even have noticed the absence of women, much less felt them to be a threat; the erotics of male bonds were quite capable of filling emotional gaps, as colleagues, friends, confidants, or lovers.[38] In doing so, scientific producers and practitioners mobilized masculine discourses for authenticating lines of authority and expertise.

SCIENTIFIC LABOR AND GENDERED SPACES

Historians of science are particularly attuned to the practices of physical and intellectual labor that produce knowledge. Such practices vary between sites of research—including scientific observations, experiments in laboratories and field locales, museum collecting, and theoretical modeling—each with their own rules of engagement.[39]

[35] Margaret Rossiter, "Which Science? Which Women?" *Osiris* 12 (1997): 169–85.

[36] Paul Lucier, "The Professional and the Scientist in Nineteenth-Century America," *Isis* 100 (2009): 699–732, on 704.

[37] Jamie Cohen-Cole has written about the convivial masculinity of the social scientists and intellectuals who attended conferences in exotic natural settings; Cohen-Cole, "The Creative American: Cold War Salons, Social Science, and the Cure for Modern Society," *Isis* 100 (2009): 256–60.

[38] On this conundrum and its entanglements, see Elizabeth A. Wilson's fascinating study of male relationships in the postwar artificial intelligence movement: Wilson, "'Would I Had Him with Me Always': Affects of Longing in Early Artificial Intelligence," *Isis* 100 (2009): 839–47.

[39] Pamela Long, *Openness, Secrecy, Authorship: Technical Arts and the Culture of Knowledge from Antiquity to the Renaissance* (Baltimore, 2003); Robert Kohler, *Lords of the Fly: Drosophila Genetics and the Experimental Life* (Chicago, 1994); Nicholas Jardine, James Secord, and Emma Spary, eds., *Cultures of Natural History* (Cambridge, 1996); David Kaiser, *Drawing Theories Apart: The Dispersion of Feynman Diagrams in Postwar Physics* (Chicago, 2005); Lorraine Daston and Elizabeth Lunbeck,

Erecting Sex:
Hermaphrodites and the Medieval Science of Surgery

*by Leah DeVun**

ABSTRACT

This essay focuses on "hermaphrodites" and the emerging profession of surgery in thirteenth- and fourteenth-century Europe. During this period, surgeons made novel claims about their authority to regulate sexual difference by surgically "correcting" errant sexual anatomies. Their theories about sex, I argue, drew upon both ancient roots and contemporary conflicts to conceptualize sexual difference in ways that influenced Western Europe for centuries thereafter. I argue that a close examination of medieval surgical texts complicates orthodox narratives in the broader history of sex and sexuality: medieval theorists approached sex in sophisticated and varied manners that belie any simple opposition of modern and premodern paradigms. In addition, because surgical treatments of hermaphrodites in the Middle Ages prefigure in many ways the treatment of atypical sex (a condition now called, controversially, intersex or disorders/differences of sex development) in the modern world, I suggest that the writings of medieval surgeons have the potential to provide new perspectives on our current debates about surgery and sexual difference.

In early fourteenth-century Catalonia, a man named Guillem Castelló of Castelló d'Empúries tried to have his marriage annulled on the grounds that his wife, Berengaria, was unable to have sex. To verify his claims before the court, Guillem sought the expertise of a surgeon, Vesianus Pelegrini, who conducted a thorough gynecological examination of Berengaria and came to a startling conclusion. As he saw it, Berengaria was not a woman at all. Instead, she had

*Department of History, Rutgers University, 111 Van Dyck Hall, 16 Seminary Place, New Brunswick, NJ 08901; leah.devun@rutgers.edu.

I am grateful for a Franklin Research Grant from the American Philosophical Society, as well as for financial support from the Institute for Medical Humanities at the University of Texas Medical Branch, the Melbern G. Glasscock Center for Humanities Research at Texas A&M University, the David Geffen School of Medicine and Louise M. Darling Biomedical Library at the University of California, Los Angeles (UCLA), and the Program in History and Philosophy of Science at Stanford University. Two librarians at UCLA, Teresa G. Johnson and Russell Johnson, deserve special mention. I would also like to thank Daniel Bornstein, Caroline Walker Bynum, Tina Chronopoulos, Mary Doyno, Kathryn Jasper, Kenneth Kipnis, Susan L'Engle, Paula Findlen, Kathleen P. Long, John Hilary Martin, Cary J. Nederman, Christof Rolker, James Rosenheim, Londa Schiebinger, Robert A. Nye, Erika Lorraine Milam, *Osiris*'s anonymous reviewers, and the other contributors to this volume for their helpful suggestions.

a male penis and testicles like a man and is so narrow that she can barely urinate through an opening that she has in a fissure that she has in the vulva, [which] lies beneath her penis. She has a flap stretched between her thighs like the wings of a bat, which covers the fissure in the vulva whenever she draws her knees toward her head. She has more the aspect of a man than a woman, and there is no way in which Guillem or any other man can lie with her, nor can she render her conjugal debt, nor conceive nor bear a child.[1]

Berengaria was an individual with neither typical male nor female anatomy—a hermaphrodite, as she would have been called in the late Middle Ages.

The selection of Pelegrini to evaluate Berengaria's sex was not an arbitrary one. By the early fourteenth century, male medical experts were frequently called upon by civil and ecclesiastical courts to testify in cases related to marriage and inheritance.[2] In the Middle Ages, one's status as male or female determined not only whom one could marry but whether one could claim a vast array of male privileges, including inheriting property, offering witness testimony in court, working in certain industries, and entering the priesthood. An examination such as Berengaria's was thus a pivotal moment that determined the legal, social, and religious fates of the examinee. Pelegrini's skills as a surgeon were especially appropriate: just a few decades before, an elite group of surgeons had begun to circulate instructions for determining the masculinity and femininity of hermaphrodites, certifying their sex, and offering "cures" that brought their bodies into conformity with standard expectations of male or female anatomy.

In this article, I focus on the emergence of this surgical science of sex in Italy and France, chiefly during the thirteenth and fourteenth centuries, when surgeons made novel claims about sexual difference and their authority to regulate it. Their theories drew upon both ancient roots and contemporary conflicts to conceptualize sexual difference in ways that influenced Western Europe for centuries after. Beyond this, I argue that a close examination of medieval surgical texts complicates orthodox narratives in the broader history of sex and sexuality: medieval theorists approached sex in sophisticated and varied manners that belie any simple opposition of premodern and modern paradigms. In addition, because surgical treatments of hermaphrodites in the Middle Ages prefigure in many ways the treatment of atypical sex in the modern Western world (a condition now called, controversially, intersex or disorders/differences of sex development [DSD]), I argue that the writings of medieval surgeons have the power to provide new perspectives on our current debates about surgery and sexual difference.[3] Modern disputes are strikingly similar to those in the distant past,

---

[1] Bernat Sunyer, *Llibre,* in Arxiu Històric Provincial Girona, fons de Perelada, fol. 52v: "habebat virgam virilem et testiculos ad modum hominis et est arta ita quod vix potest mingere per quoddam foramen parvum quod habet in fissura vulve quam habet suptus dictam virgam virilem et habet pellem inter crura sua stensam ad modum alarum vespertilionis que cohoperit fissuram dicte vulve tociens quociens ipsa Berengaria vertit genua sua versus faciem suam et quod habet formam virilem plus quam muliebrem, ita quod dictus Guillelmus nec alter homo posse iacere secum nec habere rem carnaliter nec ipsa posset reddere ullatenus debitum coniugalis nec concipere nec infantare." Michael R. McVaugh describes the case in his *Medicine before the Plague* (Cambridge, 1993), 206.

[2] Katharine Park, *Secrets of Women: Gender, Generation, and the Origins of Human Dissection* (New York, 2006), 97.

[3] I use the term "hermaphrodite" to encompass variations and ambiguities in sexual morphology, gender presentation, and sexual practice, in keeping with the historical terminology of the period. The category of hermaphroditism overlaps a great deal with what is now called intersex or DSD; however, many differences also exist. For an overview of intersex studies, see Lisa Downing, Iain Morland, and Nikki Sullivan, *Fuckology: Critical Essays on John Money's Diagnostic Concepts* (Chicago, 2015); Elizabeth Reis, *Bodies in Doubt: An American History of Intersex* (Baltimore, 2009); Morgan Holmes,

de-exceptionalizing our present "fuss" about the body; yet, in other ways, medieval categories are foreign to us, suggesting that systems of sex, gender, and sexuality have highly variable meanings across time.[4] Those differences throw into stark relief positions we often take for granted about how bodies should look and behave.

THE MASCULINIZATION OF MEDICAL PRACTICE

The backdrop of this story is the emergence of an industrious group of male surgeons, who organized to advance their professional status in the late Middle Ages, particularly in Italy and France. They developed a new genre of surgical manuals to standardize and disseminate ideas about the proper form of the body and, along the way, proposed novel ways of thinking about sexual difference. Through these texts, trained male surgeons appointed themselves as experts who could evaluate the male and female qualities of patients and counter their deficiencies through surgery. This developing science of surgery grew out of a professional dispute about bodies, gender, and authority that, in turn, shaped the nature of sex itself.

Until the thirteenth century, aspiring surgeons trained by apprenticeship, learning by oral and practical instruction from more experienced surgeons, rather than in formal schools or with standardized texts.[5] The field of surgery lacked the prestige of internal "medicine," which was a separate discipline focused on the balance of the four fluids of the body (known as the humors), and which was taught in European universities.[6] But surgeons—who focused on the exterior of the body—competed in a rough-and-tumble marketplace alongside barber-surgeons, empirics, and other practitioners—male and female—many of whom were informally trained. Yet by the mid-thirteenth century, a group of ambitious surgeons began to call for greater respect for their vocation, writing surgical textbooks in Latin that were "more elaborate, more highly organized, and more physiological in character" than anything that had appeared before in Greek or Arabic.[7] The authors cited well-known medical authorities and claimed that their science, like medicine, was based on reason, expert training, and classical medical theory.[8] The manuals provided the first systematic approach to surgery in the West, distinguishing those who wrote and read them from their less educated competitors,

---

ed., *Critical Intersex* (Aldershot, 2009); Iain Morland, ed., "Intersex and After," special issue, *GLQ* 15 (2009); Katrina Karkazis, *Fixing Sex: Intersex, Medical Authority, and Lived Experience* (Durham, N.C., 2008); Anne Fausto-Sterling, *Sexing the Body: Gender Politics and the Construction of Sexuality* (New York, 2000); Suzanne Kessler, *Lessons from the Intersexed* (New Brunswick, N.J., 1998); Alice Domurat Dreger, *Hermaphrodites and the Medical Invention of Sex* (Cambridge, Mass., 1998); Dreger, Cheryl Chase, Aron Sousa, Philip A. Gruppuso, and Joel Frader, "Changing the Nomenclature/ Taxonomy for Intersex: A Scientific and Clinical Rationale," *J. Pediatr. Endocrinol. Metab.* 18 (2005): 729–33.

[4] On "fuss," see Caroline Walker Bynum, "Why All the Fuss about the Body? A Medievalist's Perspective," *Crit. Inq.* 22 (1995): 1–33.

[5] Michael R. McVaugh, *The Rational Surgery of the Middle Ages* (Florence, 2006), 13–87, esp. 34–46.

[6] Nancy G. Siraisi, *Medieval and Early Renaissance Medicine: An Introduction to Knowledge and Practice* (Chicago, 1990), 153–86. Universities did not, however, exercise a monopoly over the training and licensing of physicians; for apprenticeships in learned medicine, see Cornelius O'Boyle, "Surgical Texts and Social Contexts: Physicians and Surgeons in Paris, c. 1270 to 1340," in *Practical Medicine from Salerno to the Black Death*, ed. Luis García-Ballester, Roger French, Jon Arrizabalaga, and Andrew Cunningham (Cambridge, 1994), 159–60.

[7] This is Michael R. McVaugh's assessment in his introduction to Guy de Chauliac, *Inventarium sive chirurgia magna*, ed. McVaugh, 2 vols. (Leiden, 1997), 1:ix.

[8] McVaugh, *Rational Surgery* (cit. n. 5), 35–41, 54–8.

whom they denounced as ignorant and even dangerous.[9] Economics played a role, too; surgeons hoped to enhance not only their reputations but also their clientele, in some cases putting themselves in direct competition with physicians.[10] Members of this group of elite surgeons, including Bruno Longobucco, Guglielmo da Saliceto, Lanfranco da Milano, and Guy de Chauliac, became among the most authoritative and widely read practitioners in medieval history. Under their aegis, surgery developed from a craft restricted to treating growths, wounds, and other injuries to a profession expansive in its applications, as well as in its social and academic pretensions.[11]

At this same point in time, medieval health care was undergoing a broader process of masculinization. The surgeons' new claim to expert, book-based knowledge was part of a growing shift in literacy and professionalization that was transforming medical care throughout Western Europe. As the historian Monica Green has shown, beginning in the High Middle Ages, medical knowledge became increasingly literate rather than oral, a change that favored the success of male medical professionals, who were much more likely to be book-learned than women.[12] Newly adopted licensing regulations further contributed to professionalization, and they too worked to the advantage of men. Restrictive laws in Spain, France, Italy, England, and parts of Germany meant that medical practice was becoming increasingly reserved for the "right" kind of males, that is, those who were licensed and authorized.[13] In this new world of organized guilds and specialized training, courts prosecuted female and other illicit healers for practicing medicine without a license.[14] Health care was thus a contested site in the late Middle Ages, pitting men against women, and men against other men, all jockeying for competitive advantage. The surgeon that I focus on in this article, Lanfranco da Milano, found himself at the center of just such a struggle when university-authorized physicians in Paris backed legislation to limit the practice of surgeons and render surgery legally inferior to medicine.[15] Lanfranco wrote his book as an impassioned defense of his livelihood, arguing that surgery was based on rationality and written texts—two decidedly masculine justifications, given that men's greater capacity for reason and literacy distinguished them from "vile and presumptuous women" who approached the body with "neither art nor understanding."[16]

---

[9] Siraisi, *Medieval and Early Renaissance Medicine* (cit n. 6), 153–86; McVaugh, *Rational Surgery* (cit. n. 5), 51–2; Peter Murray Jones, "John of Arderne and the Mediterranean Tradition of Scholastic Surgery," in García-Ballester et al., *Practical Medicine* (cit. n. 6), 290–9. It is difficult to know to what degree these changes affected the practice of surgery, but the status of surgeons was certainly transformed. See Monica H. Green, *Making Women's Medicine Masculine: The Rise of Male Authority in Pre-Modern Gynaecology* (Oxford, 2008), 6.

[10] McVaugh, *Rational Surgery* (cit. n. 5), 175.

[11] The surgeons were not completely successful in their goals to institutionalize surgical instruction in the universities, however. See ibid., 229–66.

[12] Green, *Making Women's Medicine Masculine* (cit. n. 9), viii–xiv. She argues that this transition predates the orthodox narrative, which places the professionalization of gynecology and the takeover of woman-centered medicine by men in the early modern period. This takeover accelerated in the eighteenth and nineteenth centuries; see Jean Donnison, *Midwives and Medical Men: A History of Inter-Professional Rivals and Women's Rights* (London, 1977); Ornella Moscucci, *The Science of Woman: Gynaecology and Gender in England, 1800–1929* (Cambridge, 1990).

[13] I rely on Green's account of the growing regard for text-based, theoretical medicine in late medieval Europe; Green, *Making Women's Medicine Masculine* (cit. n. 9), xiii–9, 74–117.

[14] Ibid., 1–9.

[15] McVaugh, *Rational Surgery* (cit. n. 5), 38–9.

[16] Quoted from Bruno Longobucco, *Surgery*, in Green, *Making Women's Medicine Masculine* (cit. n. 9), 14; Lanfranco made similar statements in his *Great Surgery*, as did Guy de Chauliac and Henri de Mondeville; see McVaugh, *Rational Surgery* (cit. n. 5), 40–1.

By the fourteenth century, men were also extending the anatomical boundaries of their expertise by expanding into women's and reproductive medicine.[17] While surgeons previously avoided the intimate details of women's bodies because of concerns about propriety, the new surgical manuals began to describe the anatomy of each body part, including male and female genitals.[18] It was in this context that surgeons first broached the topic of patients whose sex was neither conclusively male nor female. Surgeons argued that hermaphroditism demanded intensive inspection of the sexual organs, culminating in a pronouncement on whether the patient was masculine enough to warrant male status, or too feminine to qualify. If the patient was deemed female, the surgeons recommended amputations of masculine-looking genitals to return them to a more "natural" form. These interventions corrected nature's "mistakes" and prevented patients from exercising sexual privileges reserved for men, as I discuss below.

It is striking that a new science of sex emerged to police anatomical masculinity at precisely the moment that male medical practitioners were so intent on policing the masculinity of their own profession. This intellectual and political climate no doubt shaped the surgeons' perceptions of what sex was and how it was to be judged. Surgeons focused, first, on how sex manifested itself in the external genitalia (the surface of the body and the primary site of surgical expertise) and, second, on how those genitalia might be corrected through surgery (the surgeons' primary mode of manipulating that surface). For surgeons, locating sex in the exterior flesh was not just an epistemological but also a professional move. The traditional Hippocratic/Galenic understanding of sexual difference—popularized by the late antique physician Galen and the ascendant model for much of the Middle Ages—viewed sex as a spectrum that encompassed masculine men, feminine women, and many shades in between, including hermaphrodites—a perfect balance of male and female.[19] These sexes reflected variations in the "complexion," that is, the combination of heat, cold, dryness, and moisture within, which in turn affected the shape of the body. The "Aristotelian" model challenged this view when ancient texts by Aristotle became available to European readers in the thirteenth century. The Aristotelian model argued that hermaphrodites were not an intermediate sex but a case of doubled or superfluous genitals. The individual's underlying sex could be determined by carefully examining the complexion, which always indicated underlying male or female sex.[20] But both Galenic and Aristotelian approaches emphasized the role of the interior humors in establishing sex. Surgeons, for the most part, restricted themselves to the bodily exterior, probing the genitals for clues and optimizing the patient's sex by reshaping the flesh. It is worth noting that the surgeons certified the sex of their patients at the same time that they bolstered their own legitimacy as readers of the body, excluding from their membership both women and unauthorized, untrained men.[21] I argue that a very particularized notion of sex arises when it is decided by a male-dominated group of professionals fixated on

[17] Green, *Making Women's Medicine Masculine* (cit. n. 9), 71–117.

[18] Ibid., 98–111; McVaugh, *Rational Surgery* (cit. n. 5), 67–8; Park, *Secrets of Women* (cit. n. 2), 93.

[19] Katharine Park and Lorraine J. Daston, "The Hermaphrodite and the Orders of Nature: Sexual Ambiguity in Early Modern France," *GLQ* 1 (1995): 419–38, on 420–2; Joan Cadden, *Meanings of Sex Difference in the Middle Ages* (Cambridge, 1993), 198–202; for the significance of Thomas Laqueur's "one-sex" model in *Making Sex: Body and Gender from the Greeks to Freud* (Cambridge, Mass., 1990), see my comments below, 35–6.

[20] Park and Daston, "The Hermaphrodite" (cit. n. 19), 421.

[21] There is evidence that a small number of women practiced surgery, including wives and daughters of male guild members, but their numbers were decreasing by the beginning of the fifteenth century,

masculinity and its attendant privileges. The surgeons' impulse to interrogate and correct the body's sex in medieval Europe was inseparable from surgery's development
as a profession—and the gendered conflict that accompanied it.

CLOSING HERMAPHRODITES

Lanfranco da Milano—that same surgeon caught up in the fight over legitimate medical practice in Paris—devoted a chapter of his *Great Surgery* (1296) to the problem of
hermaphroditism. It appeared among a miscellany of ailments affecting the abdomen
and genitals, including hernias, hemorrhoids, kidney stones, and vaginal growths. The
chapter, entitled "On the Closure of a Hermaphrodite, and on the Added Pannicle of
a Woman," explained that

> [a] hermaphrodite is one who has each perfect sex in such a way that [the hermaphrodite]
> is able to be active and passive. Some of these have one [sex] that is fully formed, the
> other not fully formed, and some have neither fully formed. On the contrary, they have in
> the orifice of the vulva some added flesh, which is sometimes soft, fleshy, of a small and
> weak character, other times of a strong and sinewy character. The fleshy piece is removed
> swiftly with cutting instruments, and those parts left behind with light cauterization; the
> natural flesh must always be taken care of by means of iron, or through a ligature with
> thread until all superfluity is taken off.[22]

This passage begins Lanfranco's attempt to categorize hermaphrodites, individuals
who possessed more than one set of genitals exhibiting various states of what he called
"perfect" anatomical development. According to Lanfranco's instructions, surgeons
should cut away the added fleshy piece, which he described as superfluous. But "if the
flesh is truly hard and strong and sinewy," Lanfranco continued, "so that it resembles
the male penis and most of all if it becomes erect upon touching a woman, in no way
touch that with the iron nor think to treat it with anything."[23] Behind these instructions
lay a serious and potentially life-altering process of discernment—and one that was
utterly dependent on Lanfranco's notions of masculinity.

Whether a patient was in possession of a penis or merely excessive vulvar tissue
hinged on the size of the genitals: a penis, Lanfranco argued, was "truly hard and
strong and sinewy" rather than merely "small and weak." The organ's potential sexual
function was also extremely significant. An organ that became aroused in response to
female touch apparently met the minimum criterion for masculinity.[24] But how did the
surgeon know if a patient's penis responded to a woman? It is possible that he merely

---

and women in general did not make up more than 1.5 percent of medical practitioners. See Green,
Making Women's Medicine Masculine (cit. n. 9), 120–1.

[22] Lanfranco da Milano, *Lanfranci maioris*, in *Cyrurgia Guidonis de Cauliaco: et Cyrurgia Bruni,
Theodorici, Rogerij, Rolandij, Bertapalie, Lanfranci* (Venice, 1498), 198va: "Hermaphroditus est ille
qui habet utrumque sexum perfectum: ita quod agere potest et pati: quorum aliqui habent vnum perfectum alium imperfectum: aliqui nec vtrumque perfectum: immo habent in orificio vulue aliquid
carnis addite: que aliquando est mollis carnea parue et debilis tenacitatis: aliquando fortis et neruosa.
Carnea vero remouetur de leui cum instrumentis incidentibus: et cum leui cauterizatione residui carnem
semper cauendo naturalem a ferro: vel cum ligatione cum filo: quod quotidie plus stringatur donec tota
superfluitas auferatur."

[23] Ibid.: "Si vero dura sit et fortis: et neruosa: ita quod virgae virili assimiletur: et maxime si tangendo
mulierem erigitur: illam nullo modo ferro tangas: nec cum aliquo curare cogites."

[24] For another view of sexual function as a determinant of masculinity, see Beth Linker and Whitney Laemmli, "Half a Man: The Symbolism and Science of Paraplegic Impotence in World War II
America," in this volume.

asked the patient about his/her sexual desire. Yet this odd test of a hermaphrodite's erection mirrored similar tests of impotence in legal disputes over unconsummated marriages, in which surgeons were deeply involved.[25] These examinations were no dispassionate observations of a man's anatomy. Examiners—who often included female prostitutes—touched, stroked, prodded, and otherwise helped along the individual's physical response to arousal, a response that constituted manhood.[26] The size of the male genitals played a role here, too: one physician remarked on the small penis of an examinee, which was no larger than that of a "two-year-old boy."[27] Historian Jacqueline Murray has argued that the unfortunate litigant who failed such an impotence test was considered "less than a 'real man,'" dismissed by the court as lacking in hardness, substance, and maleness.[28] The subject's sex was to some extent legally indeterminate until the examiners either helped him achieve an erection and male status or judged him incapable. Lanfranco's warning that the surgeon should "in no way" touch a hermaphrodite's organ with any surgical tool suggests that similar criteria—a phallus that was hard, strong, sinewy, and capable of erection when prompted by a woman—were adequate to establish physical integrity and masculinity in such a way that surgical intervention was not only unnecessary but unthinkable.

Lanfranco recommended a different treatment for "male" hermaphrodites with partially formed vulvas and vaginal orifices:

> There are also certain men who have beyond their testicles two additional "saddlebags" and in the middle a pit resembling the vulva of women. In certain of them there is in this pit an opening through which they pass urine. If it was [there] like that, abandon the cure. But if it was not a deep pit, touch that place with a hot iron little by little, and then apply butter and expect the occurrence of the cautery scab. If you followed your plan, you grow back the skin. But if not, repeat with the cauterizer until that place has become well filled in and returns to the natural form.[29]

The male patients in question had bodies with genital "pits," which Lanfranco likened to female vulvas. He advised against closing the pit surgically if the urethra appeared within the pit rather than at the tip of the penis. In these cases, surgery would interfere

---

[25] Danielle Jacquart and Claude Thomasset, *Sexuality and Medicine in the Middle Ages*, trans. Matthew Adamsom (Oxford, 1988), 171–2; Bronach Kane, *Impotence and Virginity in the Late Medieval Ecclesiastical Court of York* (York, 2008), 5–37.

[26] Jeremy Goldberg, "John Skathelok's Dick: Voyeurism and 'Pornography' in Late Medieval England," in *Medieval Obscenities*, ed. Nicola McDonald (York, 2006), 105–23; Thomas G. Benedek and Janet Kubinec, "The Evaluation of Impotence by Sexual Congress and Alternatives Thereto in Divorce Proceedings," *Trans. Stud. Coll. Phys. Philadelphia* 4 (1982): 1333–53.

[27] Jacquart and Thomasset, *Sexuality and Medicine* (cit. n. 25), 171.

[28] Jacqueline Murray, "Hiding behind the Universal Man: Male Sexuality in the Middle Ages," in *Handbook of Medieval Sexuality*, ed. Vern Bullough and James Brundage (New York, 1996), 123–52, on 139; Kane, *Impotence and Virginity* (cit. n. 25), 25. Such evaluations also resonate with Nathan Ha's study of the Canadian sexologist Kurt Freund, who used penile plethysmography to quantify sexual arousal in suspected homosexual men. Medical experts at the time linked such measurements of sexual orientation to gender identity. See Ha, "Detecting and Teaching Desire: Phallometry, Freund, and Behaviorist Sexology," in this volume.

[29] Lanfranco da Milano, *Lanfranci maioris* (cit. n. 22), 198va: "Sunt etiam quidam viri qui post testiculos habent additamenta duo lateralia: et in medio foueam quasi vuluam mulierum: in quam in quibusdam est foramen: per quod emittunt vrinam: quod si sic fuerit, curam dimittas. Si vero profunda non fuerit illa loca ferro tangas calido paulatine: deinde butirum appone: et escare casum expecta. Quod si tuum fueris consecutus propositum cutem regenera. Sin autem cauterium itera donec locus optime fuerit repletus et ad formam redierit naturalem."

with the elimination of urine and threaten the life of the patient.[30] But, barring this complication, the surgeon could close the vulvar pit through repeated cauterizations and, as a result, return the body to its "natural form." A man with a vagina-like orifice was at odds with what Lanfranco deemed the natural shape of the male body—that is, closed, impenetrable, and integral. This was in contrast to the body of a woman, which had long been judged in classical and medieval thought to be lacking in physical stability, characterized by uncontrollable leaking and oozing, and subject to penetration.[31] The bodies of "men" with saddlebags failed to properly form the masculine corporeal boundaries that divided the interior of the body from its exterior. Yet surgery could eliminate the hermaphrodite's porosity, remove the blemish of feminine genitals, and render the patient decisively male.

Finally, in the same chapter on hermaphrodites, Lanfranco turned to the problem of women with penis-like genitals. According to Lanfranco, a woman with a "muscular growth" in the orifice of the vulva prompted disgust.[32] Medieval medical professionals were well aware of the existence of the clitoris, although they disagreed on its name and function (the early modern period—which supposedly "discovered" or "rediscovered" the clitoris—was hardly any more precise).[33] Lanfranco chose the word "pannicle"—a general term for a membrane, or sometimes a swelling or tumor—to name the female hermaphrodite's enlarged genitals. His inclusion of the descriptor "muscular" suggests that he had in mind a part of the genitals with substantial erectile tissue, most likely the clitoris. Sometimes, Lanfranco wrote, this muscular flesh was outsized to the extent that it hung down from the body. He worried that a woman with such imposing genitals would be loved less by men, and so he advised the surgeon to amputate the "superfluity of that skin, [and] afterwards, cauterize it with gold until you reduce it to the natural form."[34] Lanfranco's discussion reveals a striking concept of the natural: through each intervention, the surgeon's craft returns the body to its "natural form." Lanfranco also claimed in the same chapter that "nature erred seriously" in the creation of hermaphroditic genitals.[35] For Lanfranco, atypical genitals

[30] A point also made by the tenth-century Arab surgeon Al-Zahrāwī. See his discussion of hermaphroditism in his surgical manual: Albucasis, *On Surgery and Instruments*, trans. M. S. Spink and G. L. Lewis (Berkeley and Los Angeles, 1973), 454.

[31] Anne Carson, "Putting Her in Her Place: Woman, Dirt, and Desire," in *Before Sexuality: The Construction of Erotic Experience in the Ancient Greek World*, ed. Froma I. Zeitlin, John J. Winkler, and David M. Halperin (Princeton, N.J., 1990), 135–69; Sarah Alison Miller, *Medieval Monstrosity and the Female Body* (New York, 2010).

[32] Lanfranco da Milano, *Lanfranci maioris* (cit. n. 22), 198va: "Quibusdam etiam accidit mulieribus quod panniculus quidam lacertosus: qui est in orificio vulue: adeo augmentatur: quod multum dependeat: ita quod mulierem afficit tedio." This section immediately follows a discussion of uterine and vaginal obstructions that prevented what Lanfranco viewed as optimal female sexuality, that is, passive sexual intercourse with a man and the successful conception of a child. For this reason, we might surmise that Lanfranco viewed "female" hermaphroditism, at least in part, as a malady that impeded sexual entry into the female body.

[33] Karma Lochrie, *Heterosyncrasies: Female Sexuality When Normal Wasn't* (Minneapolis, 2005), 76–89; Katharine Park, "The Rediscovery of the Clitoris: French Medicine and the Tribade, 1570–1620," in *The Body in Parts: Fantasies of Corporeality in Early Modern Europe*, ed. David Hillman and Carla Mazzio (New York, 1997), 170–93; Valerie Traub, *The Renaissance of Lesbianism in Early Modern England* (Cambridge, 2002), 87–93.

[34] Lanfranco da Milano, *Lanfranci maioris* (cit. n. 22), 198va–b: "et a viro multo minus amatur: quam sic cures superfluum illius incide pellicule: postea cum auro calido cauteriza, donec ad formam naturalem reducas."

[35] Ibid., 198va: "Nam illud peccatum est in forma vbi natura fortiter errauit."

were "unnatural," but surgery was "natural," aiding the flesh to return to the form that nature intended.

Such sentiments fit well with surgeons' characterizations of themselves as the help-mates of nature, assisting her operations and correcting her mistakes, for they often imagined nature as a feminine force that created, ordered, and, occasionally, failed. The newly rediscovered writings of Aristotle seemed to support this confidence in humanity's ability to mimic and even improve upon nature. In the *Physics*, Aristotle distinguished between two types of human art, one that imitated nature and another that "perfect[ed] that which nature [could not] complete."[36] In the *Metaphysics*, he added that sometimes health was brought about by nature and sometimes by human skill.[37] Medieval philosophers and alchemists echoed his conclusions, writing that humans could help nature achieve greater perfection, and medieval surgeons made similar re-marks.[38] When speaking of grave wounds to the brain and spine, Lanfranco explained, "nothing is impossible for powerful nature, most of all when a good physician assists her with helpful things, and she is aided to increase her effort."[39] Elsewhere he noted that surgery enhanced nature's potency, even to the point of making the seemingly impossible possible.[40] Such opinions suggest that surgeons viewed themselves as na-ture's helpers, extending the efficacy of her labors. Yet some hoped to carve out a more prominent role for the surgeon, whose skills, they argued, eclipsed even those of nature. The fourteenth-century surgeon Henri de Mondeville, for instance, wrote that a sur-geon was neither nature's servant nor assistant but her master: "Nature works neither alone nor with the aid of the professional," he claimed. "It is the professional alone who works."[41] This bold assertion demonstrates how the surgeon's dominance was tied to his professionalism—that masculine mastery of textual and theoretical knowledge that set him above not only the unlearned female medical practitioner, but nature herself.

HERMAPHRODITES AND THE HISTORY OF SURGERY

Peter Murray Jones has characterized Lanfranco's text as the "culmination of the genre" of Latin scholastic surgery in its most realized form.[42] Lanfranco's innovative vocabulary—the "saddlebags" and "muscular" erectile tissue of hermaphrodites—is not apparent elsewhere in the surgical literature and may have stemmed from his expe-rience with actual cases in the field.[43] Yet Lanfranco, it turns out, was far from the first

---

[36] Aristotle, *Physica* 2.8.199a.

[37] Aristotle, *Metaphysics* 7.9.1034a8; see also Aristotle, *Physica* 2.4.194a21–23.

[38] Umberto Eco, *Art and Beauty in the Middle Ages* (New Haven, Conn., 1986), 93–4; William New-man, "Technology and Alchemical Debate in the Late Middle Ages," *Isis* 80 (1989): 423–45.

[39] Lanfranco da Milano, *Lanfranci maioris* (cit. n. 22), 178ra: "Forti namque nature nihil est im-possibile: maxime cum per bonum medicum ex rebus eam iuuantibus: et ad intentionem valentibus adiuuatur."

[40] Ibid.: "Quoniam natura cum bonis iuuaminibus facit multotiens que medico impossibilia vide-bantur."

[41] Quoted in Marie-Christine Pouchelle, *The Body and Surgery in the Middle Ages* (Cambridge, 1990), 39.

[42] Jones identifies Lanfranco and Henri de Mondeville as the height of medieval surgery in his "John of Arderne" (cit. n. 9), 290. The influential surgeon Guy de Chauliac cites Lanfranco more than any other European surgeon; see McVaugh's introduction to Chauliac's *Inventarium* (cit. n. 7), 1:xiii.

[43] Green makes a similar observation about Lanfranco in *Making Women's Medicine Masculine* (cit. n. 9), 99–100. Surgeons sometimes included empirical observations alongside theoretical material from authoritative sources; see McVaugh, *Rational Surgery* (cit. n. 5), 37–8.

to recommend the surgical amputation of hermaphroditic genitals. He almost certainly adopted some of his ideas from a long tradition that stretched back at least to Byzantine Greece. But Lanfranco was no parrot of earlier authorities: he engaged actively with those texts and elaborated upon them, suggesting that he sought out preexisting knowledge on this subject because he thought it meaningful and even urgent.[44]

Opinions on the surgical correction of hermaphrodites seem to have entered the West through the writings of Muslim medical luminaries such as Al-Majūsī, Al-Zahrāwī, and Ibn Sīnā (also known as Haly Abbas, Albucasis, and Avicenna), who were all advocates of such surgeries, and who provided a fairly consistent discussion of hermaphrodites in their writings.[45] The tenth-century Arab surgeon Al-Zahrāwī, for example, classified hermaphrodites on the basis of their genital appearance, arriving at two categories of male and one category of female hermaphrodites. "You cut away the superfluous pieces of flesh until the mark of them is eliminated; then cure them with the usual treatment for the remaining wounds," he wrote in his *On Surgery and Instruments*, which was translated into Latin in the twelfth century, after which it became a standard textbook in Europe.[46] It is also possible, Al-Zahrāwī cautioned, for the female organ to grow in size until it is "above the order of nature, and its appearance is shamefully ugly; and in certain women it is so large that it becomes erect just like [the organs of] men and it attains to coitus. You must grasp the growth with your hand or a hook and cut it off."[47] Here, Al-Zahrāwī judged large female genitals to be a cause for alarm. He assumed they had a natural size; excessive genitals were outside the natural order and an unsightly deformity that elicited shame and led their possessor to usurp the masculine role, sexually penetrating another woman's body.

Latin European surgeons echoed the interpretations of Al-Zahrāwī, Al-Majūsī, and Ibn Sīnā (who offered similar opinions) in their own analyses of hermaphrodites.[48] The Paduan surgeon Bruno Longobucco was among the first surgical writers in Western Europe to borrow extensively from Arabic material that had become available in Latin translation.[49] He described hermaphroditic anatomies as "unnatural" and

[44] On the citation of authoritative ancient texts, see Caroline Walker Bynum, *Metamorphosis and Identity* (New York, 2001), 26.

[45] They based their conclusions on the earlier recommendations of Paul of Aegina, a seventh-century Alexandrian Greek physician whose medical compendium *Epitomê Iatrikê* became a heavily cited source in Muslim texts. For the influence of Aegina in Arabic medicine, see Monica H. Green, "The Transmission of Ancient Theories of Female Physiology and Disease through the Early Middle Ages" (PhD diss., Princeton Univ., 1985), 79–80. With respect to enlarged female genitals, Aegina was likely dependent on Soranus of Ephesus, a second-century adherent of the Methodist school of medicine. On Soranus, see Rino Radicchi, ed., *La "Gynaecia" di Muscione: manuale per le ostetriche e le mamme del VI sec. d. C.*, (Pisa, 1970), 190; Ann Ellis Hanson and Monica H. Green, "Soranus of Ephesus: Methodicorum Princeps," *Aufstieg und Niedergang der Römischen Welt* 37 (1994): 968–1075; Paul Burguière, Danielle Gourevitch, and Yves Malinas, *Soranos d'Éphèse: Maladies des femmes*, vol. 1, *Texte établi, traduit, et commenté* (Paris, 1988), vii–xlvi.

[46] Albucasis, *Cyrurgia Albucasis cum cauterijs et alijs instrumentis*, in *Cyrurgia parua Guidonis* (Venice, 1501), 23vb: "incidas carnes additas donec effugiat impressio earum, deinde cura eas curatione reliquorum vulnerum." For his reliance on Aegina, see Albucasis, *On Surgery and Instruments* (cit. n. 30), 454.

[47] Albucasis, *Cyrurgia Albucasis* (cit. n. 46), 23vb: "Tentigo fortasse additur super rem naturalem donec sedatur et turpis sit aspectus eius: et quandoque magnificatur in quibusdam mulieribus adeo donec expanditur sicut in viris et peruenit usque ad coitum. Oportet ergo vt teneas superfluitatem tentiginis manu aut cum vncino et incidas: et non vltimes in incisione precipue in profundo radicis vt non accidat fluxus sanguinis."

[48] Haly Abbas, *Liber totius medicine* (Lyon, 1523), 282v (*Practica* IX, liiii); Avicenna, *Liber canonis medicinae* (Venice, 1527), bk. 3, tr. 2, fen 20, chap. 43.

[49] McVaugh, *Rational Surgery* (cit. n. 5), 17, 25–7.

"disgraceful," and he advocated the removal of such "leftover, superfluous parts" by means of cutting.[50] The influential French surgeon Guy de Chauliac agreed with this position, adding that large female genitals were a source of unhappiness and even harm to patients.[51] The Italian Guglielmo da Saliceto, another member of this circle of learned surgeons and the teacher of Lanfranco da Milano, provided an example of a related surgery in his *Summa of Preserving and Curing* (ca. 1285). The conditions he called "ragadias" and "furfurs" resulted from abrasions to the female genitals during sexual intercourse; they caused pain and bleeding, as well as the development of fleshy protrusions. Guglielmo's discussion is complex and touched upon a variety of growths, but his central concern was that a protrusion might become erect like a male penis, allowing a woman to have sex with other women, as men do.[52] Here, a woman's genitals were disturbing not just because they resembled a man's, confusing categories of sex anatomy, but because they facilitated sex with other women, confusing sexual roles. For Guglielmo, a masculine piece of flesh carried with it a corresponding masculine sexual appetite for women. This flesh split its bearer into disjunctive parts: its possessor was a woman, and yet the libido it engendered was decidedly male (since she desired as men do). Women were thought in the Middle Ages to be the more sexually voracious of the two sexes, and yet here it was the acquisition of a phallus that activated her lust and (mis)directed it toward other women.[53] Sexual responsiveness to women was in some cases the critical sign of male sex—for Lanfranco it meant an escape from the surgeon's knife. But here an organ that became erect in response to a woman did not establish maleness. It is possible that the surgeons deemed different criteria significant in different contexts; sexual arousal was certainly crucial in determining the sex of "male" hermaphrodites. In the case of Guglielmo's ragadias and furfurs, however, such a response was insufficient. Perhaps it was the experience of passive sexual intercourse that disqualified such patients from maleness? Perhaps the possession of a penetrable vagina? Guglielmo gives us too little information to be sure. In any case, this diversity of criteria for masculinity suggests that there was no stable and definitive set of anatomical characteristics or somatic experiences that defined male or female sex.[54] Guglielmo recommended here the excision of those masculine anatomical parts that were on the body but, remarkably, not of it.

Educated men such as Guglielmo and Lanfranco were no doubt privy to the intellectual conceit among medieval theologians and natural philosophers that each part of the body had a "final cause," or *telos*, a singular purpose for which it was designed, not just by nature, but by God.[55] If the *telos* of the male penis was to ejaculate semen into the womb of a female, then no other use of it could be judged natural.[56] Reports

[50] S. P. Hall, ed., "The *Cyrurgia magna* of Brunus Longoburgensis: A Critical Edition" (DPhil thesis, Oxford Univ., 1957), 290–1: "Hermafrodita, ut dicit Haly, est passio innaturalis et turpissima ualde uiris. . . . Et modus curationis est ut seces illas carnes superfluas sectione qua non remaneat ex eis aliquid."

[51] Guy de Chauliac, *Inventarium* (cit. n. 7), 1:388.

[52] Guglielmo da Saliceto, *Summa conservationis et curationis* (Venice, 1502), 62rb–va; Helen Rodnite Lemay, "Human Sexuality in Twelfth- through Fifteenth-Century Scientific Writings," in *Sexual Practices and the Medieval Church*, ed. Vern L. Bullough and James A. Brundage (Buffalo, N.Y., 1982), 196; Lemay, "William of Saliceto on Human Sexuality," *Viator* 12 (1981): 165–81, on 179.

[53] On female sexual appetite, see Cadden, *Meanings of Sex Difference* (cit. n. 19), 178.

[54] See also Erika Lorraine Milam and Robert A. Nye's discussion in the introduction to this volume, 6–7.

[55] See, e.g., Aristotle, *Physica* 2.3; *Metaphysics* 1.3.

[56] Mark D. Jordan, *The Invention of Sodomy in Christian Theology* (Chicago, 1997), 132.

varied on the purpose of the external female sexual organ (Galen thought it to shield the interior parts in some way), but the idea that it should sexually penetrate another body was out of the question: nonprocreative sex—that is, sodomy—was against "nature's order in the manner of generation, which [was] the starting point for the whole of nature."[57] The physical shapes of men and women, as well as their corresponding contributions to sexual reproduction, underpinned the existence of the two sexes and their intertwined and mutual need. To violate the proper use of a body part was to invert and even injure nature, which was the guarantor not only of the order of the cosmos but also of the existence of the creator, God. The sex of an individual and the appropriate use of his or her genitals represented an important part of ordering nature and maintaining the divinely instituted relations of human society within a Christian context. According to this model, order was embedded in the very flesh of the human body, which surgeons were called upon to reshape in the interest of societal—and even divine—imperatives.

TOWARD A SCIENCE OF MASCULINITY

We should perhaps not be surprised that medieval surgeons evaluated hermaphrodites on the basis of prevailing notions of masculinity and femininity. European surgeons were influenced by their Arabic forebears' analyses of hermaphroditism, but they also injected concerns about the proper characteristics of the sex organs drawn from Latin European medieval and classical sources. Lanfranco's discussion, for instance, capitalized on well-known medieval stereotypes about the genitals: while men's sexual organs were strong and vigorous, women's were soft and weak. If hard or obstructive flesh marred a woman's genitals, it should be surgically removed so that the body could return to its natural, soft, and open form. This characterization of the genitals was common in medieval texts, and it ultimately derived from ancient conceptions of sexual difference. The word *nervus* (tendon or sinew) was a common Roman term for penis, and the description of the penis as "sinewy" was repeated by a number of medieval interpreters of the body. Some writers, including Galen, viewed the penis as a tendon or collection of tendons.[58] According to Isidore of Seville's *Etymologies*, an encyclopedic work that became an influential source for medical knowledge during the Middle Ages, man [*vir*] received his name from the Latin word for vigor or power [*vis*], from which also derived the word strength, *virtus*, an attribute that extended from man's anatomy to his actions.[59] Lanfranco's portrayal of the male genitals as strong and *nervosus* was no doubt linked to traditional vocabulary that depicted men's bodies as vigorous, hard, and sinewy. Women's bodies, in contrast, were fleshy, yielding, and penetrable. The word for woman, *mulier*, was, according to Isidore of Seville, derived from *mollior*, or "softer," and such softness—both physical and behavioral—was a distinctive feature of femininity and effeminacy.[60] Isidore's *Etymologies* imagined that the origins of words—however fallacious or mythical those origins often were—illuminated underlying truths about the nature of things. For him and his readers, a distinction between masculinity and femininity rooted in linguistics became

---

[57] Lochrie, *Heterosyncrasies* (cit. n. 33), 76; Albert the Great, *Summa de bono*, as quoted in Jordan, *Invention of Sodomy* (cit. n. 56), 127.

[58] James Noel Adams, *The Latin Sexual Vocabulary* (Baltimore, 1990), 38.

[59] Isidore of Seville, *Etymologiae*, 11.2.17.

[60] Ibid., 18–9.

fused with one based in anatomy. An individual's proper expression of sex therefore had heavy consequences; it affected the flow of power between men and women, the complementary relationship that defined humanity.[61] Genitals too reflected these expectations: firm and vigorous masculinity or soft and yielding femininity suffused the flesh, especially the intimate parts that stood in for the sex of the whole.

These terms for men and women and their corresponding physical attributes show the extent to which such characteristics were mutually exclusive. According to Isidore, a man could not be soft and yielding in the manner of a woman without sacrificing his manhood: a man is soft [*mollis*] when he "disgraces the vigor of his sex with his enervated body, and is softened [*emollire*] like a woman [*mulier*]."[62] This usage of *mollis* was in practice at least since the classical period, when the word had a constellation of meanings that included sexual passivity, impotence, sterility, hermaphroditism, castration, and a general lack of manly vigor. The famous spring in Hallicarnassus—where the mythical nymph Salmacis joined with the god Hermaphroditus, an event recorded in Ovid's poem *Metamorphoses*—was reputed to make men who entered the waters *mollis*, *impudicus* (shameless), or *obscenus* (lewd), all terms that connoted passive sexual effeminacy.[63]

This equation of softness and effeminacy became widespread in the Middle Ages. Thomas of Aquinas pointed to a "softness of complexion" as the cause of lust in women and "phlegmatic" men (i.e., those with overabundance of the humor phlegm, and hence an unbalanced and feminized body); he followed the ancients in their linkage of physical softness, sexual impropriety, and moral weakness.[64] Effeminacy had not only undesirable sexual consequences but also broader societal ones that could threaten a community's very survival: a lifestyle of feminine vanity, torpor, and decadence "unmans" potential warriors, as the twelfth-century poet Peter of Eboli made clear: "For this soil gives birth to effeminate men; Restful shade unmans those born to the mirror / And reared on a soft couch amidst tender roses."[65] This line of reasoning also appeared in natural philosophical and medical literature of the period. Michael Scot's *Book of Physiognomy* described soft, effeminate flesh as indicative of a "weak, lustful man" while hard flesh signified a strong and brave one.[66] Henri de Mondeville identified eunuchs, phlegmatic men, and effeminate men among those "soft" parts of the body of society that represented weakness.[67] Medieval cultural imperatives demanded that men express firmness, strength, and stability in their demeanors and bodies, qualities that determined not just individual destiny but the collective fate of society.

[61] Ibid., 19.

[62] Ibid., 10.M., 179–80.

[63] M. Robinson, "Salmacis and Hermaphroditus: When Two Become One: (Ovid, Met. 4.285–388)," *Cl. Quart.*, n.s., 49 (1999): 212–23, on 212–3.

[64] Thomas Aquinas, *Summa Theologiae, Latin Text and English Translation, Introductions, Notes, Appendices and Glossaries*, 60 vols. (New York, 1963), 44:20–2 (2a2ae, q. 156, art. 2): "sicut e contrario contingit quod aliquis non persistat in eo quod consiliatum est, ex eo quod debiliter inhaeret propter mollitiem complexionis, ut de mulieribus dictum est, quod etiam videtur in phlegmaticis."

[65] Quoted in Luke Demaitre, "Skin and the City: Cosmetic Medicine as an Urban Concern," in *Between Text and Patient: The Medical Enterprise in Medieval and Modern Europe*, ed. Florence Eliza Glaze and Brian K. Nance (Florence, 2011), 97–120, on 113.

[66] Michael Scot, *Liber phisionomiae* (Venice, 1477), cap. lxxxvi (no page numbers): "Caro mollis ubique corporis significat hominem debilem: luxuriosum: ex facili timidum. . . . Cuius caro est dura uel aspera significat hominem fortem: audacem, duri ingenii."

[67] Pouchelle, *Body and Surgery* (cit. n. 41), 112–3.

Beyond their varying degrees of softness and hardness, men and women differed in the locations of their most important sexual organs. For Isidore, the physical fount of male eroticism was the loins, while a woman's was the "umbilicus," the navel or internal center of the body, rather than any external sex organ.[68] Some medieval theorists adopted this anatomical distinction between men and women, locating the primary root of female sexual appetite in the interior.[69] Lanfranco's surgical manual also identified the interior body as the chief site of female sexuality—in particular, he focused on an anatomically open pathway to the uterus, which was able to receive male semen. The geography of male and female sexual desire further distinguished their anatomies: eroticism centered on an exterior, projecting organ was incompatible with femininity. If a female hermaphrodite's external genitals responded to sexual arousal and sported a muscular erection, then the organ was unsuitably feminine and needed to be surgically removed or altered. Lanfranco's discussion of large female genitalia also tapped into medieval ideas about the proper size of the members. According to this view, body parts had to be free from physiognomic distortion, which signified moral, as well as physical, ugliness. Features that were too large or too numerous were associated in the Middle Ages with monsters and demons, a decidedly undesirable group whose hyperbolic parts were thought to signify—among other negative traits—uncontrollable sexual appetites.[70]

We see then that hermaphrodites' bodies were evaluated in ways that highlighted the divergent functions of men and women in society, as well as the confluence of the linguistic, anatomical, and social attributes that constituted masculinity or femininity. In the case of "male" hermaphrodites, Lanfranco focused on questions of sexual arousal rather than those more closely tied to reproduction or aesthetics. Surgical interventions that altered the male's external appearance—for instance, the closure of feminine vaginal pits—did not mention explicitly the attractiveness of the male body or its potential to repulse female sexual partners. But the hermaphrodites Lanfranco classified as women were judged by vastly different criteria. Their genitals—now deemed overly large female genitals rather than a small penis—were spared or removed based upon whether or not they gave aesthetic offense to men, not according to their potential sexual function. In fact, when the sexual function of such women was mentioned in other surgical manuals, it was in connection with concerns that her anatomy would lead her toward an inappropriate use of her genitals, that is, through penetrative sex with other women—a forbidden appropriation of male sexual prerogative. This contrast suggests that in the case of a female hermaphrodite, the surgeon strived to produce aesthetically pleasing genitals while eliminating barriers to reproduction; in the case of a male hermaphrodite, the surgeon removed any semblance of femininity while preserving the male's capacity for arousal. In each case, the surgeon altered a hermaphrodite's body to restore what he viewed as the proper function and appearance of the genitals, by either removing superfluous flesh or closing superflu-

[68] Lynda L. Coon, "Gender and the Body," in *The Cambridge History of Christianity*, vol. 3, *Early Medieval Christianities, c. 600–c. 1100,* ed. Thomas F. X. Noble and Julia M. H. Smith (Cambridge, 2006), 433–52.

[69] See, e.g., Scot, *Liber phisionomiae* (cit. n. 66), cap. i: "Luxuria uiri primam radicem habet in lumbis: et libido mulieris in umbilico." Opinions on this were not monolithic, however; Cadden notes that Hildegard of Bingen placed the loins inside the kidneys, and John of St. Amand located female desire in the vulva. See Cadden, *Meanings of Sex Difference* (cit. n. 19), 138–9, 180.

[70] Bettina Bildhauer and Robert Mills, "Introduction: Conceptualizing the Monstrous," in *The Monstrous Middle Ages,* ed. Bildhauer and Mills (Toronto, 2003), 10–23.

ous orifices. The task of these surgeons was then not only to identify which of two sexes prevailed but also to "cure" or, as Lanfranco put it, to "close" hermaphroditic bodies.

Such closures were not only physical but also metaphorical. Surgical interventions on hermaphrodites were intended to close off certain functions, preventing hermaphrodites from using their bodies for both active and passive sex. In order to avoid engaging in illicit sexual acts, the hermaphrodite could use either a passive or active genital member, but never both. A French encyclopedic text entitled *Placides and Timeus, or the Secrets of the Philosophers*, written in the late thirteenth century, explicitly brought together this closure of genital possibilities with the surgical closure of the body. The author wrote:

> The ancient judges established that no one should allow [hermaphrodites] to use both natures, but that one should put to them rather what nature they would want, to do it or to suffer [it to be done to them]. And when they have taken the one or the other, that is to say the nature of the man or the woman, if someone finds them practicing that which they had denied, then they must be punished bodily, because it is against nature to use both. And such men, who wish to use both natures, must not be tolerated among the people, and so the ancients judged that if they chose to be passive, then one had to cut off their testicles and then close up the skin of the penis in the front, and if they chose to be active, that is to say to use the nature of a man, then one closed the orifice, and if by malice they unclose it, than they are punished bodily for doing so.[71]

We should take note of the intriguing uses of "nature" in the text. "Nature" was a standard term for the sex organs in classical literature, a convention adopted in turn by medieval authors. James Noel Adams argues that the genitals' excretory functions, which were sometimes called the "necessities of nature," lent to the genitals first the name "parts of nature" and later merely "nature."[72] Hermaphrodites, according to *Placides and Timeus*, were not to have two natures. In this context, "nature" referred to the genitals, requiring either the male or female organ of a hermaphrodite to be rendered unusable by surgery so that he or she could not "use both natures." But nature indicated more than just anatomy. In *Placides and Timeus*, an individual's nature was also a complicated intersection of his/her expressed object of sexual desire and his/her role within a schema of passivity/activity. Therefore, hermaphrodites had to specify "what nature they would want, to do it or to suffer [it to be done to them]"; an individual could not opt to have active and passive sex alternately.[73] According to the historian Ruth Mazo Karras, sex in the Middle Ages was chiefly conceived of as

---

[71] Claude Thomasset, ed., *Placides et Timéo, ou, Li secrés as philosophes* (Geneva, 1980), 153–4: "Les anchiens justichiers establirent que nuls ne laissast tels hommes user de .II. natures, mais c'on les meist enchois a quois de prendre quelle nature qu'il vorroient, de faire ou de souffrir. Et quant il aroient prins l'un ou l'autre, c'est assavoir le nature de l'omme ou de le femme, qui les trouveroit ouvrans de celle qu'il avoient renoÿe qu'ilz fussent punis du corps, car c'est contre nature d'user de deux. Et tels hommes, qui de deux natures veulent user, ne devroient estre souffers entre gens, et si jugerent les anchiens que, s'i eslisoient a souffrir, que on leur copast les testicules et si closist on le pel de le vergue par devant, et, s'i eslisoient a faire, c'est assavoir d'user de nature d'omme, que on closist l'orefice, et se ilz par malice le destoupoient, que il en fussent pugnis du corps."

[72] Adams, *Latin Sexual Vocabulary* (cit. n. 58), 59–60.

[73] Although some scholars have viewed similar sources as evidence that hermaphrodites exercised self-determination in matters of sex during the medieval period, I argue that a mandate for hermaphrodites to choose one fixed sex under coercion, enforced by surgery and in accordance with the rules of opposite-sex sexual attraction, hardly constitutes self-determination. See my comments below, 35–6.

something one did to another, and categories of activity and passivity differentiated the vastly different experiences of the sexual act that the two partners were thought to have.[74] Thus, the exclusivity of one's passive or active role in intercourse was central to the ordering of sexual relations. Finally, nature was in *Placides and Timeus* a set of laws: "it is against nature to use both" genitals or sexual roles. Nature dictates the proper use of body parts, each of which has a naturally (and even divinely) ordained purpose for which it was made. This understanding of nature in the regulation of hermaphrodites is significant. As we have seen, surgeons were deeply invested in upholding nature's design by correcting her failings. The passage suggests that much more was at stake than preserving the hardness and softness of the flesh; surgeons were responsible for hewing human bodies to a transcendent order.

There is some indication that medieval thinkers also looked to factors beyond the genitals to establish sexual difference. Hermaphroditism was of concern to practitioners of both civil and canon—or church—law, particularly after the eleventh century, a period of reform and renewed interest in synthesis and classification. These legal scholars categorized hermaphrodites by determining whether masculine or feminine traits predominated within a particular individual. The canon lawyer Huguccio offered just such a consideration in his *Summa* (1188):

> As to a hermaphrodite, if he has a beard and always wants to engage in manly activities and not in those of women, and if he always seeks the company of men and not of women, it is a sign that the masculine sex predominates in him and then he can be a witness where a woman is not allowed, namely with regard to a last will and testament, and he also can be ordained a priest. If he however lacks a beard and always wants to be with women and be involved in feminine works, the judgment is that the feminine sex predominates in her and then she should not be admitted to giving any witness wherever women are not admitted, namely at a last will and testament, neither can she be ordained.[75]

This opinion highlights the gap between male and female privileges. Only a man was able to serve as a priest or bear witness. Interestingly, Huguccio relied on hair—specifically, the ability to grow a beard—as a primary basis for sex assignment. A similar attentiveness to hair also appeared in medieval illustrations of male and female hermaphrodites, in which the presence of facial hair and a "male" hairstyle stood in for other bodily indications of sex. In an illustrated copy of *Mandeville's Travels*, for instance, "male" and "female" hermaphrodites displayed their identical genitals; instead, hair provided a central clue to their sex (see fig. 1). For Huguccio, a preference for manly company and activities also indicated the power of social roles and networks in defining masculinity. By his account, however, there could be no men who enjoyed

---

[74] Ruth Mazo Karras, *Sexuality in Medieval Europe: Doing unto Others* (New York, 2005), 3–5.

[75] Huguccio, *Summa*, MS Vat. 2280, fol. 140v: "si quidem habet barbam et semper vult exercere virilia et non feminea et semper vult conversare cum viris et non cum feminis signum est quod virilis sexus in eo prevalet et tunc potest esse testis ubi mulier non admittitur scilicet in testamento et in ultimis voluntatibus tunc etiam ordinari potest. Si vero caret barba et semper vult esse cum feminis et exercere feminea opera iudicium est quod feminini [*sic*] sexus in eo prevalet et tunc non admittitur ad testimonium ubi femina non admittitur, scilicet in testamento sed nec tunc ordinari potest quia femina ordinem non recipit." For hermaphrodites in canon law, see Maaike van der Lugt, "Sex Difference in Medieval Theology and Canon Law: A Tribute to Joan Cadden," *Med. Fem. For.* 46 (2010): 101–21; Cary J. Nederman and Jacqui True, "The Third Sex: The Idea of the Hermaphrodite in Twelfth-Century Europe," *J. Hist. Sexual.* 6 (1996): 497–517; Christof Rolker, "The Two Laws and the Three Sexes: Ambiguous Bodies in Canon Law and Roman Law (12th to 16th Centuries)," *Z. Savigny-Stiftung Rechtsgesch. kanonistische Abt.* 100 (2014): 178–222.

Figure 1. Hermaphrodites. Livre des merveilles *(Travels of Sir John Mandeville), Bibliothèque nationale de France, MS Français 2810, fol. 195v (fifteenth century).*

the company or activities of women, nor the reverse, despite many other testaments to the contrary.[76] Secondary sex characteristics and homosocial bonds clearly possessed much explanatory power, and yet Huguccio ultimately fell back on the genitals as an arbiter of sex, closing his discussion with the observation that when discerning sex, "an inspection of the genitals is frequently effective."[77] Despite this attentiveness to alternative signs of sex, in this case the intimate parts of the body once more appear to offer a final word. And, as we have seen, by the end of the Middle Ages, the most adept reader of these parts was a man with expert medical training.

Such thoughts on embodiment and sexual difference became enshrined in surgical texts in the late Middle Ages, and they in turn acquired a timeworn authority of their own. According to the 1405 statutes of the University of Bologna, surgical students began their curriculum each year with the study of Bruno of Longobucco's *Surgery*.[78] Lanfranco's manual was translated into French, English, and Dutch and remained influential for subsequent generations of surgeons. Guy de Chauliac's *Inventarium* was printed over a dozen times and edited in the sixteenth century into what was at the time a modern version, demonstrating its continued relevance.[79] Many of the structural reforms suggested by the elite medieval surgeons waned in the early modern period thanks to changing practices in licensing and education (although many were implemented

[76] Many contemporary texts attest to the existence of effeminate males and masculine females who were nonetheless male or female. See Cadden, *Meanings of Sex Difference* (cit. n. 19), 201–9; Barbara Newman, *From Virile Woman to WomanChrist* (Philadelphia, 1995); Caroline Walker Bynum, "Jesus as Mother and Abbot as Mother: Some Themes in Twelfth-Century Cistercian Writing," in *Jesus as Mother: Studies in the Spirituality of the High Middle Ages* (Berkeley and Los Angeles, 1982), 110–69.

[77] Huguccio, *Summa*, fol. 140v: "Praeterea ad talem discretionem multum valet inspectio genitalium quid si illi duo sexus equales per omnia inveniuntur in eo credo quod debeat iudicari de eo tamquam femineus sexus in eo praevalet quia verum est virilem sexum in eo non praevalere."

[78] Paul F. Grendler, *The Universities of the Italian Renaissance* (Baltimore, 2002), 322–3.

[79] McVaugh, "Introduction" (cit. n. 7), 1:xiv; McVaugh, *Rational Surgery* (cit. n. 5), 242–51.

again in the nineteenth century, during another wave of professionalization).[80] But opinions on the treatment of hermaphrodites persisted; we find recommendations to amputate hermaphroditic genitals in early modern works, including Jacques Dalé-champs's 1570 *French Surgery*, which has been credited with producing momentous changes in beliefs about same-sex sexuality.[81] In the hands of professional surgeons, the science of masculinity was just as authoritative and objective as those treatments for abscesses, fractures, and wounds that appeared on adjacent pages. Masculinity vouched for by nature and science was a hard and firm category indeed.

* * *

I would like to end where I started, with the story of Berengaria Castelló in Catalonia. As we have seen, Vesianus Pelegrini inspected Berengaria's genitals and found a preponderance of the masculine. His observations rested not only on the contours of Berengaria's body but also her potential role in sexual intercourse. "[No] man can lie with her," Pelegrini testified, suggesting the extent to which Berengaria's passive sexual performance was key to her presumed identity. It is interesting to note that Pelegrini deemed Berengaria to be male on the basis of her failure as a female (her inability to fulfill her marital debt to a man), not on the criteria for masculinity we saw advanced by Lanfranco. The surviving summary of the case gives us no further information about what Berengaria had to say about her own body and what, if any, "cure" she received from the surgeon. We also cannot know whether Pelegrini read and absorbed the lessons of the new surgical manuals, some of which were already circulating in Catalan translation in the early fourteenth century.[82] But Berengaria's brush with surgical authority at this critical juncture indicates that evaluations of hermaphrodites in surgical manuals were not mere theoretical exercises. Surgeons surely, if infrequently, interacted with patients of atypical sex, although we might imagine that other hermaphrodites never saw a surgeon or endured surgery.[83]

It is important to point out that the taxonomies of sex favored by medieval surgeons were not the only choices available. If hermaphrodites occupied a particular point on the conventional sexual spectrum, as the Hippocratic/Galenic model of sexual differ-ence held, then why further scrutinize their bodies? If hermaphrodites possessed re-dundant genitals with no bearing on their actual sex, as the Aristotelian model argued, then why remove them? The European surgeons' attempts to reconfigure sex on the body rested not only upon ancient Greek and Arabic authority but also upon medieval cultural demands that the body look and behave only in very circumscribed ways. Men were closed, stable, and firm; women, open, soft, and penetrable; their libidos centered in divergent regions of the body, in organs of divergent size. Surgeons la-

[80] McVaugh, *Rational Surgery* (cit. n. 5), 229–66.

[81] Park, "Rediscovery of the Clitoris" (cit. n. 33), 175–6.

[82] Catalan translations of both Lanfranco's *Minor Surgery* and Teodorico's *Surgery* were available at the time of Berengaria's examination in 1331; McVaugh, *Rational Surgery* (cit. n. 5), 241–2.

[83] It is impossible to quantify how many hermaphrodites received medical care. Surgery on another fourteenth-century hermaphrodite in Bern is recounted in the *Annals of the Friars Minor of Colmar* and cited in Miri Rubin, "The Person in the Form: Medieval Challenges to Bodily 'Order,'" in *Framing Medieval Bodies*, ed. Sarah Kay and Miri Rubin (Manchester, 1994), 100–22. Christof Rolker has documented additional cases of hermaphrodites who were not subject to surgical intervention in his "Der Hermaphrodit und seine Frau: Körper, Sexualität und Geschlecht im Spätmittelalter," *Hist. Z.* 297 (2013): 593–620; he argues that the requirements for masculinity in medieval marriage law were exceedingly low. See also Rolker, "The Two Laws" (cit. n. 75).

bored to produce genitals that remained safely within the bounds of the natural and the aesthetically inoffensive, striving to facilitate their functions within overlapping social and theological frameworks. The texts also suggest that surgeons saw themselves as engaged in a moral enterprise—the elevation of sexual morality—that overrode other moral obligations to patients, whose consent to these invasive examinations and operations is never mentioned. Furthermore, the surgeons' area of expertise influenced where they imagined the body's sex to reside. Their cuts worked upon the body's surface but did nothing to bring about changes in the internal complexion.

Lanfranco's (and other surgeons') claims were also firmly rooted in the gendered politics of late medieval health care. Surgeons found themselves enmeshed in a hard-fought battle to limit women, and those men untrained in the masculine art of learned medicine, from joining their ranks. This conflict brought the surgical manuals into existence, and with them, the conviction that the masculine hand of a professional might prevail where a woman (and even nature) could not. These and other sentiments reveal the extent to which surgeons viewed women as political and economic rivals. Perhaps it was not a stretch to imagine them as sexual rivals too, eager to usurp the penetrative role in intercourse if their anatomy so allowed. Although surgeons maintained that gender roles and bodily shapes were "natural," they devoted considerable energy to articulating precisely what the sexes were and should remain, and how they might be effectively contained within such categories.

This surgical science of sex also suggests that much more was at play than the "one-sex model" popularized by Thomas Laqueur's influential *Making Sex: Body and Gender from the Greeks to Freud*. According to Laqueur, until the eighteenth century medical thinkers believed that male and female sexual organs were the same anatomical structure: the vagina was an inverted penis that appeared internally or externally, depending on the heat of the individual's complexion.[84] With respect to hermaphrodites, Laqueur argues, enlarged clitorises were unambiguously female anatomies that did not count in establishing male status; the internal phallic/vaginal structure had to descend to mark maleness.[85] Some medieval surgeons do in fact mention these correspondences between male and female body structures, but they also subscribed to a more nuanced view of sex that considered factors beyond the presence or absence of a descended vagina/penis.[86] Lanfranco's writings clearly demonstrate how surgeons evaluated the size and function of a phallic/clitoral structure to determine a patient's sex and potential therapy. In these cases, the clitoris and penis were separated by a fuzzy line that counted crucially in establishing sex. Laqueur writes that there was no essential sex in the premodern world, but "neither were there two sexes juxtaposed in various proportions."[87] But this is just what we find in Lanfranco's "men" with penises, vulvar "saddlebags," and vaginal pits alongside a testicular structure. It seems that medieval understandings of sexual difference cannot be reduced to a simple one-sex model, as Joan Cadden, Katherine Park, and Robert A. Nye have also argued, and we must give voice to the full range of complicated, and even contradictory, ideas

[84] Laqueur, *Making Sex* (cit. n. 19), 25–62.

[85] Ibid., 137–8.

[86] Katharine Park, "Cadden, Laqueur, and the 'One-Sex Body,'" *Med. Fem. For.* 46 (2010): 96–100, on 98.

[87] Laqueur, *Making Sex* (cit. n. 19), 124.

that medieval theorists of the body found significant.[88] The surgeons' approach to hermaphroditic bodies also suggests that the premodern world offered far less freedom to choose one's sex than has been previously imagined. Scholarship, following Michel Foucault, has sometimes hailed the period as refreshingly fluid in its approach to sex and sexuality, prior to the biomedical regulatory practices that demanded from the body a single "true sex" beginning in the eighteenth century.[89] While the self-reported desires and proclivities of hermaphrodites figured into evaluations of sex in the Middle Ages, a narrative focused on self-determination takes too little account of the expert surgeon, whose interrogations and operations eradicated from the body sexual and anatomical possibilities.

For those familiar with our own controversies over sex and genitals in the twenty-first century, this medieval science of sex may seem surprisingly familiar. Modern understandings and treatments of intersex are similarly dominated by gendered assumptions about bodies. Studies by Suzanne Kessler, Alice Dreger, and Katrina Karkazis, among others, have documented surgical procedures performed in the twentieth and twenty-first centuries to "normalize" intersex patients, reshaping genitals deemed too small to be believably male, or too large to be female.[90] Stereotypical notions about active/male and passive/female sexuality have continued to shape surgeons' understandings of what constitutes functionally or aesthetically appealing bodies.[91] Also notable is the growing market in "designer vaginas," voluntary cosmetic genital surgeries that trade in related assumptions about the softness and smallness of a female genital ideal.[92] In both the medieval and modern worlds, an individual's apparent failure to achieve masculinity or femininity and fulfill his/her obligations is a medical problem that can be cured at least in part by surgery. And surgery brings about a much more radical transformation than a simple rearrangement of the body's surface. It enforces symmetry between the constituent parts of the self, validating or erasing certain desires, assigning the medical and legal status that confers admission to a human community, and reinforcing gendered power dynamics. Despite the medical advances and technologies of recent decades, our contemporary conventions of masculinity and femininity remain strikingly parallel to past debates, even if they are expressed in (sometimes) divergent terms.

For when we hold up the modern in the light of the medieval, we also see a range of historically contingent ideas about masculine and feminine bodies. No one talks much now about male porosity or superfluous labial folds, which were among the problems that preoccupied Lanfranco. A penis with a variably positioned urethra, on the other hand, is much more bothersome to doctors now than it was to medieval practitioners. Atypical genitals now vex pediatrics in the same way that they once troubled

---

[88] Cadden, *Meanings of Sex Difference* (cit. n. 19), 3; Katharine Park and Robert A. Nye, "Destiny Is Anatomy," *New Republic* 204 (Feb. 18, 1991): 53–7; Park, "Cadden" (cit. n. 86), 96–100; see also Helen King, *The One-Sex Body on Trial: The Classical and Early Modern Evidence* (Farnham, Surrey, 2013).

[89] Michel Foucault, *Herculine Barbin: Being the Recently Discovered Memoirs of a Nineteenth-Century French Hermaphrodite*, trans. Richard McDougall (New York, 1980), vii–xvii; Anne Clark Bartlett, "Foucault's 'Medievalism,'" *Mystics Quart.* 20 (1994): 10–18, on 15–6.

[90] For contemporary studies of intersex/DSD, see n. 3 above.

[91] See, e.g., Kiira Triea's story in "Power, Orgasm, and the Psychohormonal Research Unit," in *Intersex in the Age of Ethics*, ed. Alice Domurat Dreger (Frederick, Md., 1999), 140–4.

[92] Virginia Braun, "Selling the 'Perfect' Vulva," in *Cosmetic Surgery: A Feminist Primer*, ed. Cressida J. Heyes and Meredith Jones (Burlington, Vt., 2009), 133–49.

adult medicine in the Middle Ages.[93] Such divergences might also suggest that bodily morphologies are most at stake during different moments in our life cycles (a contrast that is worthy of further exploration). Perhaps the process of allowing the past and present to resonate productively will make our contemporary debates about the body seem less new and exceptional. It might also help us to pose more critical questions: What categories are upheld, challenged, or created by these surgeries? What cultural demands for activity/passivity, hardness/softness, or beauty/ugliness still, or yet again, animate our bodily choices? What unacknowledged allegiances might we have to a perceived natural or moral order, a transcendent sense of the way that things are or should be, that influences our care? The surgical texts discussed in this article may help illuminate our assumptions about masculinity and femininity, charting their history, and our roles as agents who accept and enforce their demands. Our history of the sexes is—if nothing else—a history of the contingent, the changing, and therefore the changeable.

[93] Little is known about medieval pediatrics and hermaphroditic infants, although at least one case appears in the inquisitorial record in the context of Judaizing and a suspected circumcision. See Josep Hernando i Delgado, "El procès contra el convers Nicolau Sanxo, ciutadà de Barcelona, acusat d'haver circumcidat el seu fill (1437–1438)," *Acta Hist. Archaeol. Med.* 13 (1992): 75–100. My thanks to Paola Tartakoff for bringing this case to my attention.

"Beards, Sandals, and Other Signs of Rugged Individualism":
Masculine Culture within the Computing Professions

*by Nathan Ensmenger**

ABSTRACT

Over the course of the 1960s and 1970s, male computer experts were able to successfully transform the "routine and mechanical" (and therefore feminized) activity
of computer programming into a highly valued, well-paying, and professionally
respectable discipline. They did so by constructing for themselves a distinctively
masculine identity in which individual artistic genius, personal eccentricity, anti-
authoritarian behavior, and a characteristic "dislike of activities involving human
interaction" were mobilized as sources of personal and professional authority. This
article explores the history of masculine culture and practices in computer programming, with a particular focus on the role of university computer centers as key sites
of cultural formation and dissemination.

In 1976, the MIT computer science professor Joseph Weizenbaum published *Computer Power and Human Reason*, a scathing intellectual and moral indictment of the
discipline of artificial intelligence, a field that he himself had helped to establish.
More than three decades later, his book continues to be widely read and influential,
although not perhaps for the reasons that Weizenbaum had hoped or expected. It was
not his carefully constructed philosophical arguments that attracted the attention of
most audiences but rather his lurid descriptions of what he regarded as one of the most
dangerous and disturbing phenomena associated with the emerging technology of
electronic computing: namely, the increasing prevalence of the compulsive programmer, or the "computer bum."

  In computer centers around the world, Weizenbaum argued, these computer bums,
"bright young men of disheveled appearance, often with sunken glowing eyes," could
be discovered hunched over their computer consoles, "their arms tensed and waiting
to fire their fingers, already poised to strike, at the buttons and keys on which their
attention seems to be as riveted as a gambler's on the rolling dice."[1] When not other-

  * School of Informatics and Computing, Indiana University, Bloomington, IN 47408;
nensmeng@indiana.edu.
  [1] Joseph Weizenbaum, *Computer Power and Human Reason: From Judgment to Calculation* (San
Francisco, 1976), 116.

wise transfixed by their computer screens, these compulsive programmers pored over their computer printouts "like possessed students of a cabalistic text. . . . They work until they nearly drop, twenty, thirty hours at a time. Their food, if they arrange it, is brought to them: coffee, Cokes, sandwiches. If possible, they sleep on cots near the computer." But such interludes in the real world were few and far between, and the computer bums never wandered far from their machines. They existed almost entirely in an electronic universe of their own creation, isolated from material concerns and conventional social interactions, haunting the sheltered cloisters of the computer center. "Their rumpled clothes, their unwashed and unshaven faces, and their uncombed hair all testify that they are oblivious to their bodies and to the world in which they move. They exist, at least when so engaged, only through and for the computers."[2]

For Weizenbaum, the disheveled figure of the computer bum represented the embodiment of the dehumanizing effects of pursuing computer power as an end rather than a means: deceived by the illusion of omniscience associated with mastery of this powerful technology, these wasted young men were not scientists uncovering new truths about the universe, or engineers building useful products to benefit society, but mere junkies in search of a fix. That such myopic and socially maladjusted tinkerers were being accorded such a prominent and influential role in the construction of the essential structures of the modern information society was, for Weizenbaum, the dangerous and disturbing consequence of a reckless computational imperative. These were not the type of people he wanted to be entrusted with the technological keys to the increasingly computerized kingdom.

Although Weizenbaum's *Computer Power and Human Reason* was early, authoritative, and persuasive (among its many admirers was his fellow MIT professor Sherry Turkle, who would later extend his arguments in her even more popular and influential *The Second Self*), his was not the only, or even the first, mainstream account of the compulsive programmer phenomenon.[3] At the same time that Weizenbaum was deriding the obsessive tendencies of the computer bum, a powerful counternarrative was emerging in which such single-minded focus was lauded as desirable, possibly even heroic. In this interpretation, the glowing screens in the computer centers represented not retreat from the world, but mastery over it.

The best exemplar of this alternative portrayal of the computer bum was actually published four years prior to *Computer Power and Human Reason*. In a rollicking essay in *Rolling Stone* magazine provocatively entitled "Spacewar: Fanatic Life and Symbolic Death among the Computer Bums," Stewart Brand had heralded the arrival of the electronic digital computer as "good news, maybe the best since psychedelics." Where Weizenbaum perceived in computerization the realization of impersonal, bureaucratic and authoritarian imperatives, Brand saw revolutionary potential and the empowerment of individuals. Via the computer, revolutionary citizens/programmers could appropriate Cold War technologies for the purposes of progressive social transformation.[4]

Although the ostensible subject of his article was Spacewar, an early video game developed by students at MIT to demonstrate the capabilities of the then-novel

---

[2] Ibid.

[3] Sherry Turkle, *The Second Self: Computers and the Human Spirit* (New York, 1984).

[4] Fred Turner, *From Counterculture to Cyberculture: Stewart Brand, the Whole Earth Network, and the Rise of Digital Utopianism* (Chicago, 2006), 116–7.

cathode ray tube display, Brand was clearly less interested in Spacewar than he was in its computer "hacker" developers.[5] While he acknowledged the double-edged bite of the "hacker" epithet (which he deemed both "a term of derision and the ultimate compliment"), Brand's representation of the Spacewar hackers was unambiguously positive. Yes, being a computer bum might reflect a "kind of fanaticism," but this was the fanaticism of the artist, the inventor, and the explorer. These "magnificent men in their flying machines" were "scouting a leading edge of technology." They were "brilliant," "revolutionary," and "servants in the human interest." To the degree that they violated the norms of conventional society, it was as the heroic outsider or iconoclast. Anticipating the Wild West metaphors that continue to be popular within the free software/open source software movements, Brand portrayed computer hackers as the "outlaws," "heretics," and "revolutionaries" of the modern era, fighting to bring computer power to the people.[6]

According to Brand, Spacewar was not just a computer game but a kind of software samizdat, the vehicle through which the subversive hacker subculture was smuggled into the network of research laboratories sponsored by the Advanced Research Projects Agency. The result was the creation of an increasingly global community of technician-radicals. Every night, "hundreds of computer technicians" in computer centers around the world engaged in an effectively out-of-body experience, "locked in life-or-death space combat computer-projected onto cathode ray tube display screens, for hours at a time, ruining their eyes, numbing their fingers in frenzied mashing of control buttons, joyously slaying their friends and wasting their employers' valuable computer time." These centers were anything but the isolated social wastelands portrayed by Weizenbaum; rather, the computer center that Brand described constituted a vibrant social space, with its own "language and character, its own legends and humor." In fact, as Brand recalled it, his evening spent with the denizens of the Stanford Artificial Intelligence Laboratory was "the most bzz-bzz- busy scene I've been around since the Merry Prankster Acid Tests."[7]

These two radically different interpretations of the same phenomenon, as portrayed by Weizenbaum and Brand, neatly capture the perplexed, ambivalent, and conflicted attitudes toward computer programming—and more specifically, computer programmers—that characterized the early decades of electronic computing.

Computer programming was, from its very origins, a mongrel discipline. Originally envisioned as low-status clerical work, programming soon acquired a reputation as being one of the most complex, arcane, and esoteric of technical disciplines. Although associated with the emerging discipline of computer science, the majority of programmers had no academic training and did not see themselves as scientists. (And, as the indignant and dismayed response of Weizenbaum to the computer bums clearly illustrates, many computer scientists did not always know what to do with programmers.) Programmers clearly built things, but they generally did not regard what they did as engineering. They most often described their work and expertise using vague analo-

[5] Stewart Brand, "SPACEWAR: Fanatic Life and Symbolic Death among the Computer Bums," *Rolling Stone*, 7 December 1972.

[6] Mitch Kapor and John Perry Barlow, "Across the Electronic Frontier," Electronic Frontier Foundation, 10 July 1990, https://w2.eff.org/Misc/Publications/John_Perry_Barlow/HTML/eff.html (accessed 5 August 2014); Jonathan J. Rusch, "Cyberspace and the 'Devil's Hatband,'" *Seattle Univ. Law Rev.* 24 (2000): 577–98.

[7] Brand, "SPACEWAR," (cit. n. 5), 51.

gies and mixed metaphors: to many of its practitioners, programming was simultaneously art and science, high tech and black magic, work and play.[8]

If the discipline itself was opaque and incomprehensible to outsiders, so too were its practitioners. The colorful sobriquets invented to describe them—"wizards," "gurus," "computer boys," the "high priests of the new technology"—reflected the curious mix of wonder, respect, suspicion, and contempt with which they were regarded by their contemporaries. On the one hand, the technical skills that they possessed were clearly powerful, perhaps even dangerous; on the other, their odd practices (and sometimes personal appearance) and seeming disregard for conventional social norms and authority figures made them bizarre if fascinating characters. To many, they appeared to be as much a subculture as an occupational group. Indeed, many popular accounts of programmers emphasized their innate and idiosyncratic genius. "Excellent developers, like excellent musicians and artists, are born, not made," declared one industry observer, and "the number of such developers is a fixed (and tiny) percentage of the population."[9]

From a contemporary perspective, of course, the association of computer programming ability with a particular personality type is familiar to the point of being cliché. Today we would call such individuals not computer bums but computer hackers or, even more likely, computer nerds. Indeed, within a decade of the publication of *Computer Power and Human Reason*, the computer nerd would became a stock character in the repertoire of American popular culture, his defining characteristics (white, male, middle-class, uncomfortable in his body, and awkward around women) well established in popular histories of computing such as Tracy Kidder's Pulitzer Prize–winning *Soul of a New Machine* (1981) and Steve Levy's *Hackers* (1984), as well as the 1983 Hollywood blockbuster *WarGames*.[10] During the boom years of the personal computer and Internet revolutions, the business and popular press embraced the nerd identity as key to success in the new economy. Each carefully constructed "origin story" of a self-respecting high-tech entrepreneur reads as a minor variation on a formula. The "lonely-nerd-turned-accidental-billionaire" narrative has assumed the mantle of Great American Success Story, as exemplified in the hit PBS documentary *Triumph of the Nerds* (1996) and the Academy Award–winning *The Social Network* (2010).[11]

Indeed, in much of popular culture, the character of the computer nerd has become so hegemonic that it threatens to erase other cultural representations of scientists and engineers. In the work of the best-selling science fiction writer Neal Stephenson, for example, Isaac Newton and his contemporaries in the Royal Society are represented as early incarnations of the hacker mentality whose mannerisms and motivations are largely indistinguishable from those of the modern open source software community.[12] In the popular genre of steampunk fiction, the Industrial Revolution is reimagined as

[8] Nathan Ensmenger, *The Computer Boys Take Over: Computers, Programmers, and the Politics of Technical Expertise* (Cambridge, Mass., 2010); Wendy Hui Kyong Chun, *Programmed Visions: Software and Memory* (Cambridge, Mass., 2011).

[9] Bruce Webster, "The Real Software Crisis," *Byte* 21 (1996): 218.

[10] Tracy Kidder, *The Soul of a New Machine* (New York, 1981); Steven Levy, *Hackers: Heroes of the Computer Revolution* (Garden City, N.Y., 1984); *WarGames*, directed by Wolfgang Petersen (Los Angeles, 1983), DVD.

[11] *Triumph of the Nerds*, directed by Paul Sen (New York, 2002), DVD; *The Social Network*, directed by David Fincher (Culver City, Calif., 2010), DVD.

[12] Neal Stephenson, *Quicksilver* (New York, 2003).

an abortive first attempt at the computer revolution, with Charles Babbage standing in for an early Alan Turing.[13] The perceived connection between computer "nerdery" and mild forms of autism has stimulated retrospective diagnoses of bookish intellectuals and scientific figures from the fictional Doctor Frankenstein to Newton, Darwin, and Einstein, suggesting a line of descent leading directly to the contemporary computer nerd.[14] The remarkable genius and accomplishments of Thomas Edison are now compared to those of Steve Jobs, and not the other way around.[15]

Like the 1970s-era computer bum, with whom he shares certain characteristics, the contemporary computer nerd is defined primarily by his consuming obsession with technology, his lack of conventional social skills, and inattention to his physical health and appearance. Though images of both "bums" and "nerds" were more stereotypical than representative, they are historically significant for the role they played as weapons and resources in the ongoing process of the social construction of the computer professions. The contested debate about the identity, expertise, and authority of computer programmers shaped many of the technical, managerial, and professional developments in electronic computing for the first several decades of the electronic computer era.[16] The disparate responses of Weizenbaum and Brand to the character of the computer bum are both reflections of this debate and contributions to it; they were not simply describing what they thought computer programmers were like but arguing for a particular vision of what they ought to be.

In this essay, I explore the history of the most iconic and invariable attribute of the computer nerd stereotype: namely, that he is a "he." This is not, of course, to suggest that women do not program computers; in fact, women played an unusually prominent role in the history of computer programming, especially in its earliest decades. And yet computer programming today is both male dominated and hypermasculine. Even in an era in which even the most traditionally masculine disciplines, such as mathematics, physics, and engineering, have opened up opportunities for women, female participation in computing remains dismally low. The number of women studying computer science (as a percentage of total enrollments) has actually decreased over recent decades, and representations of female nerds in popular film, fiction, and history are virtually nonexistent. The notorious misogyny of certain subcultures of the computing community is well documented, as is the discouraging effect that this has on female participation.[17]

To argue that a discipline is dominated by males is not necessarily to suggest that it embodies uniquely masculine characteristics. There are structural, legal, or historical reasons why certain occupations are dominated by men that have little to do with whether the work involved is essentially masculine. In the case of computer programming, however, the dominant assumption is that there are certain intellectual and emotional characteristics that are associated with computer programming ability—logical, detached, narrowly focused—that also just happen to be more prevalent in males. The belief that males are much more likely to be antisocial, antisensual, and

[13] William Gibson and Bruce Sterling, *The Difference Engine* (New York, 1991).

[14] Benjamin Nugent, *American Nerd: The Story of My People* (New York, 2008).

[15] Walter Isaacson, *Steve Jobs* (New York, 2011).

[16] Ensmenger, *Computer Boys* (cit. n. 8).

[17] Thomas Misa, ed., *Gender Codes: Why Women Are Leaving Computing* (Hoboken, N.J., 2010); Jane Margolis and Allan Fisher, *Unlocking the Clubhouse: Women in Computing* (Cambridge, Mass., 2002).

attracted to the "hard mastery" of arcane technology pervades even the academic literature, most notably the influential work of Sherry Turkle, who provided the principal
psychoanalytic framework through which the (male) nerd personality has been interpreted.[18] More recently, the perceived association between the "programmer personality" and mild forms of autism (to the point that Asperger's was sometimes referred to
as the "geek disorder" or "Silicon Valley Syndrome") has reinforced the notion that
there is a natural, historical, and inevitable connection between male forms of sociability and cognition and virtuoso computer programming ability.[19]

In my historical analysis of the masculinization of computer programming, I will
focus on three distinct but related themes. The first is that, contrary to conventional
wisdom, the computer industry was initially open to women, who were extraordinarily
well represented in computer programming. In fact, at its origins, computer programming was a largely feminized occupation. The male computer nerd, far from being a
natural or essential form of the computer user, was in many respects a response within
the early computing community to uncertainties about the occupational status and
gender identity of the computer programmer and, by extension, about the reputation
of the computer industry itself.[20] A remarkable demographic shift occurred in programming over the course of the 1960s and early 1970s, a shift that can be explained
not only in terms of the professionalization of the discipline but also by reference to
very specific structural mechanisms, such as the use of psychometric testing in corporate hiring processes. In this respect, the history of computer programming provides
novel insights into the structural factors through which the gendering of institutions
and practices occurs.[21]

The second intriguing feature of the history of masculinity in the programming
professions has to do with the significance of specific sites of practice. Place matters, even in the history of a technological genre that claims to make place irrelevant.
In this case, it was the university computer labs, the sheltered, unsupervised, and
subsidized environments in which the burgeoning computer hacker culture became
inextricably linked with the cultural practices of adolescent masculinity. Later, as
the locus of hacker activity shifted from the university mainframe to the household
personal computer, these practices were re-created in other masculine spaces, such as
bedrooms, basements, and dormitories. They persist today in the form of the corporate
"campuses" (complete with "play areas," "nap rooms," and even "tree houses") of innumerable tech firms and start-ups.[22]

The final feature of this history concerns the ways in which male programmers
mobilized multiple, and sometimes even competing, visions of masculine identity.

---

[18] Turkle, *Second Self* (cit. n. 3); Turkle, *Life on the Screen: Identity in the Age of the Internet* (New York, 1995).

[19] Jordynn Jack, *Autism and Gender: From Refrigerator Mothers to Computer Geeks* (Urbana, Ill., 2014).

[20] Michael Mahoney, "Boy's Toys and Women's Work: Feminism Engages Software," in *Feminism in Twentieth-Century Science, Technology, and Medicine*, ed. Angela Creager, Elizabeth Lunbeck, and Londa Schiebinger (Chicago, 2001), 169–85.

[21] Margaret Rossiter, *Women Scientists in America: Before Affirmative Action, 1940–1972* (Baltimore, 1995); Ruth Oldenziel, *Making Technology Masculine: Men, Women and Modern Machines in America, 1870–1945* (Amsterdam, 1999).

[22] Katherine Losse, *The Boy Kings* (New York, 2012); Jim Edwards, "We're Jealous of This Start-up's Hammock-Filled Treehouse Office," *Fortune*, 14 December 2012; Ariel Schrag, "The Ping-Pong Theory of Tech Sexism," https://medium.com/matter/the-ping-pong-theory-of-tech-world-sexism -c2053c10c06c (accessed 8 August 2014).

Computer programmers might be predominantly male, but the masculinity of the computer nerd is hardly that of the police officer or the football player—or even that of the engineer or scientist. In fact, there was no single, unified masculine ideal that male computer programmers drew upon to establish their authority or elevate their status. Some embraced the asceticism of the "compulsive programmer," while others found it repellent. Weizenbaum clearly deplored the lack of adult male "professionalism" displayed by the "bright boys" of the computer lab; for others, acting the role of the perpetually adolescent "whiz kid" was a useful professional resource. In contrast to the isolated programming nerd, the recent emergence of the frat-boy culture of "brogramming" in certain high-tech companies constitutes still another alternative form of masculinity at play in the computer industry.[23] Despite the variety of forms that it assumed, however, many computer programmers embraced masculinity as a powerful resource for establishing their professional identity and authority.

INFORMATION FACTORIES AND FEMINIZED LABOR

The first computer programmers were women. This is well-established historical fact and has been much celebrated in recent years by scholars both looking to uncover what Judy Wajcman has called the "hidden history" of women in technology and seeking to engage in contemporary debates about declining female enrollments in computer science programs.[24] These are important issues, but such treatments tend to represent the first female programmers as trailblazers carving out a role for women in a traditionally male-dominated field. As I have argued extensively elsewhere, however, the presence of women in computer programming is not just a historical anomaly or a reflection of a temporary wartime shortage of men; rather, computer programming was a feminized occupation from its origins.[25] The use of low-wage, low-skilled female programming labor was integral to the design of early electronic computation systems. For the leaders of many of the pioneering computer projects, the assumption was that the process of "coding" a computer was largely rote and mechanical—and therefore work that could be best be assigned to women. Or, to borrow a relevant metaphor from computer programming itself, the presence of women in early computing was a feature, not a bug.

The realization that women were essential, not incidental, to the invention of computer programming turns on its head the conventional interpretation of its subsequent history, at least in terms of its gender dynamics. Programming was not born male, but rather had to be made masculine. It behooves us, therefore, to provide a brief outline of the feminized origins of the discipline.

The gendered nature of early computer work can be seen clearly in the US Army Electronic Numerical Integrator and Computer (ENIAC) project, arguably the most visible and influential of the pioneering wartime electronic computer development efforts. For the male leaders of the ENIAC, all of whom were scientists, engineers, or military officers (and, more often than not, all three), the important challenges as-

[23] Marie Hicks, "De-Brogramming the History of Computing," *IEEE Ann. Hist. Comput.* 35 (2013): 86–8.
[24] Judy Wajcman, "Reflections on Gender and Technology Studies: In What State Is the Art?" *Soc. Stud. Sci.* 30 (2000): 447–64.
[25] Nathan Ensmenger, "Making Programming Masculine," in Misa, *Gender Codes* (cit. n. 17), 115–41.

sociated with the development of a working electronic computer system involved the design and construction of the actual computer. The subsequent operation of the computer was considered to be relatively trivial, and therefore work that could be successfully delegated exclusively to women. The expectation was that these women would replicate, in the electronic computer, the elaborate "plans of computation" that were already being performed by human computers in existing large-scale computational efforts. At the time, the ENIAC managers imagined the electronic computer as "nothing more than an automated form of hand computation," and therefore it seemed obvious that the same people who had directed the activities of female "computers" could also be trusted to "set up" and monitor the operations of their electronic equivalent.[26]

The sexual division of labor established at the ENIAC project provided a model for subsequent computer development projects. The very first written manual on computer programming, published in 1947 by Herman Goldstine and John von Neumann (and based on their experience with the ENIAC), carefully distinguished between the work of the "planner," who did the intellectual labor of analyzing a problem and deciding on a mathematical approach to its solution, and the "coder," who was responsible only for transcribing the thoughts of the planner and mechanically translating this solution into a form that the computer could understand.[27] The work of the "coder" was low-status, largely invisible, and therefore generally performed by women.[28]

Of course, the use of female labor to perform routine tasks was not peculiar to the ENIAC, or to electronic computing as a whole. The combination of mechanization, division of labor, and a reliance on low-skilled (or at least low-wage) workers is the essence of industrialization, and in the United States at least, women were the first factory workers.[29] This was especially true of the "information factories" that emerged in the post–Civil War period. From the multidivision firm to the modern nation-state, a growing number of information-centric organizations were made possible not only by innovations in information technology, such as typewriters, tabulating machines, mechanical calculators, and vertical filing cabinets, but also by the mass mobilization of low-wage, low-skilled female labor.[30] In fact, by the beginning of the twentieth century, women dominated the clerical occupations. The reinvention of the electronic computer as a business machine in the postwar period, driven by office technology firms such as IBM, Remington Rand, Burroughs, and NCR, assured that the gendered division of labor that existed in most business data-processing departments was simply mapped onto the new technology of electronic computing. This too would be an office technology designed by men but used by women.

In any case, the association of computer programming with routine clerical work meant that, although computer programming in the 1950s was not a job performed exclusively (or even predominantly) by women, it was nevertheless gendered female. The assumed characteristics of programming work—routine, repetitive, and highly

[26] David Alan Grier, *When Computers Were Human* (Princeton, N.J., 2005).

[27] Herman Goldstine and John von Neumann, *Planning and Coding of Problems for an Electronic Computing Instrument* (Princeton, N.J., 1947).

[28] See, e.g., Jennifer Light, "When Computers Were Women," *Tech. & Cult.* 40 (1999): 455–83; and Nellwyn Thomas, "Selling the First Computer: The Legacy of the ENIAC's Publicity" (manuscript, University of Pennsylvania).

[29] Thomas Dublin, *Transforming Women's Work* (Ithaca, N.Y., 1994).

[30] Margery Davies, *Woman's Place Is at the Typewriter: Office Work and Office Workers, 1870–1930* (Philadelphia, 1982); Sharon Hartman Strom, *Beyond the Typewriter: Gender, Class, and the Origins of Modern American Office Work, 1900–1930* (Urbana, Ill., 1992).

amenable to mechanization (or so it was believed, or at least hoped, by many computer managers at the time)—meant that it was work more likely to be assigned to women than to men. So deeply entrenched was this association that in her book *Recoding Gender*, Janet Abbate quotes one female programmer who recalls being astonished even at the suggestion that the situation could be otherwise: "It never occurred to any of us that computer programming would eventually become something that was thought of as a men's field."[31]

In practice, however, the planner/coder distinction quickly broke down, and the work of the (female) coders became entangled with the intellectual operations originally carried out by the (male) planners.[32] This proved true even at the pioneering ENIAC project, where the tasks that the programmers performed turned out to be unexpectedly difficult, requiring the development of creative new techniques, further blurring boundaries between computer design and operation, hardware and software, and men's and women's work.[33] As more and more powerful computers were developed, the significance of software became even more apparent. By the early 1960s, companies like IBM and Remington Rand UNIVAC were manufacturing relatively low-cost electronic computers that were economically competitive with earlier forms of data-processing technology. But while the computers themselves could be mass-produced, the software systems that made them useful had to be custom developed.[34] Not only were there many more organizations in need of programmers, but the types of problems that these programmers were being called upon to solve were increasingly varied and complex. Whereas the first generation of experimental electronic computers was largely used for scientific purposes, commercial machines were designed for business applications. The task of devising an algorithm capable of solving a differential equation, as challenging as that might be, paled in comparison to the complexity involved in constructing a computerized accounting system. The optimistic assumption of many computer department managers that programming was simply a matter of having a low-status "coder" implement the plan sketched out by a "planner" was revealed to be simplistically naive.

This newfound appreciation for computer programmers, combined with an increasing demand for their services, was accompanied by an equally dramatic rise in their salaries. Estimates from the mid-1960s suggested that although there were already 100,000 programmers working in the United States alone, there was an immediate demand for as many as 500,000 more.[35] One of the leading industry analysts, in a 1967 article on the "persistent personnel problem" in programming, predicted that salaries for programmers would rise 40–50 percent over the course of the next four to five years.[36] "Competition for programmers has driven salaries up so fast," warned *Fortune* in 1967, "that programming has become probably the country's highest paid technological occupation. . . . Even so, some companies can't find experienced programmers

[31] Paula Hawthorne, as quoted in Janet Abbate, *Recoding Gender* (Cambridge, Mass., 2012), 1.

[32] Jean Jennings Bartik, *Pioneer Programmer* (Kirksville, Miss., 2012); W. Barkley Fritz, "The Women of ENIAC," *IEEE Ann. Hist. Comput.* 18 (1996):13–20.

[33] Thomas Haigh, Mark Priestley, and Crispin Rope, "Los Alamos Bets on ENIAC: Nuclear Monte Carlo Simulations, 1947–1948," *IEEE Ann. Hist. Comput.* 36 (2014): 42–63.

[34] Martin Campbell-Kelly, *From Airline Reservations to Sonic the Hedgehog: A History of the Software Industry* (Cambridge, Mass., 2003).

[35] Stanley Englebardt, "Wanted: 500,000 Men to Feed Computers," *Popular Sci.,* January 1965.

[36] Richard Canning, "The Persistent Personnel Problem," *EDP Analyzer* 5 (1967): 1–14.

at any price."[37] A talented programmer not only could command a high salary but also possessed an unusual degree of autonomy and mobility.[38]

The elevation of both the status and pay scale of computer programmers attracted a growing number of men to the occupation. Some of these men drifted in from disciplines with intellectual affinities to computing, such as mathematics, philosophy, or electrical engineering. Others entered via corporate computerization efforts and had backgrounds in traditional business specialties such as accounting.[39] In either case, these recent converts to computing brought with them the traditions, practices, and status hierarchies of their former disciplines, often attempting to re-create them in their newly discovered discipline. For these aspiring male programming professionals, the lingering association of computer programming with the feminized activities of "coding," corporate data processing, and other forms of clerical work was a source of perpetual career anxiety.[40]

One strategy for dealing with this occupational insecurity was to emphasize the degree of skill required to be a successful programmer. If the problem with programming, at least from an occupational status perspective, was that it was considered to be straightforward and mechanical, then the solution was to reframe the occupation as being active, creative, and unpredictable. Given the growing scope of software projects in this period and the limitations of existing hardware, this reframing was not difficult to accomplish. Consider, for example, the work involved with writing a program for an IBM 650 computer (the first of the truly mass-produced computers, often referred to as the Model T of electronic computing). The main memory of the 650 was a rotating metal drum covered in magnetic oxides. Not only did the programmer have to analyze a complex business process and construct a program that automated its solution (using a limited and cryptic machine code instruction set), but, because magnetic drum memory was so slow relative to the 650's central processor, he also had to optimize the order and timing of critical operations to coincide with the exact moment that the desired data had rotated under the read head.[41] The difference between a program that ought, in theory, to work and one that actually functioned with an acceptable degree of performance was often a function of the singular skills and abilities of an individual programmer. No wonder they were so rare—and so valuable.

Getting a computer program to work properly under such conditions clearly required a great deal of skill, but what kind of skill was it? It wasn't exactly math, and certainly not a science, and most programmers did not consider what they did to be proper engineering.[42] More often, they described their work as a form of directed tinkering, a highly specialized form of puzzle solving that required not only skill and experience but also innate genius (fig. 1). According to John Backus, the IBM researcher most famous for developing the FORTRAN programming language, programming in the 1950s was "a black art, a private arcane matter" in which "each problem required a unique beginning at square one, and the success of a program depended primar-

[37] Gene Bylinsky, "Help Wanted," *Fortune* 75 (1967): 141–68.

[38] John Thompson, "Why Is Everyone Leaving?" *Data Management* (1969): 25–7.

[39] Thomas Haigh, "Masculinity and the Machine Man," in Misa, *Gender Codes* (cit. n. 17), 51–72.

[40] Edsger Dijkstra, "The Humble Programmer," *Comm. ACM* 15 (1972): 859–66; RAND Symposium, "Is It Overhaul or Trade-in Time: Part I," *Datamation* 5 (1959): 24–33.

[41] Mark Halpern, "Turning into Silicon: Further Episodes from Programming's Early Days," *IEEE Ann. Hist. Comput.* 14 (1992): 61–9.

[42] C. A. R. Hoare, "Programming: Sorcery or Science?" *IEEE Software* 1 (1984): 5–16.

Are **YOU** the man

to command electronic giants?

From the recent advance of electronic digital computers has emerged an exciting new job—creating instructions that enable these giant computers to perform logical operations for a variety of tasks in business, science and government.

You could be eligible for a position in computer programming. Because it is a new and dynamic field, there are no rigid qualifications. Do you enjoy algebra, geometry or other logical operations? Can you do musical composition or arrangement? Do you have an orderly mind that enjoys such games as chess, bridge or anagrams . . . finally, do you have a lively imagination?

If you do, *you* can qualify. You will receive training (at full pay) and work at IBM's Engineering Laboratories—among the most modern in the world. For more information, write to: G. W. Woodsum, Dept. 203, International Business Machines Corp., Research Laboratory, Poughkeepsie, N. Y.

DATA PROCESSING
ELECTRIC TYPEWRITERS
TIME EQUIPMENT
MILITARY PRODUCTS

INTERNATIONAL
BUSINESS MACHINES
CORPORATION

Figure 1. IBM Advertisement (New York Times, 13 May 1956, 157).

ily on the programmer's private techniques and inventions."[43] While Backus did not intend this description to be complimentary—as an aspiring computer scientist he saw this reliance on individual ability and local knowledge to be demeaning—many other programmers saw this emphasis on personal creativity and esoteric skill as the source of their professional authority. To be a devotee of a dark art, a high priest, or a sorcerer (all popular metaphors used to describe programming in this period) was to be privileged, elite, master of one's own domain.[44] It was certainly preferable to being characterized as a glorified clerical worker or a "mere" technician.

Anecdotal accounts of the unique genius of individual programmers were reinforced by an emerging empirical literature on programmer performance. In the late 1960s, the IBM Corporation commissioned a study (still widely cited today, despite its

[43] John Backus, "Programming in America in the 1950s—Some Personal Impressions," in *A History of Computing in the Twentieth Century: A Collection of Essays*, ed. Nicholas Metropolis (New York, 1980): 125–35.

[44] David Freedman, "Computer Magic," in *Proceedings of the 11th Annual Computer Personnel Research Conference* (New York, 1973), 1–9.

serious methodological shortcomings) that suggested that a truly talented programmer was at least twenty-six times more productive than his merely average colleague.[45] These exceptionally gifted "super-programmers" were "worth an army of programmers of lesser average calibre" argued one participant at a 1968 NATO conference on software engineering.[46] The conclusion drawn by many corporate managers was that "the major managerial task" they faced was finding and keeping "the right people": "With the right people, all problems vanish."[47] It would be hard to find a more compelling endorsement of the professional power conveyed by the possession of individual expertise. A skilled programmer was effectively irreplaceable.

At first glance it might seem that this focus on individual skill would provide equal opportunities for both men and women in the programming professions, and to a certain degree that is true. The literature from this period is full of anecdotal evidence about the former secretary or fashion model who turned out to be an excellent programmer (along with the male mathematician who did not).[48] Women did continue to be hired as programmers in relatively high numbers, and through the beginning of the 1970s computer programming was regarded as unusually open to female participation, at least by the dismal contemporary standards set by comparable technical professions.[49] But there were also clearly masculine associations in the language and metaphors used to describe the distinctive and temperamental character ascribed to programming professionals. Tinkering, for example, has long been gendered as a masculine approach to technology use, one in which keeping "close to the machine" was privileged over all other considerations.[50] When *Cosmopolitan* magazine published an article encouraging young women to pursue careers as "computer girls," the Computer Services Corporation, one of the largest employers of contract programmers in this period, published its own "humorous" defense of the inherent masculinity of their discipline.[51] In the advertisements from this period, women were often used as a visual proxy for low-skilled, low-wage labor. For example, if a computer manufacturer wanted to signal that its latest high-level programming language was easy to use, it would portray it being used by a female programmer—or, even more pointedly, a female secretary.[52] Such high-level languages were dismissed as "sissy stuff" by "real programmers" who preferred the "heroic" work of binary programming.[53]

---

[45] Hal Sackman, W. J. Erickson, and E. E. Grant, "Exploratory Experimental Studies Comparing Online and Offline Programming Performance," *Comm. ACM* 11 (1968): 3–11.

[46] Edward David, quoted in Peter Naur, Brian Randall, and John Buxton, eds., *Software Engineering: Proceedings of the NATO Conferences* (New York, 1976), 32.

[47] Robert Gordon, "Personnel Selection," in *Data Processing . . . Practically Speaking*, ed. Fred Gruenberger and Stanley Naftaly (Los Angeles, 1967), 79–88.

[48] Brendan Gill and Andy Logan, "Talk of the Town," *New Yorker*, 5 January 1957, 18–9; H. A. Rhee, *Office Automation in Social Perspective: The Progress and Social Implications of Electronic Data Processing* (Oxford, 1968).

[49] Lois Mandel, "The Computer Girls," *Cosmopolitan*, April 1967, 52–6; Richard Canning, "Issues in Programming Management," *EDP Analyzer* 12 (1974): 1–14; Bruce Gilchrist and Richard Weber, "Enumerating Full-Time Programmers," *Comm. ACM* 17 (1974): 592–3.

[50] Tine Kleif and Wendy Faulkner, "'I'm No Athlete but I Can Make This Thing Dance!' Men's Pleasures in Technology," *Sci. Tech. Hum. Val.* 28 (2003): 296–325; Ruth Oldenziel, "Boys and Their Toys: The Fisher Body Craftsman's Guild, 1930–1968, and the Making of a Male Technical Domain," *Tech. & Cult.* 38 (1997): 60–96.

[51] Mandel, "Computer Girls" (cit. n. 49); Computer Sciences Corporation, "In Case You Missed Our First Test . . . ," *Datamation* 13 (1967): 149.

[52] Ensmenger, "Making Programming Masculine" (cit. n. 25).

[53] Richard Hamming, *The Art of Doing Science and Engineering* (Australia, 2005).

Even the softer comparisons of computer programming to literary production—in his classic software engineering textbook, *The Mythical Man-Month*, Frederick Brooks famously compared programming to poetry—invoked traditionally masculine identities.[54] And the organizational role of "super-programmer," "hot shot," or "whiz kid" was likely more comfortable for men than for women.

By the end of the 1960s, a stereotype of the programming guru had emerged that was distinctively masculine. As the computer personnel consultant Richard Brandon described in a 1968 Association of Computing Machinery conference, the programmer type was "excessively independent," even to the point of mild paranoia. He was "often egocentric, slightly neurotic, and he borders upon a limited schizophrenia. The incidence of beards, sandals, and other symptoms of rugged individualism or nonconformity are notably greater among this demographic group." Tales about programmers and their peculiarities "are legion," Brandon argued, and "do not bear repeating here."[55]

Why such stories were legion is an open question. There were some structural reasons why programmers in this period might have been perceived as scruffy and antisocial mavericks, at least by their white-collar coworkers: for a variety of technical and economic reasons, programmers would often work odd hours and overnight shifts, meaning that on the occasions when they were visible to other employees, they were often unshaven and bedraggled (fig. 2).[56]

Possibly more significant was what Brandon described as the "Darwinian selection mechanisms" of computer industry hiring practices. By this he meant the industry reliance on psychometric testing, specifically aptitude tests and personality profiles, for the purposes of identifying trainees who possessed the "right stuff" to be skilled programmers. Such tests, which were used by more than two-thirds of all employers in this period, tended to filter for candidates who preferred to "work more with machines than with people."[57] After all, the widespread perception that computer programming was an innate and idiosyncratic ability, although conducive to the status and job security of individual programmers, provided little by way of practical guidance for an industry that suddenly found itself in need of hundreds of thousands of skilled professionals. It was one thing to recognize, as did G. T. Hunter of the IBM Corporation, the need for programmers "who were above average in training and ability" but another to specify what kind of training, and what kind of abilities.[58] Prior to the late 1960s, there were no formal academic programs in computer science, and even after such programs were established, they never provided more than a small

[54] Frederick Brooks, *The Mythical Man-Month and Other Essays on Software Engineering* (New York, 1982).

[55] Richard Brandon, "The Problem in Perspective," in *Proceedings of the 1968 23rd ACM National Conference* (New York, 1968), 332–4; Theodore Willoughby, "Are Programmers Paranoid?" in *Proceedings of the Tenth SIGCPR Conference on Computer Personnel Research* (New York, 1962), 47–54.

[56] These unusual hours often posed particular barriers to women, as many employers in this period had explicit rules against women being on the premises after hours. See Gerald Weinberg, *The Psychology of Computer Programming* (New York, 1971).

[57] W. J. McNamara. "The Selection of Computer Personnel: Past, Present, Future," in *Proceedings of the Fifth SIGCPR Conference on Computer Personnel Research* (New York, 1967), 52–6; Dallis Perry and William Cannon, "Vocational Interests of Female Computer Programmers," *J. Appl. Psychol.* 52 (1968), 31–5.

[58] G. Truman Hunter, "Manpower Requirements by Computer Manufacturers," in *Proceedings of the First Conference on Training Personnel for the Computing Machine Field*, ed. Arvid Jacobson (Detroit, 1955), 14–8.

"We're expecting vistors today so
shave, comb your hair, wash up, polish your
shoes and stay out of sight . . ."

Figure 2. Datamation *9 (1963): 42.*

fraction of the programmers required by industry. Employers struggled with the difficult task of identifying the special "twinkle in the eye" or "indefinable enthusiasm"
that separated the genuinely skilled programmer from his or her merely average colleague.[59] If the primary selection mechanism that they used to identify programming
talent associated programming ability with a "detached" personality (read antisocial,
mathematically inclined, and male), then it is no wonder that antisocial, mathematically inclined males became overrepresented in the programmer population, which
in turn reinforced the original perception that programmers ought to be antisocial,
mathematically inclined, and male.

Whatever the reasons for its origins, the association of masculine personality characteristics with innate and intuitive programming ability helped create an occupational culture in which female programmers were seen as exceptional or marginal.
Like Edwin Boring's women Experimentalists, described elsewhere in this volume,
only by behaving less "female" could they be perceived as being acceptable.[60] Many
women still did continue to be hired as programmers and other computer specialists, but they did so in an environment that was becoming increasingly normalized as
masculine, and in which the selection mechanisms privileged male candidates. Even
today, companies such as Google and Microsoft are notorious for their reliance on

[59] Datamation Report, "The Computer Personnel Research Group," *Datamation* 9 (1963): 38–9.
[60] See Alexandra Rutherford, "Maintaining Masculinity in Mid-Twentieth-Century American Psychology: Edwin Boring, Scientific Eminence, and the 'Woman Problem,'" in this volume.

confrontational interview techniques in which logic and math puzzles play a prominent role—despite the substantial evidence that such techniques are severely gender biased.[61]

IDENTITY CRISIS?

Of course, defining oneself in terms of esoteric genius or "rugged individualism" was not the only way to establish a professional identity. While many programmers continued to relish their role as technological savants, others pursued more mainstream approaches to establishing a professional monopoly of competence. These more corporate or academically oriented aspiring computer professionals, the majority of them male, worked to establish professional societies, publish academic journals, develop credentialing programs, and lobby employers and governments for recognition and legitimacy. In doing so, they mobilized a different set of masculine resources and rhetorics.[62]

As Margaret Rossiter and others have demonstrated, professionalization generally implies masculinization.[63] Consider, for example, the Data Processing Management Association (DPMA), which in the early 1960s established the Certified Data Processor (CDP) program, which was modeled after the widely recognized Certified Public Accountant (CPA). In the case of the CDP program, the masculine bias of professional standards was particularly apparent: the requirement of formal educational credentials, a minimum of three years of industry experience, and the possession of "high character qualifications" (the specifics of this requirement were vague, and rarely enforced, but appeared to involve letters of recommendation from other established "professionals") privileged not only males but males with an established commitment to a corporate managerial culture. The majority of CDP holders were middle managers, an organizational role that was often explicitly denied to women in this period, or at the very least was implicitly associated with masculine characteristics.[64] The more computer professionals were seen as not only technical experts but also potential corporate managers, the more women were excluded. The man in the gray flannel suit might have occupied the opposite extreme from the bearded, sandal-wearing, programming guru, but they were sitting on the same spectrum of masculinity.

The principal alternative to the business-oriented DPMA was the Association of Computing Machinery (ACM), which was founded in 1947 as an outgrowth of an academic conference, and which continued afterward to focus on the concerns of professional academics. As might be expected from an explicitly academically oriented professional society, the ACM was even more stringent in its educational requirements. In 1965, a period when the ratio of male to female college undergraduates was close

---

[61] William Poundstone, *Are You Smart Enough to Work at Google? Trick Questions, Zen-like Riddles, Insanely Difficult Puzzles, and Other Devious Interviewing Techniques You Need to Know to Get a Job Anywhere in the New Economy* (New York, 2012).

[62] Nathan Ensmenger, "The 'Question of Professionalism' in the Computer Fields," *IEEE Ann. Hist. Comput.* 4 (2001): 56–73.

[63] Margaret Rossiter, *Women Scientists in America: Struggles and Strategies to 1940* (Baltimore, 1982); Jeffrey Hearn, "Notes on Patriarchy, Professionalization and the Semi-Professions," *Sociology* 16 (1982): 184–202.

[64] Marie Hicks, "Meritocracy and Feminization in Conflict: Computerization in the British Government," in Misa, *Gender Codes* (cit. n. 17), 95–114.

to 2:1, it imposed a strict four-year degree requirement for its members.[65] The ACM was also notorious for its disdain for business-oriented programmers and in turn was castigated by many working programmers as "dominated by, and catering to, Ph.D. mathematicians."[66] There were even fewer female PhD mathematicians than there were women with undergraduate degrees.[67] To the extent that belonging to the ACM or possessing a computer science degree was considered an essential component of being a "professional" programmer, programming was assuming an increasingly masculine identity. A survey from the late 1970s showed that fewer than 10 percent of ACM members were women.[68]

The ACM was also responsible for setting the agenda for the emerging discipline of computer science. A comprehensive scholarly history of academic computer science has yet to be written, but for the purposes of this essay it is sufficient to note only that (*a*) the institutionalization of computer science as an academic discipline was well under way by the late 1960s and (*b*) it involved a turn toward the theoretical, the mathematical, and the abstract.[69] This latter agenda sometimes alienated computing practitioners and industry employers, who criticized the computer scientists for being "too busy teaching simon-pure courses in their struggle for academic recognition," but the pursuit of theory and abstraction were effective strategies within the academy, and the ACM quickly became dominated by those who perceived their professional identity in terms of the academic research scientist.[70] This identity was less accessible to women and other minorities, whose participation rates in both academic computer science and academically oriented professional societies were lower than their rate of participation in the computer industry more generally.[71]

It is important to note that although the academic discipline of computer science was indeed masculine, it was masculine in ways that were typical of most of academia in this period. The traditional masculinity of the academic professions had little to do with the uniquely gendered nature of computing in the corporate world. To the degree that computer scientists were decried as "eggheads" divorced from the needs and realities of the "real world," it was in terms of the traditional critique of academics as "ivory-tower" types that had little to do with the nascent masculinity of the "computer cowboy" or "whiz kid." In fact, in many respects the academic computer science persona was cultivated in direct opposition to the emerging stereotypes of the computer programmer as an intuitive genius. What the academic computer scientist wanted was to establish his discipline on a firm foundation of theoretical knowledge.[72]

[65] Claudia Goldin, Lawrence Katz, and Ilyana Kuziemko, "The Homecoming of American College Women: The Reversal of the College Gender Gap," *J. Econ. Perspect.* 20 (2006): 133–56.

[66] Editorial, "The Cost of Professionalism," *Datamation* 9 (1963): 23.

[67] Margaret Anne Marie Murray, *Women Becoming Mathematicians* (Cambridge, Mass., 2001).

[68] Thomas D'Auria, "ACM Membership Profile Report," *Comm. ACM* 20 (1977): 688–92.

[69] Michael Mahoney, "Computer Science: The Search for a Mathematical Theory," in *Science in the Twentieth Century*, ed. John Krige and Dominique Pestre (Amsterdam, 1997), 617–34.

[70] Quote from Harold Sackman, "Conference on Personnel Research," *Datamation* 14 (1968): 74–6, 81, on 76; Saul Gass, "ACM Class Structure" (letter to editor), *Comm. ACM* 2 (1959): 4; Anthony Oettinger, "On ACM's Responsibility" (president's letter to ACM membership), *Comm. ACM* 9 (1966): 545–6; "Why Are Business Users Turned Off by ACM?" (1974), CBI 23, "George Glaser Papers, 1960–1989," Box 1, Folder 3, Archives of the Charles Babbage Institute, University of Minnesota, Minneapolis.

[71] D'Auria, "ACM Membership" (cit. n. 68).

[72] Dijkstra, "Humble Programmer" (cit. n. 40).

The long-standing association of computer programming with individual aptitude, machine-specific techniques, and arcane knowledge was anathema to the computer scientist. It was, after all, an MIT professor of computer science who launched the first major attack against the burgeoning phenomenon of the "computer bum." These compulsive and unsystematic tinkerers, no matter how brilliant, represented everything that the rigorous and conscientious computer scientist was not. That the emergence of the pathologically undisciplined "computer bum" was a direct consequence of the academic institutionalization of computer science is therefore a particularly delicious irony.

COMPUTER LABS AS SOCIAL SPACES

The "computer bum" of the late 1970s superficially resembled his corporate cousin, the "computer boy." He too possessed a skill that was innate, idiosyncratic, and individual—to the point of being as much a personality type as an aptitude. He too was scruffy and unkempt, antisocial, and out of sync with the prevailing organizational norms of professional behavior. And finally, he too was "more interested in machines than in people," and in mastering technology for pleasure rather than in the pursuit of some larger purpose. But although the computer bum represented the extreme end of a spectrum that had already been defined in the corporate setting, this particular extreme could only be achieved outside the corporation. The computer bum of the late 1970s was the product of a distinctive combination of technology and place, a combination that was specific to the research university but which developed outside of, or perhaps parallel to, academic programs in computer science. Without the computer lab, the computer bum would not have existed. In these unconventional and unruly places, where the already gendered stereotypes associated with computer culture would become inextricably linked to adolescent masculinity, bright young students were allowed almost unlimited—and largely unsupervised—access to cutting-edge experimental electronic digital technologies. The norms, ethos, and practices established in the university computer centers of the 1970s formed the basis for the emergent computer hobbyist culture of the 1980s (and beyond) and would be perpetuated and re-created in similarly masculine spaces, from the bedrooms of pimply teenage computer hackers to the couches and erstwhile dormitories of innumerable Internet start-ups, to the studiously informal work spaces/playgrounds of corporate campuses at Apple, Microsoft, and Google, where free sodas and foosball tables are seen as being as essential to the production of software as product labs and computer workstations.[73]

The computer center was a social and technological space unique to the Cold War research university, although its origins predate the advent of the electronic digital computer. Beginning in the early 1930s, several major research universities had established, often in collaboration with equipment manufacturers, computational service bureaus aimed at providing computational support for scientific researchers.[74] It would be these computer centers that built (or later purchased) most of the early electronic computers, and in many cases, the first formal academic training in elec-

[73] Eric Raymond, "A Brief History of Hackerdom," in Raymond, *The Cathedral and the Bazaar* (Sebastopol, Calif., 2001), 19–64; Tim Jordan and Paul Taylor, "A Sociology of Hackers," *Sociol. Rev.* 46 (1998): 757–80.

[74] Grier, *When Computers Were Human* (cit. n. 26).

tronic computing was provided through these centers, rather than via traditional departments.[75]

Even after the establishment of independent computer science departments, a separation of computer operations from computer science research was typical of most universities. In part this represented the logic of capital: it was difficult and expensive to purchase and operate a large-scale computer facility (a situation that remains true today), and so it made sense for universities to centralize computing and distribute the costs across multiple departments.[76] But it was also true that the nascent discipline of computer science was not particularly interested in controlling its own computing resources. In fact, computer scientists worked hard to distance themselves from the "service" connotations of the computer center and, indeed, from any association with actual computers.[77] After all, one of the strongest objections made to the establishment of their discipline in the first place was that what they did was not science at all, but technology. It was in their professional interest to focus on the computer as a logical abstraction rather than an embodied technology. The last thing that research-oriented computer scientists wanted to be associated with were the "mere technicians" who tended the machinery, which explains both the continued existence of the autonomous computer center and the great antipathy academic researchers had for the activities of the "computer bums" with whom these centers were increasingly identified.[78]

In its physical configuration, the academic computer center closely resembled its nearest cousin, the corporate data-processing department: the size, expense, and power requirements of computers in this period demanded the construction of dedicated computer rooms with raised floors, reinforced cooling systems, and securely locked doors. But whereas in the corporate context the enforced segregation of computer equipment and personnel served to reinforce the elite and privileged status of the computer experts—the literature from this period is replete with references to "high priests" of computing carefully controlling access to the "air-conditioned holy of holies" of the computer room—the marginal location of the computer center encouraged experimentation and exploration.[79] In this sheltered but unsupervised environment, the links between electronic computing and the culture and practices of adolescent masculinity would be firmly established. During the day, the university computer centers were run by staff technicians in the service of faculty research projects. At night, however, the computer centers were turned over to the use of undergraduates, either explicitly or with the implied consent of the faculty and administration. It was the after-hours activities of unofficial computer enthusiasts that would establish the distinctive computer "hacker" identity.[80]

[75] William Aspray, "Was Early Entry a Competitive Advantage? US Universities That Entered Computing in the 1940s," *IEEE Ann. Hist. Comput.* 22 (2000): 42–87.

[76] William Aspray and B. O. Williams, "Arming American Scientists: NSF and the Provision of Scientific Computing Facilities for Universities, 1950–1973." *IEEE Ann. Hist. Comput.* 16 (1994): 60–74.

[77] Atsushi Akera, *Calculating a Natural World: Scientists, Engineers, and Computers during the Rise of U.S. Cold War Research* (Cambridge, Mass., 2007).

[78] Michael Mahoney, "What Makes Computer Science a Science?" in *Science in the Context of Application*, ed. Martin Carrier and Alfred Nordmann (Dordrecht, 2010), 389–408.

[79] L. R. Fiock, "Seven Deadly Dangers in EDP," *Harvard Bus. Rev.* 40 (1962): 88–96; Anthony Chandor, *Choosing and Keeping Computer Staff* (London, 1976); Backus, "Programming in America" (cit. n. 43).

[80] Levy, *Hackers* (cit. n. 10).

The association between the social architecture of the computer center and the expression of the computer bum personality was first made public by the psychologist Lucy Zabarenko and her colleague Ellen Williams at the 1971 ACM Conference on Personnel Research. In doing their empirical research on programmer education, Zabarenko and Williams had noticed a "special cultural phenomenon" peculiar to the university computer center—a culture so unusual that they thought it worthy of further study by anthropologists.[81] There was something "especially compelling" about the nature of computer programming, they argued, that absorbed its practitioners to such a degree that they lost their sense of time and place. In their quest to "get [time] on" the machine, the inhabitants of the computer center stayed up late at night, slept all day, and lost all interest in their other academic work. Their obsession would cause them to neglect their bodies, to the point that "many of these men appeared poorly nourished and all were thin," subsisting as they did "mainly on coffee and carbohydrates." These practices, originating from necessity, soon became part of the "invariant custom" of the "computer bum," who increasingly associated only with others of his kind, making it a point "to be informally dressed, elaborately unaware of time, and constantly underfed." For Zabarenko and Williams, the computer bum was an unsavory character, one who threatened, rather than encouraged, the advancement of computer technology. "Can we teach young children computer skills," they worried, "without also transmitting the beliefs and values of the computer center?" They believed the pervasive presence of the disheveled computer bums in the computer center would deter more "normal" programmers.

We have already seen how Stewart Brand, just a year after the publication of Zabarenko and Williams's report, provided a radically different assessment of the relative virtues and vices of the computer bum culture. But Zabarenko, Williams, and Brand (and, just a few years later, Joseph Weizenbaum) were in surprising agreement about the nature and causes of the phenomenon. What made the computer bum possible was not simply the availability of computer technology, but the combination of technology, culture, and environment. This was a combination peculiar to the university computer center. It did not exist within the corporate data-processing department, despite their apparent similarities.

Three features of the academic computer center significantly contributed to the formation of its unique culture. To begin with, the computer center was an isolated, and therefore largely unsupervised, environment where students had an unprecedented degree of access to the equipment. In the corporate setting, even the most ardent computer enthusiasts were limited in their ability to engage directly with the machine. Rarely if ever was this access individual or unmediated. After hours in the university computer center, it was possible to exercise what came to be known as the "hands-on imperative" (a practice that would later be elevated to the status of central tenet of the "hacker ethic" in a popular and sensational account of the history of the computer center at MIT, revealingly entitled *Hackers: Heroes of the Computer Revolution*).[82] Even if direct access to the machine was officially forbidden, motivated and creative student programmers could usually find a way. At MIT, for example, the long-standing tradition of "lock hacking" proved a useful resource to a new generation of aspiring

[81] Lucy Zabarenko and Ellen Williams, "The Computer Center as a Subculture," *Council on Anthropology and Education Newsletter* 2 (1971): 5–8.
[82] Levy, *Hackers* (cit. n. 10).

"computer hackers."[83] In an era of mainframe computers that occupied an entire room, this was as close as you could get to the experience of a "personal" computer. It is no wonder that computer centers tended to attract the type of individual who found one-on-one interactions with a computer particularly compelling.

Second, the students who frequented the computer center were sheltered from the economic realities—and consequences—of their actions. In the corporate world, computer time was expensive and therefore carefully rationed and monitored. In addition, corporate programmers were being paid for their work and as such were accountable to managers, budgets, and schedules. Student programmers, on the other hand, were largely free to pursue their own interests, agendas, and aesthetics.[84] This last was especially significant: while industry employers had long complained that graduates of computer science programs had only learned to write "trick programs" rather than real applications, the codes that the computer bums obsessed over did not generally serve a pedagogical purpose and were rarely associated with their academic studies; in fact, the very best of the bums were notorious for not completing their course work, even when it related directly to their computer science curriculum.[85] Quite a number failed out of university—but nevertheless continued to frequent the university computer labs. In an era in which many academic computer centers were saturated with grant money (largely from the Department of Defense), a skilled computer bum could pick up enough work to support his habit almost indefinitely. And, in stark contrast to the present era, the work that went on in the computer center was not intended to kick-start a commercial project. The goal of becoming the next Steve Jobs or Mark Zuckerberg would not become the dominant obsession of the aspiring computer nerd until a later generation.

When the bums in the computer centers did write code, it was often to solve trivial puzzles or to tinker with programs that had already been written. The goal was not so much to accomplish an objective but to produce code that was beautiful, elegant, humorous, or otherwise aesthetically appealing. For example, one popular challenge was to attempt to solve a given problem in as few instructions as possible. Programmers would spend hours, even days, eliminating (or "bumming," as it was called) a single line of code. Whether the resulting program ran quickly or efficiently, or even solved some useful or interesting problem, was irrelevant. The goal was simply to please oneself (as Stanford Professor John McCarthy described it, his students "got the same kind of primal thrill from 'maximizing code' as fanatic skiers got from swooshing frantically down a hill") or, more frequently, to impress one's peers.[86] A truly elegant program listing would be "bummed to the fewest lines so artfully that the author's peers would look at it and almost melt with awe."[87] These listings would be passed around the computer center to be shared, admired, envied. Trimming code served as a form of masculine competition, a means of both demonstrating mastery over the machine and establishing dominance within the community hierarchy.[88]

---

[83] Ibid.

[84] Roy Rosenzweig, "Wizards, Bureaucrats, Warriors, and Hackers: Writing the History of the Internet," *Amer. Hist. Rev.* 103 (1998): 1530–52; Sam Williams, *Free as in Freedom: Richard Stallman's Crusade for Free Software* (Sebastopol, Calif., 2002).

[85] Richard Hamming, "One Man's View of Computer Science," *J. ACM* 16 (1969): 3–12.

[86] McCarthy, quoted in Levy, *Hackers* (cit. n. 10), 13.

[87] Levy, *Hackers* (cit. n. 10), 32.

[88] Ibid.

This brings us to the last of the three features of the university computer center that made it so distinct and significant, and which was noted by all of its observers, whether with admiration or disdain: despite the stereotype of the computer person as individualistic and "disinterested in people," the computer center was a profoundly social space. To be sure, the computer bums came to the computer center to indulge their fascination with the machine, and it was in part the machine that kept them glued to their screens and keyboards. But they were more than simply working alone, together. In practice, computer centers were abuzz with conversation and other forms of social interaction.

In fact, in his 1971 analysis of *The Psychology of Computer Programmers*, Gerald Weinberg argued that the sociability of the computer lab was the key to effective learning and innovation in computer programming. In his study of the sociology of computer labs, he found that even small perturbations in the social and spatial networks of the center (e.g., the relocation of the soda machine) proved disruptive to learning and productivity. Programmers learned through conversation, by watching one another code, and by telling one another stories over Chinese food at three in the morning. Even practical jokes and pranks could serve a purpose: Stewart Brand, for example, relates the story of an MIT hacker who wrote a program called "The Unknown Glitch," "which at random intervals would wake up, print out I AM THE UNKNOWN GLITCH. CATCH ME IF YOU CAN, and then it would relocate itself somewhere else in core memory, set a clock interrupt, and go back to sleep."[89] Searching for the glitch was at once a form of collective entertainment, a lesson in computer architecture, and a rite of passage. Although in the sheltered womb of the computer room the computer bums might be isolated from the outside world, they were in intense interaction with one another.

The incorporation of video display units into computer terminal technology, which began in the 1960s, created new opportunities for socialization within the computer center. Hackers could now demonstrate their programs to others more readily and tinker with the computer's graphical capabilities. Among other things, they could develop competitive games such as Pong and Spacewar and then play against one another.[90] The virtual violence of the computer video game, at this point available only within the confines of the university computer center, provided the link between the abstract and disembodied activities of the computer hacker and more traditional, physical forms of masculine competition.[91] Finally, these graphical displays could be used to display pornography. The earliest documented computer "girlie pics" date from the mid-1950s, but no doubt these were the first of many. In fact, one widely distributed digital scan of a 1960s-era Playboy pinup, the so-called Lena image, became a reference image for researchers in computer graphics and has been reproduced and/or cited in hundreds of academic papers.[92] Looking at "girlies" on computer screens (as opposed to, e.g., pursuing them in real life) might be a pathetically compensatory

[89] Brand, "SPACEWAR" (cit. n. 5).

[90] Henry Lowood, "Videogames in Computer Space: The Complex History of Pong," *IEEE Ann. Hist. Comput.* 31 (2009): 5–19.

[91] Gitte Jantzen and Jans F. Jensen, "Powerplay—Power, Violence and Gender in Video Games," *AI & Soc.* 7 (1993): 368–85; J. Jansz, "The Emotional Appeal of Violent Video Games for Adolescent Males," *Comm. Theory* 15 (2005): 219–41.

[92] Jaime Hutchinson, "Culture, Communication, and an Information Age Madonna," *IEEE Professional Comm. Soc. News.* 45 (2001): 1–7.

and adolescent masculinity, but it was masculinity nonetheless.[93] And at the very least, sharing such images with your friends in the computer lab created yet another opportunity for male sociability.

Compared to other places where young men would go to prove themselves—say, for example, the lofty peaks pursued by the Victorian mountaineers—the computer center might seem a vastly inferior alternative. And yet, as Michael S. Reidy describes in this volume, even in environments that were inherently dangerous, risk was socially constructed (by limiting one's food supply, avoiding the beaten paths, pushing the body's physiological limits).[94] In a similar manner, by engaging in marathon coding sessions, surviving for days on Cokes and junk food, and otherwise denying themselves, computer programmers could also engage in manly demonstration. In *Second Self*, Turkle describes a practice known as "sport death," in which computer programmers challenged one another to push the limits of sleep deprivation. As one of her MIT hackers described it, the "essence of sport death is to see how far you can push things, to see how much you can get away with. . . . I generally wait until I have to put in my maximum effort and then just totally burn-out."[95] And in fact, although Turkle does not seem to be aware of this, the concept of sport death was imported into the MIT computer center by a geology student, who had picked it up from rock climbers and parachutists at Yosemite.[96] The physical risks of computer programming might have been artificial and contrived, but they were nevertheless a form of masculine competition and display. In a rare moment of self-reflection about the gendered nature of such practices, the same hacker who described the phenomenon to Turkle noted, seemingly as an afterthought, that "women are not so into sport death."[97]

To the degree that the computer center was a social environment, however, it was almost exclusively a homosocial environment. Again, this is a stark contrast to the corporate computing experience. Although the stereotype of the bearded, besandaled computer programmer was well in place by the late 1960s, in actual practice women were still very much present in most corporate computer departments. There were certain rare circumstances in which women were explicitly excluded from the sanctum sanctorum of the corporate computer center (generally after hours, and then ostensibly to protect their personal safety), but in most corporations women represented at least 25–30 percent of all computer personnel.[98] If we include computer operators and keypunch operators (by then the most feminized of computer specialties), then the representation of women would be even higher. Not so in the university computer centers, particularly during the overnight hours—which is when most of the interesting action occurred. At these moments, the computer center was effectively males only.[99]

In part this was simply a reflection of the demographics of the student population—at

[93] Ben Edwards, "The Never-Before-Told Story of the World's First Computer Art (It's a Sexy Dame)," *The Atlantic*, 24 January 2013, http://www.theatlantic.com/technology/archive/2013/01/the-never-before-told-story-of-the-worlds-first-computer-art-its-a-sexy-dame/267439/ (accessed 5 August 2014).

[94] Michael S. Reidy, "Mountaineering, Masculinity, and the Male Body in Mid-Victorian England," in this volume.

[95] Turkle, *Second Self* (cit. n. 3), 194.

[96] Pepper White, *The Idea Factory: Learning to Think at MIT* (New York, 1991), 299.

[97] Sherry Turkle, "Computational Reticence: Why Women Fear the Intimate Machine," in *Technology and Women's Voices: Keeping in Touch*, ed. Cheris Kramarae (New York, 1988), 41–60, on 45.

[98] Sherry Turkle, "Advanced Programmers, Women Employment Seen Rising," *Datamation* 10 (1964): 69; Richard Canning, "Issues in Programming Management," *EDP Analyzer* 12 (1974): 1–14.

[99] Levy, *Hackers* (cit. n. 10), 75.

some of the earliest universities to develop computer centers, such as Princeton and Columbia, women were not even able to enroll until the late 1960s (or even later). But even as female enrollments in formal academic computer science programs increased, their participation in the informal computer center culture did not. The male camaraderie defined by inside jokes, competitive pranks, video game marathons, and all-night code fests simply was unfriendly to a more mixed-gender social environment, a fact noted by many women who cited the male-dominated culture of the computer center as an obstacle to their continued participation in computing.[100] As Douglas Thomas has suggested, university computer centers, and the hacker culture that emerged out of them, are examples of what Anthony Rotundo, in his history of American masculinity, has called "boy culture."[101] In such cultures, both affection and mastery is expressed through "friendly play," "rough hostility," and "affection through mayhem."[102] In the absence of opportunities for physical conflict, hackers turned to pranks, trash talk, and other forms of emotional aggression as a means of establishing masculine identity.

The Duke of Wellington famously ascribed his victory at Waterloo to the manly virtues acquired by his officers on the playing fields of Eton. Similarly, we might argue that the start-up culture of Silicon Valley was conceived in the computer labs of Stanford and MIT. These academic computing centers served as key sites of play and learning, central nodes in the informal networks of knowledge exchange that defined computing practice in this period, and obligatory passage points for the emerging hacker community. The historical exclusion of women from these environments, and the continuing gender specificity of their more modern equivalents, is therefore of profound and lasting significance. Reforming the culture means re-creating the spaces and places in which that culture is reified and transmitted.

TRIUMPH OF THE NERDS?

By the end of the 1970s, when Joseph Weizenbaum first published his scathing critique of the computer bum, the unique combination of technology, culture, place, and practices that had created this phenomenon was already coming to an end. The expensive mainframe computers that had justified the existence of the computer center were being replaced by smaller personal computers. As the computer centers were reconstituted and reconfigured (socially, technologically, and institutionally) as classroom-oriented "computer labs," they lost some of their sense of mystery, seclusion, and sacredness. But many of the norms and practices that had been established in the computer centers had become so thoroughly integrated with hacker culture that they endured long after their original reasons for being disappeared. Life in the new computer labs continued to be nocturnal, despite the fact that there was no longer any real competition for computer time during daylight hours. The all-night coding sessions continued, reinterpreted as a rite of passage and a cultural marker rather than a structural necessity. And more often than not, these sessions were happening not in a university computer lab, but in the homes and bedrooms of the latest incarnation of Weizenbaum's compulsive programmer, the so-called computer hacker.

[100] Karen Frenkel, "Women and Computing," *Comm. ACM* 33 (1990): 34–46.

[101] Douglas Thomas, *Hacker Culture* (Minneapolis, 2002).

[102] Anthony Rotundo, *American Manhood: Transformations in Masculinity from the Revolution to the Modern Era* (New York, 1994). Quoted in Thomas, *Hacker Culture* (cit. n. 101), 19.

While the computer hacker bears some resemblance to the computer bum (indeed, Stewart Brand had deployed the two terms almost synonymously), the two figures are not identical. The computer bum was intimately associated with the university computer center (the only place where there were computing resources to bum); the computer hacker increasingly had access to his own machine, often in the privacy of his bedroom. Absent from hacker culture were the mediating influences of employers, faculty advisors, or professionally minded colleagues. Whereas the computer bum might have been a pathetic, wasted figure, the computer hacker was tinged by an element of danger. Prior to 1983, the word "hacker" appeared only infrequently in the literature; within a year, hacker was a household term. And in almost every case, the concept of the hacker was associated, if only indirectly, with the emerging problem of computer crime.[103] The most notable instance of this was the 1983 film *WarGames*, in which a young computer genius in pursuit of a video game brings the world to the brink of nuclear annihilation. But real-world examples of computer hacking, most notably the exploits of the "414 gang," who infiltrated computers at Los Alamos National Laboratory, the Sloan Kettering Cancer Center, and the Security Pacific Bank, suggested that "computer security" (as the problem would eventually come to be known) was a serious and growing threat to technological, economic, and even national stability. Hackers, unlike bums, were potentially malicious. Suddenly the notion of an obsessed, socially maladjusted young man armed with a powerful computing device didn't seem quite so harmless.

It is beyond the scope of this article to do more than sketch a brief outline of the computer hacker. Like the earlier stereotypes of the singularly gifted computer genius, the hacker is young, white, male, and focused on the computer to the exclusion of other interests. Youth (and maleness) had always been a defining feature of the popular conception of the computer expert (although the "boys" in "computer boys" was more often an expression of derision than a demographic reality), but the computer hacker, as constructed by sensationalist media of the 1980s, was almost by definition an adolescent male. It was a rare article on the growing incidence of hacker-driven computer crime that did not mention, generally in the first sentence, the age of the alleged perpetrator, and the younger the better.[104] Of the two central protagonists of journalist Steve Levy's *Hackers* (first published in 1984), one is a physically underdeveloped freshman at MIT, and the other is a twelve-year-old boy who happens to wander into the MIT computer lab. And while the statistics might have shown that a substantial number of computer programmers were neither young nor male, movies like *WarGames* provided a visual guide to what a "real" computer hacker looked like.[105] Even the dominant psychological explanation of the hacker mindset, which was Freudian, explicitly excluded women![106]

The adolescent male hacker introduced yet another layer of masculine identities

[103] Nathan Ensmenger, "From Whiz Kids to Cybercriminals: Emerging Narratives of Risk in Computer Security," *IEEE Ann. Hist. Comput.* (forthcoming).

[104] See, e.g., Paul Ciotti, "The Hacker's World," *Los Angeles Times*, 14 November 1988, which opens with the subheading "Young and gifted, they aren't afraid of getting emotionally involved with their computers," or Michael Schrage, "Teen-Computer Break-Ins: High-Tech Rite of Passage," *Washington Post*, 21 August 1983, among many others.

[105] As the movie poster for *WarGames* (cit. n. 10) makes clear, it is a young man who is the master of the machine. The only female character of any significance in the film stands behind him, looking over his shoulder in awe and admiration.

[106] Turkle, "Computational Reticence" (cit. n. 97).

and practices to the increasingly male-dominated computing subculture, this time borrowed from ham radio and hobby electronics. These were already activities dominated by men. In fact, as Susan Douglas and Kristin Haring have convincingly demonstrated, many of the characteristics and practices that are commonly assumed to have originated in 1980s hacker culture were actually well defined a half-century earlier by ham radio operators, for similar reasons and via similar processes.[107] Douglas in particular ties these practices to the late nineteenth- and twentieth-century "crisis of masculinity," in which young men, struggling to define themselves in a white-collar information economy in which physical strength and courage were largely irrelevant, turned to the mastery of technology as a means of demonstrating their fitness and potential. By engaging in ritualized forms of competition—in the case of early amateur radio operators this meant playing pranks on commercial and military operators, for personal computer enthusiasts "hacking" into computer systems—these young men could participate in controlled (and often socially approved, or at least condoned) acts of juvenile rebellion. The same skills and abilities that lent an edge of danger to the computer hacker or the phone "phreaker" were also those that could land him a high-paying job—or, by the late 1980s, turn him into a personal computer industry millionaire. Many parents were willing to risk the vaguely defined legal consequences associated with adolescent hacking in exchange for the opportunity for their sons to become the next Bill Gates or Steve Jobs.

What is most significant about hacker culture is that it was hegemonic. Although most computer programmers, then and now, did not consider themselves to be hackers, it became increasingly difficult for them to distance themselves from the connotations associated with popular representations of hacker culture. This appears to have been a particular problem for female programmers. Figure 3 shows a time series representing the percentage of women enrolled in undergraduate computer programs. As you will note, the graph demonstrates the notorious "bump" that occurs in the early 1980s: prior to 1983–4, female enrollments in computer science had been gradually increasing; in the years following, enrollments have, on the whole, continuously declined. There are many explanations for this decline, including a failure of the STEM education pipeline, a lack of female role models, and institutionalized sexism in higher education.[108] When we overlay the figures on female computer science enrollments with a representation of the rising number of media mentions of the word "hacker" (as measured by Google Ngram) we can visualize an alternative explanation.[109] The two graphs form an almost perfect inverse of one another. As the hacker stereotype came to dominate the popular image of what computer programmers do and who they are, they marked computing as an almost exclusively male domain. As Tove Hapnes and Knut Sorenson have argued in their study of Norwegian hacker culture, for many women in computing, the concept of the computer hacker became a metaphor "for all the things they did not like about computer science: the style of work, the infatuation with computers leading to neglect of normal non-study relations, and the concentration on

---

[107] Susan Douglas, *Inventing American Broadcasting, 1899–1922* (Baltimore, 1987); Kristen Haring, *Ham Radio's Technical Culture* (Cambridge, Mass., 2006).

[108] Caroline Clark Hayes, "Computer Science: The Incredible Shrinking Woman," in Misa, *Gender Codes* (cit. n. 17), 265–74.

[109] Google Ngram Viewer, https://books.google.com/ngrams/graph?content=hacker&year_start =1966&year_end=2000&corpus=15&smoothing=5&share=&direct_url=t1%3B%2Chacker%3B %2Cc0 (accessed 4 August 2014)

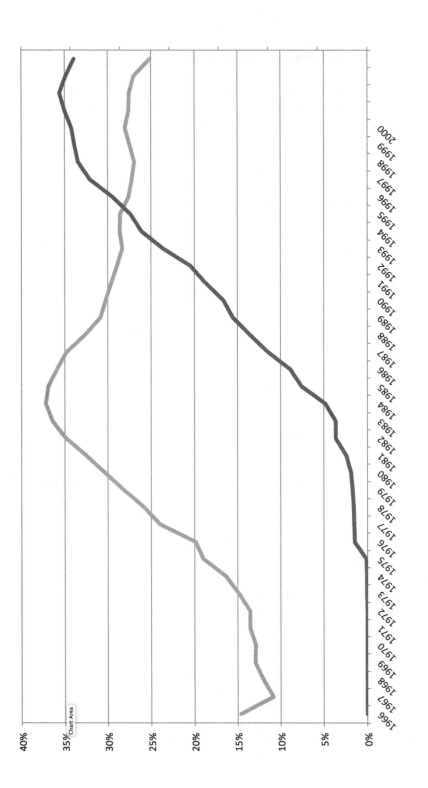

Figure 3. Female enrollments in undergraduate computer science climbed steadily until 1982, when they suddenly started declining. For years the explanation for this has been something of a mystery. The dramatic rise in media representations of the computer hacker might be the explanation. The data on female enrollments are from the National Science Foundation. The references to the word "hacker" come from the Google Ngram viewer (National Science Foundation, Science and Engineering Degrees: 1966–2006 [Arlington, Va., 2008]; Google Ngram viewer, https://books.google.com/ngrams/ [accessed 4 August 2014]).

problems with no obvious relation to the outside world."[110] The same origin myths and "triumph of the nerd" fairy tales that male hackers find comforting and empowering are, for many of their female counterparts, profoundly limiting narratives.[111]

It is important to note that many male programmers are also uncomfortable with the hacker stereotype. Not only is it not an identity available to every man (as Ron Eglash has described, it is particularly problematic for African-Americans), but it also precludes other forms of masculine and professional persona.[112] For those who aspired to be computer scientists or software engineers, the character and habits of the hacker were an embarrassment. But such is the nature of hegemony; by the middle of the 1980s, even for those programmers who aspired to more conventional professional identities, the emerging stereotype of the undisciplined computer bum provided a necessary foil to position oneself against. To be a professional computer scientist or software engineer was to not be a hacker, maverick, or bum. The existence of such amateurs was nevertheless assumed, or at least asserted, in the rhetorical construction of one's chosen disciplinary agenda.

CONCLUSIONS

In Turkle's now-classic analysis of computer culture, she provided a psychoanalytical interpretation of the compulsive computer user. The obsessive computer user, according to Turkle, was a kind of paradox, a "loser" who saw himself as elite. Overwhelmed by the complexity of real-world social interactions, he retreated into the controlled, predictable microworld of the electronic computer. In the chapter that deals most directly with hacker culture and mentality, Turkle focuses specifically on the failure of the hacker to come to terms with his own masculinity. The chapter opens with a description of the annual "ugliest man on campus" competition, in which male MIT students flaunted "their pimples, their pasty complexions, their knobby knees, their thin, underdeveloped bodies." Turkle focuses on the contrast between the self-perception of the MIT students ("Everyone knows that engineers are ugly. . . . [That] to be at MIT is to be a tool, a nerd, a person without a body") and what they imagine to be a more ideal form of masculinity ("To be at Harvard is to be a gentleman, to be sexy, to be desired"). In Turkle's influential interpretation, the computer hacker is defined in large part by his lack of masculine identity. Hackers are good with machines for the same reasons that they are unsuccessful with people (and, in particular, women). However, in the context of their interaction with "the intimate machine," their tendency to be antisocial, antisensual, and overly focused on control may be transformed from liabilities into assets.[113]

To argue that computer hackers have constructed for themselves a "world without women" is not to suggest that they are not deeply invested in their own masculine

[110] Tove Hapnes and Knut H. Sorensen, "Competition and Collaboration in Male Shaping of Computing: A Study of a Norwegian Hacker Culture," in *The Gender-Technology Relation: Contemporary Theory and Research*, ed. Keith Grint and Rosalind Gill (London, 1995), 174–91.

[111] Robert Cringely, *Accidental Empires: How the Boys of Silicon Valley Make Their Millions, Battle Foreign Competition, and Still Can't Get a Date* (Reading, Mass., 1992); Alison Adam, "Hacking into Hacking: Gender and the Hacker Phenomenon," *ACM SIGCAS Computers and Society* 33 (2003): 3.

[112] Ron Eglash, "Race, Sex, and Nerds: From Black Geeks to Asian American Hipsters," *Social Text* 2 (2002): 49–64.

[113] Turkle, "Computational Reticence" (cit. n. 97).

identity, however.[114] As I have attempted to demonstrate, the practices of "bumming," pranking, and other forms of technical display that originated in the university computer labs of the 1970s form the basis for a rich culture of masculinity within computing communities. Some of the most conspicuous features of this masculinity, and in particular the association of computer programming with the "computer nerd" personality type, are not so much a reflection of the essentially gendered nature of the activity (or, as Turkle suggests, the uniquely "intimate" nature of the technology) but are instead the by-product of attempts by early programmers to elevate the status of their discipline. In a wide variety of periods and contexts, from the corporation to the academy to the computer center, male programmers have mobilized masculinity as a means of pursuing professional status and autonomy. Many male programmers saw the role of the eccentric and exceptional computer genius as a desirable alternative to that of a lowly, routinized, and feminized "coder." Although there were some downsides to being categorized as a "whiz kid" or a "computer boy," most particularly the stigma of being narrowly focused, antisocial, and corporate unfriendly, this identity nevertheless provided programmers with many of the perceived benefits of professionalization: the establishment of barriers to entry to the discipline, the possession of a "monopoly of competence," and mastery over an esoteric body of knowledge.[115] In fact, one might argue, contra Turkle, that computer programmers, rather than being insufficiently masculine, have elevated the performance of masculinity to an extreme.

[114] David Noble, *A World without Women: The Christian Clerical Culture of Western Science* (New York, 1992). See also Erika Lorraine Milam, "Men in Groups: Anthropology and Aggression, 1965–75" (in this volume), on the self-conscious performance of ritualized competition among academic and philanthropic "men in groups" during the 1960s and 1970s.

[115] Magali Sarfatti Larson, *The Rise of Professionalism: A Sociological Analysis* (Berkeley and Los Angeles, 1977).

Men in Groups:
Anthropology and Aggression, 1965–84

by Erika Lorraine Milam*

ABSTRACT

By the late 1950s, Harry Frank Guggenheim was concerned with understanding why some charismatic leaders fought for freedom, while others sought power and domination. He believed that best-selling books on ethological approaches to animal and human behavior, especially those by playwright and screenwriter Robert Ardrey, promised a key to this dilemma, and he created a foundation that would fund research addressing problems of violence, aggression, and dominance. Under the directorship of Rutgers University professors Robin Fox and Lionel Tiger, the Harry Frank Guggenheim Foundation fostered scientific investigations into the biological basis of human nature. This essay analyzes their discussions of aggression as fundamental to the behavior of men in groups in order to elucidate the private and professional dimensions of masculine networks of US philanthropic and academic authority in the late 1960s and 1970s.

Multimillionaire Harry Frank Guggenheim read Robert Ardrey's *Territorial Imperative* in November 1967 and found it "of fascinating interest."[1] Guggenheim had long been concerned with improving "man's relation to man" but had trouble deciding how best to proceed. In 1959, together with a small cadre of close friends, he began a conversation with University of Michigan professor of psychology Paul Fitts. A few years later, Fitts assembled a group of scientists to tackle the problem at a 1964 symposium entitled "Strategies of Dominance and Social Power," which was held at Henry Ford's former home, Fair Lane, by then part of the university's Dearborn campus. For both Guggenheim and Fitts, one of the most promising lines of inquiry lay in analyzing

---

* Department of History, 136 Dickinson Hall, Princeton University, Princeton, NJ 08544-1017; emilam@princeton.edu.

I owe a huge debt to Robert A. Nye, my stalwart coeditor, for his advice and friendship and for introducing me to fabulous restaurants in cities across the country. The feedback from participants at the *Masculinities in Science/Sciences of Masculinity* workshop vastly improved an early draft, as did later readings by Michael Gordin, Sarah Milov, Karen Rader, Alex Rutherford, participants in Princeton's History of Science Program Seminar, and two anonymous reviewers. Charles Greifenstein at the American Philosophical Society and Karen Colvard at the Harry Frank Guggenheim Foundation granted me access and guidance through their collections. Thank you to the APS Library for permission to quote from their archived collections. I am additionally grateful to Robin Fox and Lionel Tiger for sharing their memories of these events and for their clemency should our interpretations differ.

[1] Robert Ardrey, *The Territorial Imperative: A Personal Inquiry into the Animal Origins of Property and Nations* (New York, 1966); Harry Frank Guggenheim (HFG) to Henry Allen Moe, 10 November 1967, Folder: Guggenheim, Harry Frank #5, Mss.B.M722, Henry Allen Moe Papers, American Philosophical Society, Philadelphia, Pa. (hereafter cited as "Moe Papers").

the origins, development, and mechanisms of dominance in order to discover ways of controlling its expression. Fitts hoped Guggenheim would fund a research center at Michigan, but Guggenheim remained skeptical of the institutional stagnation he felt would inevitably characterize any university-based center.[2] In the end, it didn't matter. Fitts died less than a year after the symposium at Fair Lane, mere months after Guggenheim formalized arrangements for his philanthropic foundation "to promote the development of knowledge concerning, and the application of such knowledge to the improvement of, man's relation to man for scientific and charitable purposes."[3]

The Harry Frank Guggenheim Foundation (HFGF) came to play a significant role in fostering research on the biological basis of human nature in the service of understanding the domination of some men by others. Guggenheim believed that in order for his new venture to succeed, it was "imperative" to "enlist the interest not only of top flight men in the field, but the right men" and to find "a first class person to head up the project."[4] In the final years of his life, Guggenheim struck up an unlikely friendship with Ardrey, a playwright and screenwriter who in the early 1960s turned his attention to nonfiction with the wildly successful *African Genesis*, describing the evolutionary origins of humanity.[5] Together with Konrad Lorenz and Desmond Morris, Ardrey was often characterized in the popular press as advancing a vision of man as nothing but an animal. In the United States, Lorenz may have been best known for his popular science writings—*King Solomon's Ring* and *On Aggression*—but his authority as an expert on animal behavior was girded by his position as director of the Max Planck Institute for Behavioral Physiology in Seewiesen, Germany.[6] After Morris earned his DPhil at Oxford under Nikolaas Tinbergen, Lorenz's close scientific collaborator and friend, he moved to London to head the Granada TV and Film Unit at the Zoological Society of London (producing the popular television show *Zootime*) before accepting an appointment as curator of mammals at the Zoological Society, where he penned *The Naked Ape*, an international best seller.[7] Unlike Lorenz and Morris, Ardrey had little training in the sciences, but he maintained a devoted readership nevertheless.[8] Ardrey personally recommended social anthropologists Robin Fox and Lionel Tiger as candidates for directing the fellowship program at the HFGF. After Guggenheim's death and some ensuing debate among members of the board, the foundation hired them as co–research directors of the fellowship program in 1972.

As midcareer scientists fascinated by questions of aggression and human behavior, Fox and Tiger constituted a logical choice. They had published the first in a series

[2] HFG to "All Hands—Man's Relation to Man Project," 16 December 1968, Folder: Guggenheim, Harry Frank, Foundation #2, Moe Papers.

[3] "30 March 1965, State of New York, Department of State, James E. Allen, Jr. Commissioner of Education of the State of New York," Folder: Guggenheim, Harry Frank, Foundation #1, Moe Papers.

[4] HFG to Moe, 8 January 1964, Folder: Guggenheim, Harry Frank #2, Moe Papers.

[5] Robert Ardrey, *African Genesis: A Personal Investigation into the Animal Origins and Nature of Man* (New York, 1961).

[6] Konrad Lorenz, *On Aggression*, trans. Marjorie Kerr Wilson (New York, 1966). On Lorenz's public persona, see Tania Munz, "'My Goose Child Martina': The Multiple Uses of Geese in the Writings of Konrad Lorenz," *Hist. Stud. Nat. Sci.* 41 (2011): 405–46.

[7] Desmond Morris, *The Naked Ape: A Zoologist's Study of the Human Animal* (New York, 1967). On the intellectual ecologies of this community, see Richard W. Burkhardt Jr., *Patterns of Behavior: Konrad Lorenz, Niko Tinbergen, and the Founding of Ethology* (Chicago, 2005).

[8] Nadine Weidman, "Popularizing the Ancestry of Man: Robert Ardrey and the Killer Instinct," *Isis* 102 (2011): 269–99; Erika Lorraine Milam, "Making Males Aggressive and Females Coy: Gender across the Animal-Human Boundary," *Signs* 37 (2012): 935–59.

of collaborative efforts in 1966—"The Zoological Perspective in Social Science."[9] Social scientists ought to pay more attention to recent advances in the biological sciences, they argued, especially those emerging from the study of animal behavior. They posited that, by understanding humans as cultural animals ("with an as yet insufficiently explored repertoire of genetically programmed behavioural predispositions"), social scientists should take seriously recent insights from ethology, paleoanthropology, and genetics. These factors constrained the variability of human social action that typically occupied the research efforts of sociologists and anthropologists.[10] They hoped the object of study in the social sciences would remain the same but become more nuanced as a result of such evolutionary reasoning. Of particular concern to Tiger was the question of homosocial association—how and why groups of men function the way they do.[11] Fox, for his part, had been rethinking notions of kinship in human societies. He emphasized the mother-child bond as the primary basis for understanding kinship patterns in human and nonhuman primates.[12] When we look at the collective body of their work, a sexual division of labor comes into stark relief—men hunted and women reproduced.

Throughout the 1960s and early 1970s the effects of the women's movement became visible as women entered graduate school and professional scientific careers in greater numbers.[13] Workplaces and organizations with male homosocial networks were increasingly called into question, not only by left-leaning feminists but also by right-leaning conservatives who had, since the 1950s, been concerned with the influence of "homosexuals" in the public sphere.[14] Tiger and Fox thus sought to defend the normality of homosocial association between adult men against the perceived threat of feminists and Freudians, basing their research on ostensibly universal behaviors of primates and humans. Their arguments functioned to both exclude women and safeguard a masculine preserve for the "right" men.[15]

[9] Lionel Tiger and Robin Fox, "The Zoological Perspective in Social Science," *Man*, n.s., 1 (1966): 75–81. In their copublications, Tiger's name always came first. I thus refer to them as "Tiger and Fox" when discussing joint publications, and alphabetically as "Fox and Tiger" at all other times.

[10] Ibid., 76–7, 80. Critics read their arguments as biologically determinist, yet Tiger and Fox believed culture and experience acted to modify behavior, too, and would not have self-identified as "determinists." At issue were differing convictions regarding the degree of constraint. See, e.g., Ullica Segerstråle, *Defenders of the Truth: The Battle for Science in the Sociobiology Debate and Beyond* (New York, 2001), 27–8.

[11] Lionel Tiger, *Men in Groups* (New York, 1969). On the cultures of homosociality in the sciences, see also Alexandra Rutherford, "Maintaining Masculinity in Mid-Twentieth-Century American Psychology: Edwin Boring, Scientific Eminence, and the 'Woman Problem,'" and Nathan Ensmenger, "'Beards, Sandals, and Other Signs of Rugged Individualism': Masculine Culture within the Computing Profession," both in this volume.

[12] Robin Fox, *Kinship and Marriage: An Anthropological Perspective* (Baltimore, 1967).

[13] Margaret W. Rossiter, "The Path to Liberation: Consciousness Raised, Legislation Enacted," in *Women Scientists in America: Before Affirmative Action, 1940–1972* (Baltimore, 1995), 361–82; and on the more recent decades, Rossiter, *Women Scientists in America: Forging a New World since 1972* (Baltimore, 2012).

[14] Rosabeth Kanter, *Men and Women of the Corporation* (New York, 1977). On midcentury politics over homosexuality, see David K. Johnson, *The Lavender Scare: The Cold War Persecution of Gays and Lesbians in the Federal Government* (Chicago, 2004); Margot Canaday, *The Straight State: Sexuality and Citizenship in Twentieth-Century America* (Princeton, N.J., 2009). On the science of homosexuality, see also Nathan Ha, "Detecting and Teaching Desire: Phallometry, Freund, and Behavioral Sexology," in this volume.

[15] Robert A. Nye, "Kinship, Male Bonds, and Masculinity in Comparative Perspective," *Amer. Hist. Rev.* 105 (2000): 1656–66; Nye, "Medicine and Science as Masculine 'Fields of Honor,'" *Osiris* 12 (1997): 60–79.

During their tenure as research directors at the HFGF (1972–84), Fox and Tiger weathered considerable changes in Americans' scientific search for human nature. Cultural and biological approaches to understanding "the human" increasingly diverged, dashing their hopes for a unified, more biologically based anthropology.[16] A great many cultural anthropologists, for example, refused to believe that the pair's zoological framework provided any explanatory power in interpreting human actions and behaviors. Yet at the same time, the HFGF provided monetary and therefore institutional protection from these critics. Perhaps most importantly for understanding the fate of evolutionary theories of human sociality, the foundation provided a haven for research that buttressed gendered norms in the evolutionary past of humans and promoted the idea that male and female evolutionary strategies worked necessarily at odds with one another.

This essay analyzes how discussions about dominance and aggression exemplified masculine social dynamics of the 1960s and 1970s. As a mechanism for making visible the tight-knit alliances and friendships binding some men together and excluding others, I refer to the men comprising the HFGF's board of directors as they referred to each other. Standard writing convention now suggests that authors should use the full name of their subjects at the first mention and solely the last name in all subsequent references. Among its other functions, this convention avoids perpetuating outdated tendencies to refer to female scientists, politicians, and other professionals by their first names, thereby connoting a false sense of familiarity or diminution of status.[17] The strength of the practice speaks to the continuing power of forms of address in mediating professional and personal relationships.[18] In the first half of the article, then, I use familiar names to reflect the private social circles constructed and maintained, in part through such informal forms of address, by the men about whom I write. This is simply a matter of using actors' categories. More public conversations about Fox and Tiger's research and work at the HFGF occupy the second half of the article, and I accordingly switch to the more traditional naming convention that characterized those discussions.[19]

The early history of HFGF's investment in "Man's Relation to Man" provides an intimate glimpse into how these groups of men hoped to use an evolutionary perspective to transform research on human nature and how they enacted their scientific

[16] On the diversity of anthropological approaches in the twentieth century, many of which incorporated alternative evolutionary perspectives, see Henrika Kuklick, ed., *A New Anthropology* (Oxford, 2008), and Susan Lindee and Ricardo Ventura Santos, eds., "The Biological Anthropology of Living Human Populations: World Histories, National Styles, and International Networks: Wenner-Gren Symposium Supplement 5," *Current Anthropology*, vol. 53, suppl. 5 (2012).

[17] R. Brown and M. Ford, "Address in American English," *J. Abnormal Soc. Psychol.* 62 (1961): 375–85; D. Slobin, S. Miller, and L. Porter, "Forms of Address and Social Relations in a Business Organization," *J. Personality Soc. Psychol.* 8 (1968): 289–93.

[18] Gloria Cowan and Jill Kasen, "Form of Reference: Sex Differences in Letters of Recommendation," *J. Personality Soc. Psychol.* 46 (1984): 636–45; Hilary Takiff, Diana Sanchez, and Tracie Stewart, "What's in a Name? The Status Implications of Students' Terms of Address for Male and Female Professors," *Psychol. Women Quart.* 25 (2001): 134–44. On the long-standing importance of forms of address in the sciences, see also Mario Biagioli, *Galileo, Courtier: The Practice of Science in the Culture of Absolutism* (Chicago, 1994).

[19] In my interviews with both Fox (8 November 2011) and Tiger (10 November 2011), they often referred to each other by first name but used last names when discussing each other's research. At the time of writing this essay, neither had archived their correspondence. However, Tiger's papers have now been deposited at Special Collections and University Archives, Rutgers University Archives (R-MC 117).

commitments socially by establishing their own standing in the worlds of private philanthropy and academia.[20] It also illustrates the importance of informal masculine networks of money and authority, embodied in the leadership of the HFGF, in defining the kinds of questions scientists asked and how they answered them. Understanding these dynamics requires a careful exploration of the bonds of friendship that tied together Ardrey, Guggenheim, Fox, and Tiger.

THE PHILANTHROPIC SAVANNA

In 1848, Harry Guggenheim's grandparents moved to Philadelphia, where they started a successful mining company. Harry's father, Daniel Guggenheim, eventually took over the burgeoning family business, and Daniel and his nine siblings became fixtures in East Coast philanthropic networks, creating the Solomon R. Guggenheim Museum and Foundation in New York City, the John Simon Guggenheim Memorial Foundation, and the Daniel and Florence Guggenheim Aviation Safety Center at Cornell University. Harry served in both world wars as a member of the Naval Aviation Forces, where he met James "Jimmy" Doolittle and Charles "Slim" Lindbergh. He was appointed US ambassador to Cuba from 1929 to 1933 and cofounded *Newsday* with Alicia Patterson (his third wife) in 1940.[21] Jimmy and Slim were family friends and were awarded Daniel Guggenheim Medals in 1942 and 1953, respectively, honoring their "notable achievements in the advancement of aeronautics." Jimmy retired from active military service in 1959 but remained interested in aviation safety throughout his life (a concern shared by both Harry and Slim). Slim had been an associate of the Guggenheim family at least as early as the 1920s, when he toured the country promoting aviation under the sponsorship of Daniel Guggenheim. In short, Harry was a man who moved through elite New York circles, surrounded by socially and economically powerful men whose mettle, he believed, had been tested by combat and hardened by business. He intended to spend his money on practical solutions to one of the persistent dilemmas confronting all humanity.

Harry was also strong-willed, perhaps obstinate. He admitted as much when writing to his "old and valued friend" Henry Allen Moe.[22] Henry was then president of the American Philosophical Society, would soon become interim chairman of the National Endowment for the Humanities (until the first official chairman appointed by President Lyndon Johnson could begin his duties), and believed firmly in the importance of universities in creating an intellectually healthy nation.[23] Harry apologetically wrote to Henry, "I'm afraid I am perhaps a difficult donor, and perhaps unable to accept the role of a philanthropist who calls in experts to dispense his funds. I have been the head of three Foundations, two of which I still head, not as a philanthropist but as

[20] On the mutual mapping of professional and knowledge agendas in the study of human nature, see Donna Haraway, *Primate Visions: Gender, Race, and Nature in the World of Modern Science* (New York, 1989).

[21] Under Patterson and Guggenheim, *Newsday* was a successful conservative suburban daily newspaper (Monday–Saturday) in a tabloid format; Lee Smith, "The Battle for Sunday," *New York Magazine*, 25 October 1971, 34–9.

[22] HFG to Moe, 19 April 1965, Folder: Guggenheim, Harry Frank #2, Moe Papers; Guggenheim referred to Moe as his "old and valued friend" in a letter to Paul Fitts, 21 January 1964, Folder: Guggenheim, Harry Frank #3, Moe Papers.

[23] E.g., Henry Allen Moe, "The Shortage of Scientific Personnel," *Science* 105 (1947): 195–8, and Moe, "The Power of Freedom," *Amer. Assoc. Univ. Professors* 37 (1951): 462–75.

the directing spirit, with some expertise, dispensing funds of others." He wondered whether, due to this experience, he found it "hard to turn over these funds to professionals and say, 'I want to improve man's relation to man; here are X dollars; now get to work.'" Expressing a sentiment he came to repeat often in his correspondence with friends, Harry added, "In the six years that I have been attempting to make some progress on this project I have found that the only suggestions . . . in what I consider a practicable manner were not suggested by professionals, but were the intuitive suggestions of laymen."[24] Practicality, or common sense, was a quality Harry deemed especially lacking in social scientists.

Harry and his associates expected the HFGF would award about six fellowships a year of between $5,000 and $9,000 each (in 2015 terms, between $37,000 and $68,000). Awardees were to be granted a great deal of leeway with their research projects under the assumption that as vetted men of quality they would produce top-notch results. Henry (chairman), Fitts, and G. Edward "Ed" Pendray formed the initial fellowship committee, but they hoped additionally to find a part-time director for the fellowship program.[25] Ed was another long-term associate of the Guggenheim family, having helped develop the Guggenheim Jet Propulsion Laboratory at the California Institute of Technology, among other such ventures. In the wake of Fitts's death, the nascent foundation floundered, waiting for someone to assume responsibility for the entire project or (at least) the fellowship program—part-time initially, but full-time later if they and the project were to "take-fire."[26]

In the spring of 1966, Harry wrote to his confidant Slim Lindbergh that throughout history men had abused their political power. "In pursuit of that primary urge to dominate their fellow man," he suggested, "they have decimated him and caused incalculable destruction to the accumulated works of beauty and utility that man has created." Harry further noted that in the 1960s the world still contained several of these men, who needed to be controlled lest they "continue to cause holocausts of destruction."[27] Slim believed that the quickest, most effective, and reasonable strategy for improving man's relationship to man would be to ameliorate the conditions of human life—especially through the conservation of natural resources.[28] To this Harry replied that the fundamental issue he wished to address was located not in the environment but in the "qualities in man." He asked Slim, "How can we determine the cause of this destructive rather than constructive competitive quality in man? How can we educate him so that we may divert these energies to competition that is good rather than evil?"[29] In the face of Harry's queries, Slim remained firm: "It seems to me there is good domination, and bad domination (possibly 'leadership' would be a better term to work with), and all kinds of forms in between. Again, 'good' and 'bad' vary with frameworks of reference." He staunchly continued, "I think that men who

[24] HFG to Moe, 19 April 1965, Folder: Guggenheim, Harry Frank #2, Moe Papers.

[25] Pendray also wrote popular books (including several science fiction novels under the pseudonym Gawain Edwards) and received a John Simon Guggenheim Fellowship in 1964.

[26] G. Edward Pendray, "Summary of Progress, Man's Relation to Man Project," 15 April 1961 to 1 August 1965, Folder: Guggenheim, Harry Frank Foundation: Lindbergh, Charles A., Moe Papers.

[27] HFG to General Charles A. Lindbergh, 10 May 1966, Folder: Guggenheim, Harry Frank, Foundation: Lindbergh, Charles A., Moe Papers.

[28] Lindbergh to HFG, 26 April 1966, Folder: Guggenheim, Harry Frank, Foundation: Lindbergh, Charles A., Moe Papers.

[29] HFG to Lindbergh, 10 May 1966, Folder: Guggenheim, Harry Frank, Foundation: Lindbergh, Charles A., Moe Papers.

gain great ability to dominate and exert power intertwine with their environment, in addition to being affected by hereditary characteristics. I don't believe you can separate them from institutions of their times any more than you can separate heredity and environment."[30] Harry again responded without addressing Slim's point. Their dialogue, he wrote, helped him "think through the answer to a most difficult question: 'Is there a basic quality in man that can be isolated which is the cause of strife with his fellow man?'"[31]

The following year, when Harry picked up *Territorial Imperative*, he was looking for a new perspective as well as a new man to spearhead research on "man's relation to man" with the full financial backing of his foundation. He wrote Henry, brimming with enthusiasm for the *Territorial Imperative*.[32] Henry suggested that Harry was bound to find Desmond Morris's recently released *The Naked Ape* equally worthwhile. (In fact, Harry later reported, he found it "extraordinary."[33]) The previous year, Henry had also recommended Lorenz's *On Aggression*, which Harry had consumed with equal vigor. Harry even published an editorial in *Newsday*—entitled "The Mark of Cain"—in which he called his readers' attention to the need to understand "the nature of the beast within man" and lavished praise on both Ardrey and Lorenz for their efforts to uncover man's instinctive aggression.[34] Lorenz's conception of "aggression" as key to human nature, and fundamental to our virtuous qualities (leadership and kindness) as well as our violent tendencies (dictatorship and murder), fit neatly within Harry's vision of human social relations. Harry thought these books so useful to his incipient foundation that he sent copies to his board.[35] A remarkable exchange of letters then ensued between Harry, Henry, Jimmy, Slim, and Ed, discussing the merits of a biological perspective on aggression, including the male drive to defend territory and compete with other males for social status.[36] Harry habitually excerpted and distributed letters among members of the group.[37] A letter to him thus often functioned as a letter to all.

Ardrey, impressed by the "Mark of Cain" editorial in *Newsday*, wrote to introduce himself to Mr. Guggenheim, and they began a lively correspondence. Rather than pessimistically predicting man's inevitable doom, however, Ardrey closed on a positive note: "I believe that when one regards oneself as a risen ape, the future becomes illimitable. When one regards oneself as a fallen angel, one has no future at all. What

---

[30] Lindbergh to HFG, 29 May 1966, Folder: Guggenheim, Harry Frank, Foundation #27, Moe Papers.

[31] HFG to Lindbergh, 1 June 1966, Folder: Guggenheim, Harry Frank, Foundation #27, Moe Papers.

[32] HFG to Moe, 10 November 1967, Folder: Guggenheim, Harry Frank #5, Moe Papers.

[33] HFG to Moe, 19 December 1967, Folder: Guggenheim, Harry Frank #5, Moe Papers.

[34] [Harry F. Guggenheim], "The Mark of Cain," *Newsday*, 25 September 1967, 33.

[35] HFG to Doolittle, 6 July 1966, Folder: Guggenheim, Harry Frank #6; HFG to Moe, 10 November 1967, Folder: Guggenheim, Harry Frank #5; HFG to Moe, 19 December 1967, Folder: Guggenheim, Harry Frank #5, Moe Papers. Based on this correspondence, it appears likely that Guggenheim had not yet finished *Territorial Imperative* at the time he penned his editorial.

[36] Recent scholarship draws our attention to the social and material natures of letter writing, the complex etiquette dictating the relationships between public and private correspondence, and the hierarchies of credibility that define membership in epistolary communities. See, e.g., James How, *Epistolary Spaces: English Letter-Writing from the Foundation of the Post Office to Richardson's "Clarissa"* (Aldershot, 1988); Anthony Grafton, "The Humanist as Reader," in *A History of Reading in the West*, ed. Gugliemo Cavallo and Roger Chartier (Amherst, Mass., 1999), 179–212; Lorraine Daston, "Sciences of the Archive," *Osiris* 27 (2012): 156–87.

[37] Most of the letters I discuss are thus preserved in a single archive—Henry Allen Moe's papers at the American Philosophical Society. Guggenheim carefully distinguished between private news and professional discussions, and his secretary reproduced only those sections of letters that directly addressed the question of "man's relation to man."

man needs in our time, above all else is an elation. I think that's what my ethologist friends are finding."[38] Here, it seemed, was the expert guidance for which Harry had been looking. He replied to Mr. Ardrey, informing him of his foundation and asking him to contribute an article to *Newsday* as part of a series on "The Condition of the American Spirit."[39] Ardrey responded immediately, taking the opportunity to cultivate a potential patron, and cleverly made his intellectual project about Guggenheim's: "With admirable intuition you as long ago as 1963 grasped the problem of dominance as central to the human predicament, and I have outlined the program of my work to demonstrate that I too regard it as central."[40] Yet by dating Guggenheim's interest to 1963, Ardrey also established his own priority, as *African Genesis* had been published two years earlier. Ardrey followed his compliment with a request—would Guggenheim be so kind as to send him a copy of the bibliography on dominance that Fitts had prepared after the conference?

In *African Genesis*, Ardrey had expounded paleoanthropologist Raymond Dart's idea that man was born evolutionarily when he picked up a bone or piece of rock and realized its power as a weapon. Through the group hunt, he posited, came greater quantities of meat in our diet, an enlarged brain, and cooperative hunting. Man had not fathered the weapon; "the weapon, instead, had fathered man."[41] Five years later, he stood by this argument. While preparing to write the *Territorial Imperative*, Ardrey returned to Africa, where he also came to believe that males fight for status and females mate with whichever male happens to be occupying the best territory, using antelope as his example. "The female wants her affection," Ardrey noted, "but she wants it at a good address. Whether or not our human sensibilities are offended or intrigued, it is a harsh truth that the doe is attracted and excited by the qualities of the property, not the qualities of the proprietor."[42] Although he refrained from invoking explicit parallels to human courtship, the implications hung heavily in the air. Absent was any mention of male-female affection or family structure as the basis of social order.

Even though reality rarely lived up to the iconic tropes of the 1950s, Ardrey's reframing of social structure in terms of male-male interactions represented a substantial break with traditional norms of familial masculinity. Such rhetoric, with its core American values depicted in the nuclear families of June and Ward Cleaver (*Leave It to Beaver*) or Ozzie and Harriet Nelson (*The Adventures of Ozzie and Harriet*), had emphasized the importance of reproduction to the country's democratic future.[43]

Ardrey found inspiration for part of his vision of social behavior in the work of ecologist Vero Copner Wynne-Edwards.[44] Wynne-Edwards argued that birds and other

[38] Ardrey to HFG, 18 October 1967, Folder: Guggenheim, Harry Frank #5, Moe Papers.

[39] HFG to Ardrey, 30 and 31 October 1967, Folder: Guggenheim, Harry Frank #5, Moe Papers.

[40] Ardrey to HFG, 4 November 1967, Folder: Guggenheim, Harry Frank #5, Moe Papers.

[41] Ardrey, *African Genesis* (cit. n. 5), 29.

[42] Ardrey, *Territorial Imperative* (cit. n. 1), 45–8.

[43] Elaine Tyler May, *Homeward Bound: American Families in the Cold War Era* (New York, 1988); Stephanie Coontz, *The Way We Never Were: American Families and the Nostalgia Trap* (New York, 1992); Joanne Meyerowitz, ed., *Not June Cleaver: Women and Gender in Postwar America, 1945–1960* (Philadelphia, 1994); Wendy Kline, *Building a Better Race: Gender, Sexuality, and Eugenics from the Turn of the Century to the Baby Boom* (Berkeley and Los Angeles, 2001); James Gilbert, *Men in the Middle: Searching for Masculinity in the 1950s* (Chicago, 2005); Alexandra Stern, *Eugenic Nation: Faults and Frontiers of Better Breeding in Modern America* (Berkeley and Los Angeles, 2005).

[44] Wynne-Edwards's suggestion that animal species acted to limit their own population growth for the good of the species would eventually land him in intellectual hot water; Mark Borello, *Evolutionary Restraints: The Contentious History of Group Selection* (Chicago, 2010).

animals rarely caused each other lasting physical harm during their contests for su-
periority. "Instead," he suggested, "they merely threaten with aggressive postures,
vigorous singing or displays of plumage. The forms of intimidation of rivals by birds
range all the way from the naked display of weapons to the triumph of splendor revealed in the peacock's train."[45] Humans, on the other hand, had lost the knack for
ritualized combat and too often succumbed to real killing. Yet based on his firm belief
that much social behavior was indeed ritualized, even in humans, Wynne-Edwards
defined societies as "brotherhood[s] tempered by rivalry," expressed through ritualized combat so as to preserve the longevity of the population.[46] Ardrey made use of
Wynne-Edwards's twinned concepts of brotherhood and rivalry in his explanation of
noyaus, societies whose members cohered thanks only to shared suspicion of a common enemy.

Ardrey also followed the lead of ethologists when he dismissed the idea that male
courtship was female directed in favor of arguing that males competed with each other
over territories through ritualized behavior.[47] He submitted to his readers that the true
objects of a male's thoughts during the mating season were other males—his performance mattered "in the eyes of his fellows"—and thereby discounted Charles Darwin's theory that male courtship display functioned to attract the attention of females.
Ardrey fretted that he was leaving himself open to accusations of universalizing a homosocial and potentially homosexual tendency, a position he quickly disavowed.[48] He
need not have worried. The idea that males were preoccupied with gaining the respect
of other males became one of the oft-cited conclusions of his work.

Harry found these ideas enthralling. Still, before responding, he decided he should
vet Ardrey with his trusted friends and so telephoned Henry. As Harry later recounted,
he was delighted to discover that Ardrey was "an old [John Simon] Guggenheim Fellow and so an old friend" of Henry's.[49] Henry later passed along a copy of an article
about Ardrey's experience in the theater industry to help acquaint Harry with Ardrey's
primary career in the arts.[50] In his reply to Ardrey, Harry promised to dig out a copy of
Fitts's bibliography but remarked, "the bibliography is going to be a disappointment
to you as it was to me. It consists mainly of references to power in industry." He also
forwarded copies of his complete correspondence with Ardrey to Henry and Slim—

[45] V. C. Wynne-Edwards, "Population Control in Animals," *Sci. Amer.,* August 1964, 68–74, on
71. Ethologists had long believed that male animals engaged in "ritualized" fights rather than killing
each other; see, e.g., A. David Blest, "The Concept of 'Ritualisation,'" in *Current Problems in Animal
Behaviour,* ed. W. H. Thorpe and O. L. Zangwill (Cambridge, 1961), 102–24; Julian S. Huxley, "A
Discussion on Ritualization of Behaviour in Animals and Man," *Phil. Trans. Royal Soc. London Ser.
B* 251 (1966): 249–71.

[46] Wynne-Edwards, "Population Control in Animals" (cit. n. 45), 71.

[47] Ardrey, *Territorial Imperative* (cit. n. 1), 55.

[48] At the time, occasional hostile accusations circulated that manly characters, such as Batman and
Robin, the men of *Bonanza,* and even James Bond (due to his utilitarian engagements with women),
reflected a growing homosexual tendency in American culture: Carol L. Tilley, "Seducing the Innocent:
Frederic Wertham and the Falsifications That Helped Condemn Comics," *Inform. & Cult.* 47 (2012):
383–413; Wendall Hall, "The Fag-Jag on the Boob-Tube," *Fact* 4 (1967): 16–23.

[49] HFG to Ardrey, 7 November 1967, Folder: Guggenheim, Harry Frank #5, Moe Papers. Ardrey
received a fellowship from the John Simon Guggenheim Memorial Foundation in 1937 for "Creative
Arts—Drama & Performance Art."

[50] Moe to HFG, 8 January 1968, Folder: Guggenheim, Harry Frank #5, Moe Papers; Robert Ardrey,
"Reflections on the Theatre," *Amer. Sch.,* 4 December 1967, 111–20.

writing that he anticipated "this contact with Mr. Ardrey opens up a new vista in our Man's Relation to Man Project" and asking for their thoughts and reactions.[51]

Ed Pendray wrote back first and most critically. If Ardrey were correct, which Ed doubted, his arguments seemed to call into question the utility of proceeding with the man's relation to man project. "If human behavior is basically instinct, what can ever be done (short of a long period of evolution) to modify or improve it?" Ed continued, "I still believe profoundly in the modifiability of human behavior, based on knowledge, understanding, social pressures and education. How else can we account for all the varieties of cultures already to be found in the world?"[52] Harry defended Ardrey, insisting that his studies "confirm my thesis that dominance is a basic instinct of man. Man's actions are governed by the sum of his inheritance and environmental characteristics." By changing the crucial environment for powerful men, then, he hoped "we can influence man by directing his instinct to dominate for the progress of rather than for the depravity of mankind. In the former case we had Christ, in the latter a Hitler."[53] Harry's response is remarkable for any number of reasons, not least that he appears to have internalized Slim's earlier arguments along these lines. This assertion, of the malleability of man in the face of his inherited instincts, would prove to be a sticking point among several of the inner group (as it was for many social scientists).[54]

While waiting for his other friends to respond, Mr. Guggenheim and Mr. Ardrey, now Harry and Bob, grew closer. The next time he wrote, Bob addressed Ed's criticisms (which Harry had forwarded to him along with his own response). Bob reported enthusiastically, "You couldn't have given a better answer. . . . What we now know about dominance is that in the males of all species it's an instinct. The drive is there, born in, and cannot be obliterated as it cannot be ignored. But the goals are adjustable."[55] Bob argued that, therefore, by "denying" the innate drive to dominance in all men, "we lose control over the goals." Headway on the problem of man's relation to man could be made only by accepting man's base nature. Ever the careful correspondent, however, Bob had no wish to ostracize Ed (perhaps recognizing by now that his letter would surely find its way into Ed's hands), so he added that he "retain[ed] a considerable sympathy for Mr. Pendray's question," even if it "rests on the false concept of instinct that we've been taught and is forced on us by every Ashley Montagu in the hope that we'll deny it exists."[56] Bob said that given the novelty of his ideas, he understood the resistance to his books. He then discounted the work of continental ethologists (including Lorenz and Tinbergen, on whose research he had based much of *The Territorial Imperative*), cryptically noting that "their attitudes towards

[51] HFG to Moe, 10 November 1967, Folder: Guggenheim, Harry Frank #5, Moe Papers; see also Charles R. Lindbergh to HFG, 24 April 1968, Folder: Guggenheim, Harry Frank, Foundation #13, Moe Papers.

[52] G. Edward Pendray to HFG, 10 January 1968, Folder: Guggenheim, Harry Frank #5, Moe Papers.

[53] HFG to Pendray, 8 January 1968, Folder: Guggenheim, Harry Frank #5, Moe Papers. Lindbergh's comments should be understood in the context of his deep sympathy with the American eugenics movement. See, e.g., Andrés Horacio Reggiani, "Charles Lindbergh and the Institute of Man," in *God's Eugenicist: Alexis Carrel and the Sociobiology of Decline* (New York, 2007), 85–102.

[54] E.g., M. F. Ashley Montagu, ed., *Culture: Man's Adaptive Dimension* (Oxford, 1968).

[55] Ardrey to HFG, 17 January 1968, Folder: Guggenheim, Harry Frank, Foundation: Ardrey, Robert, Moe Papers.

[56] As a cultural anthropologist, Montagu publicly dismissed Ardrey's arguments but until the later 1960s remained sympathetic to the man. See M. F. Ashley Montagu, ed., *Man and Aggression* (New York, 1968); Montagu, *The Nature of Human Aggression* (New York, 1976).

instinct were formed at the cellular level . . . a level at which nothing so far has proved demonstrable." Again, Bob hewed closely to Harry's stated position: "what you wrote is utterly correct. A genetically determined behavior pattern is a cup of determined shape. What rain falls into that cup God and man must decide. But something will fall, and something will be retained. So one d[e]termines the difference between a Christ and a Hitler." Bob was now a trusted friend, and Harry circulated his letter at once, not for approval but for information.

Harry also arranged a "stag dinner (business clothes)" at his five-story New York City town house.[57] For posterity and the HFGF records, Ed kept minutes of the occasion.[58] Henry attended, as did other New York notables, including Roger Straus (Harry's cousin and later cofounder of publishing house Farrar, Straus and Giroux), Peter Lawson-Johnston (heir to the Guggenheim Brothers business), and Dr. Malachi Fitzmaurice-Martin (a Catholic priest and recent recipient of a grant from the Harry Frank Guggenheim Foundation)—Slim and Jimmy regretfully declined. Bob took the opportunity to reinforce his allegiance to Harry and to advance his own cause by arguing that his volumes, together with those of Lorenz and Morris, were the only "worth-while books" about human aggression and dominance. When asked who might be appropriate to direct the man's relation to man project, he requested more time to think but also suggested "the most important possibility to be Dr. Robin Fox, Chairman of the Department of Anthropology at Rutgers University."[59] All told, the dinner seemed to go quite well.

The next friend to respond to Harry's query about Bob's ideas was Slim Lindbergh. Although he began his letter by stating that he largely concurred with Harry's positive assessment, his final opinion turned out to be more complicated. "I could not be in more agreement with Ardrey's statement that 'To me, the nightmare is the denial of instinct, the denigration of competition—' (I think the essential value of competition is deplorably lacking in the marvelous philosophy of Jesus, at least as it has been handed to us)."[60] Yet he shared Ed's doubts about applying this philosophy to the foundation. "When we relate dominance to competition, it seems to me we are relating a fragment to a whole, and I cannot believe this is the best way to approach improving man's relation to man." As he had become both older and more experienced, Slim continued, he also became "more aware of the limitations of man's sciences." The difficulties facing the task ahead of them were tremendous and long-standing. "How do we clarify the issues of war in Viet Nam, of riots in our cities, of our cold-war with Russia?" In sum, he closed, "man's relationship to man expands into the miracle, and here the tools of science are inadequate." The sheer magnitude of these problems belied any easy answer, even Bob's.

When Jimmy Doolittle chimed in, he pragmatically noted the different time scales required to address the problem of man's relation to man with various methods.[61] The

[57] Memo to Dr. Moe, 13 March 1968, Folder: Guggenheim, Harry Frank, Foundation #15, Moe Papers.

[58] Edward Pendray, "Summary of Meeting: The Man's Relation to Man Project," 19 March 1968, Folder: Guggenheim, Harry Frank, Foundation, 1968 #41, Moe Papers.

[59] Ibid. No one considered Bob a contender: he lived happily in Rome, possessed no administrative experience, and lacked a PhD.

[60] General Charles R. Lindbergh to HFG, 24 April 1968 [copied/sent to Moe: May 13, 1968], Folder: Guggenheim, Harry Frank, Foundation #13, Moe Papers.

[61] Doolittle to HFG, 24 October 1968, Folder: Guggenheim, Harry Frank, Foundation #14, Moe Papers.

most permanent solution operated on the longest time frame. "There is little question but that improving mankind through evolution will take a very long time indeed; measured in many millennia, or perhaps even millions of years," he wrote. "The ability to successfully change the genes—cut the meanness out and put kindness in—may come in tens or hundreds of years." In the interim, "social pressures cannot change man's instincts but may well serve to suppress the baser of them. With proper planning and implementation this effect might well occur in the relatively short time left to you and me." Like Ed, Jimmy advocated strategies that would allow a rapid change in human actions, and that meant concentrating the foundation's resources on immediately controllable elements of the human social environment.

Henry Moe took an even more direct approach by presenting his own candidate for directorship—Glenn A. Olds. At first glance, Olds was an odd choice. As a faculty member at Yale, he had taught philosophy, ethics, logic, and religion and, as an ordained Methodist minister, had directed Cornell's United Religious Work. Yet he possessed all the qualities of manly leadership Henry valued. Olds had served as president of Springfield College in Massachusetts and had been a professional boxer during his college years. In 1968, he was executive dean of the State University of New York and was widely appreciated in academic circles as a forward-thinking and energetic man. Olds was nothing if not an administrator and humanist, cast from a similar mold as Henry himself. Henry urged Harry to consider him a serious candidate, despite Bob's standing recommendations.[62]

Although members of Harry's inner circle seemed inclined to trust the fate of humanity to evolution in the long run, and genetics in the slightly less distant future, research into the social and cultural causes of violence struck them as entirely more practicable in the short term. Against this background of skepticism, Bob continued corresponding with Harry and Henry, and two names recur in his letters as suggestions for the post of research director.[63] "This pair with the improbable names," as Bob dubbed Fox and Tiger, intended to "establish an anthropological redoubt against orthodoxy." From Harry's perspective on the general inutility of most social scientists, that made them only more attractive as candidates.

When Harry replied, however, it was from the hospital, one of many visits to battle the cancer that was already eating away at his body.[64] By the time he passed away two years later, in January 1971, no decisions had been made as to the fate of his foundation. Yet Harry Frank Guggenheim's legacy lived on because of the board's desire to honor his intentions, even if they disagreed with them. Survival in the philanthropic savanna required an intuitive sense of the dominance hierarchies governing the social interactions of these men. Bob, although a relative newcomer to the group, quickly figured out an appropriate rubric for maintaining a friendship with a powerful man like Harry—a well-balanced combination of maverick bravado and studied deference.[65]

[62] Moe to HFG, 12 November 1968, Folder: Guggenheim, Harry Frank, Foundation: Ardrey, Robert, Moe Papers.

[63] Ardrey to Moe, 24 April 1968, Folder: Guggenheim, Harry Frank, Foundation: Ardrey, Robert, Moe Papers.

[64] HFG to Ardrey, 17 January 1969, Folder: Guggenheim, Harry Frank, Foundation #2, Moe Papers.

[65] The character of Ardrey's letters varied widely by correspondent. With men he identified as kindred spirits, like Clarence Ray Carpenter and Kenneth Oakley, Ardrey tended to write in more personal terms. In his letters to Farley Mowat, I would describe his phrasing as positively earthy. See Series I, Professional Correspondence, MC 190, Robert Ardrey Papers, 1955–1980, Special Collections and University Archives, Rutgers University Libraries, New Brunswick, N.J.

THE ACADEMIC JUNGLE

Fox and Tiger met in London in 1965. According to Fox, they started talking after one of Desmond Morris's seminars at the London Zoo—in front of the gibbon cages they discussed human and animal instincts and the importance of male bonding and competition in controlling human social relations. (Tiger insists there were no gibbons.) Fox "was taken immediately by this funny, smart, talkative, small but confident son of the Montreal ghetto," and they bonded "fiercely."[66] When Fox founded the Department of Anthropology at Rutgers, he hired Tiger, and there they made quite a pair.[67] Fox was lanky, charming, British, and chose to settle in the farmlands of rural New Jersey, in a house outside of Princeton. Tiger was shorter and steelier, preferring the urban sophistication of New York City. They commuted to Rutgers from different directions, literally and figuratively.

Fox and Tiger acquired their fascination with evolution from different sources. Fox's attention was caught by John Bowlby's Darwinian take on psychoanalytic theory and Michael Chance's research on the social behavior of monkeys.[68] Tiger had been reading Morris, ethologist John Crook, and primatologist John Napier, all of whom circulated through London.[69] When importing this evolutionary perspective into the social sciences, Tiger and Fox made two subsidiary scholarly arguments that are worth particular note in the context of their incipient directorship of the HFGF. First, both made academic names for themselves in the late 1960s and early 1970s by suggesting that human societies should be studied as a function of two distinct kinds of social relationships shared by all primates—the friendships uniting males and the parental love between females and their offspring. These bonds seemed to require, at least for Tiger, a sexual division of labor that worked against the claims of feminists.[70] In biological terms, male-male bonding had been crucial in driving the intellectual evolution of humanity, while female-child bonding had perpetuated the species. Fox additionally reasoned that, as a result, female-male bonding was incidental to the overall social stability of the population.[71] Both located the origins of the imbalance between male and female contributions to human evolution in the development of early human hunting practices.[72] Because of hunting in groups, men exhibited higher levels of testosterone and competitive behavior, making them ideally suited to life in

[66] Robin Fox, *Participant Observer: Memoir of a Transatlantic Life* (New Brunswick, N.J., 2004), 329–30; see also Alex Walter, "An Interview with Robin Fox," *Curr. Anthropol.* 34 (1993): 441–52. Fox's autobiography is narrated in the third person.

[67] Fox submitted his thesis in anthropology to the University of London and was awarded a PhD in 1965. Before moving to New Jersey in 1967, he lectured at the London School of Economics (LSE). Tiger earned his PhD in 1962 at LSE for a thesis in sociology and taught at the University of British Columbia for several years before joining Fox's new department of anthropology in 1969.

[68] Robert G. W. Kirk, "Between the Clinic and the Laboratory: Ethology and Pharmacology in the Work of Michael Robin Alexander Chance, c. 1946–1964," *Med. Hist.* 53 (2009): 513–36; Marga Vicedo, "The Social Nature of the Mother's Tie to Her Child: John Bowlby's Theory of Attachment in Post-War America," *Brit. J. Hist. Sci.* 44 (2001): 401–26.

[69] E.g., John Napier and N. A. Barnicot, eds., *The Primates*, Symposia of the Zoological Society of London, no. 10 (London, 1963); and O. G. Edholm, ed., *The Biology of Survival*, Symposia of the Zoological Society of London, no. 13 (London, 1964).

[70] Lionel Tiger, "Male Dominance? Yes, Alas. A Sexist Plot? No." *New York Times*, 25 October 1970, SM18.

[71] Robin Fox, "The Evolution of Sexual Behavior," *New York Times*, 24 March 1968, SM32.

[72] Contemporaneous academically oriented collections advanced a similar fascination with the idea; Richard B. Lee and Irven DeVore, eds., *Man the Hunter* (Chicago, 1968).

the corporate jungle.[73] Tiger argued that feminists, by ignoring these scientific facts, rendered their vision of social equality frustratingly difficult to achieve.[74] Second, Tiger and Fox worked self-consciously to replace a Freudian psychological conception of homosociality—which they associated with a juvenile phase of development or with homosexual tendencies—with a biological understanding of homosocial association as fully adult and normal for all men. Ardrey (who had similarly attacked Freudian psychology as so much nonsense in *Territorial Imperative*) enthusiastically agreed.[75] In his review of *Men in Groups*, he argued, "Men want, need, and must have the opportunity of exclusive association. And the all-male group, bonded by long familiarity, furnishes society with its spine."[76]

The sympathy and mutual regard worked both ways. In an interview, Tiger attributed his initial interest in biology and paleoanthropology to picking up *African Genesis*. "Sometimes a book makes a difference," he said. "I thought to myself, this, this is something."[77] They, too, struck up a correspondence, finally meeting in London when Ardrey invited him to dinner at the Savoy. Conversation naturally turned to animal behavior. Fox first encountered Ardrey through a chance recommendation from his cousin, who called *African Genesis* "quite dramatic" and "all about" Fox's "latest enthusiasms for Zoology and early man."[78] When he read it, he was impressed by the "remarkable" book's basic argument, that (in his words) "society was older than man; we did not invent it, we inherited it. This animal heritage could only be understood by putting together the knowledge of animal society (territory, mating, dominance etc) with the knowledge of primate and human evolution."[79] Fox finally met Ardrey while in Bristol for a televised joint interview and found him "instantly likeable in his no-nonsense, tell-it-like-it-is fashion. . . . He had an easy manner and a witty delivery, with a wicked line in sarcasm."[80] Fox noted, "he was immediately Bob."

Fox wrote his first book, *Kinship and Marriage*, for use in university courses, and it received little popular attention.[81] His next book was an introduction to anthropology designed to appeal to the elusive "general public" and consisted in large part of republished essays adapted, translated when necessary, and compiled for the book.[82] The last of these was "The Cultural Animal."[83] In it, Fox maintained that culture and nature were inextricably intertwined—just as early humans produced culture, it in turn produced us. But culture, in his vision, could never be expanded to the infinite possibilities of habits and traditions that we might intellectually conceive. All humans possessed "the *capacity* for culture," he continued, but were simultaneously bound by

[73] Tiger, "Male Dominance?" (cit. n. 70).

[74] Lionel Tiger and Joseph Shepher, *Women in the Kibbutz* (New York, 1975).

[75] Tiger, Fox, and Ardrey attacked a caricature of psychological theory in the 1960s; for a more nuanced vision of the field, see Jonathan Metzl, *Prozac on the Couch: Prescribing Gender in the Era of Wonder Drugs* (Durham, N.C., 2003).

[76] Robert Ardrey, "A Tiger about to Stir up a Mare's Nest," review of *Men in Groups*, by Lionel Tiger, *Life* 66 (20 June 1969): 11.

[77] Lionel Tiger, interview by Erika Lorraine Milam, 10 November 2011, New York City.

[78] Fox, *Participant Observer* (cit. n. 66), 321.

[79] Ibid., 324.

[80] Ibid., 349.

[81] Rodney Needham, "[Review of] Kinship and Marriage," *Man* 3 (1968): 324–5; Eugene A. Hammel, "[Review of] Kinship and Marriage," *Amer. Anthropol.* 70 (1968): 972–3.

[82] Robin Fox, *Encounter with Anthropology* (New York, 1973).

[83] Fox's "The Cultural Animal" was previously published in the journal *Social Science Information* (9 [1970]: 7–25).

"the *forms* of culture, the universal grammar of language and behavior."[84] Here Fox invoked Noam Chomsky's differentiation between highly variable "surface grammar" and the "deep structures" underlying all human languages as analogous to the relationship between biologically irrelevant cultural variation (shelters built from, e.g., wood, stone, or snow) and the existence of (for example) "laws about property, rules about incest and marriage, customs of taboo and avoidance, [and] methods of settling disputes with a minimum of bloodshed" found in all human cultures.[85] In his own work on human kinship patterns, Fox asserted that all social systems performed two basic functions—they defined kinship (Fox termed this "descent") and demarcated eligible mates ("alliance"). In doing so, they established traditions ensuring that members of the same kinship group would not mate, thereby preventing inbreeding without the need for a psychologically imposed "incest taboo."[86]

Fox's larger point was that evolution acted to modify human behavior, just as it had altered our anatomy.[87] A couple of years earlier, Fox wrote an article for the *New York Times* that began with the same sentiment but quickly expanded to a consideration of the evolution of differences in male and female sexual behavior.[88] Some of his lessons echoed those of Ardrey: males competed with each other for control of reproductively available females, and females fought for status within the social hierarchy to ensure the survival and health of their offspring. To succeed evolutionarily, he theorized, males had to be smart, able to defer gratification (sexual or otherwise), socially graceful and cooperative (with larger, more important males), and acceptable to females. Most important, a male "must also be tough and aggressive in order to assert his rights" within the hierarchy. Control over his emotions turned into the capacity to use tools, wield weapons, and ultimately shape his environment. Other qualities might contribute to a female's status, some of which they shared with dominant males, Fox added, "with the exception, perhaps, of bitchiness and bossiness."[89] He further observed that although harems might appear to be out of vogue in Western culture, making it difficult to gauge male status according to the number of women they oversaw, a "big man" in the office may have several women at his beck and call: "one wife, two full-time secretaries, 20 typists, and a girl who comes in to do his manicuring." Here was evidence of women as status symbols and some reassurance that "monogamous nuclear families" were merely cultural fictions, perpetuated by (among others) the *Saturday Evening Post*.

Fox attributed the most carefully elaborated example of a deep cultural structure to Tiger's research: "Whatever the overt cultural differences in male-group behavior . . . in society after society one thing stands out: men form themselves into associations from which they exclude women."[90] Tiger, for his part, hoped his *Men in Groups*, published in 1969, would reach a wide audience. He succeeded, receiving two reviews and

[84] Ibid., 21.

[85] Ibid., 15; Noam Chomsky, *Aspects of the Theory of Syntax* (Cambridge, Mass., 1965); Chomsky, *Syntactic Structures* (1957; repr., The Hague, 1965).

[86] Robin Fox, "Primate Kinship and Human Kinship," in *Biosocial Anthropology* (New York, 1975), 9–35; Fox, "Alliance and Constraint: Sexual Selection in the Evolution of Human Kinship Systems," in *Sexual Selection and Descent of Man*, ed. Bernard Campbell (Chicago, 1972), 282–331.

[87] Fox claimed inspiration from both Konrad Lorenz and Charles Darwin, especially *The Expression of Emotions in Man and Animals* (London, 1872).

[88] Robin Fox, "The Evolution of Human Sexual Behavior," *New York Times*, 24 March 1968.

[89] Ibid., 79.

[90] Fox, "Cultural Animal" (cit. n. 83), 12.

two interviews in the *New York Times* alone. The interviews described him as a "puck-ish, 33-year-old Canadian" and as "tailored in London, shod in France and automo-tively equipped by Alfa-Romeo."[91] As a representative of a new breed of hip scientific men, Tiger sought to make two points. First, like Fox, he wanted to explore the possi-bility of uniting biological and anthropological analyses of human behavior. Second, as his case study of such an approach in action, he asked, "Why do human males form all-male groups? What do they do in their groups? And, what are the groups for?"[92] He proposed, in answer, that all-male groups were ubiquitous in human societies and required some kind of initiation ceremony to demarcate members from nonmembers. These ceremonies, in turn, reflected a form of "unisexual" selection "for work, de-fense, and hunting purposes" that paralleled sexual selection for reproductive pur-poses. In this unisexual selection, male bonding and aggression were intimately linked.

"Aggression," Tiger wrote, "is both the product and cause of strong affective ties between men."[93] He further proffered that unisexual organizations, like all-male prep schools and universities, promoted conceptions of masculinity that arose out of these institutions. The net effect kept males from high-ranking families dominant and subor-dinated males from families without access to such resources; it also naturally excluded females from the upper echelons of political and social power.[94] If he were right, Tiger reasoned, then "modifying the dynamics and repercussions of the male bond may be a crucial feature of altered attitudes to power, to the value of destroying other communi-ties' people and property, and to the concept that manliness is strength rather than flex-ibility and authority rather than attentiveness to others."[95] Given his working-class back-ground, Tiger was likely attuned to the mechanisms by which class could be perpetuated through codes of masculinity defined by youthful experience with exclusive all-male institutions.[96] Yet he also despaired of changing the resulting social dynamics that char-acterized contemporary life. He argued that increasing the proportion of women in male-dominated fields would not help—men would continue to see younger women as sexual objects, older women as competitors, and neither as colleagues—while men would instinctively continue to seek alliances with other men. Tiger's reluctance to admit that the system could be changed left him open to charges of biological determinism.

Reviews of *Men in Groups* were mixed. Elaine Morgan, an early feminist critic of man-the-hunter theories of human evolution, argued that Tiger's theory of male bonding dominated discussions of human social interactions and thus obscured at-tention to equally important female-female relations.[97] Another reviewer called his theory "necessarily tentative and tenuous, based as it is on analogy, speculation and

---

[91] Quotes from, respectively, Joan Cook, "Explaining Why Men Love to Be One of the Boys," *New York Times*, 21 June 1969, 14; and Marylin Bender, "No Time for Dandies?" *New York Times*, 14 Sep-tember 1969, SM2A2.

[92] Tiger, *Men in Groups* (cit. n. 11), xiii.

[93] Ibid., 196.

[94] Ibid., 202–3.

[95] Ibid., 214.

[96] For a historical perspective, see Nye, "Medicine and Science" (cit. n. 15); Carroll Smith-Rosenberg, "Writing History: Language, Class, and Gender," in *Feminist Studies, Critical Studies*, ed. Teresa de Lauretis (Madison, Wis., 1986), 31–54. For more on all-male institutions and privilege, see Amy Milne Smith's *London Clubland: A Cultural History of Gender and Class in Late-Victorian Britain* (London, 2011).

[97] Elaine Morgan, *Descent of Woman* (New York, 1972), 190–3; Erika Lorraine Milam, "Dunking the Tarzanists: Elaine Morgan and the Aquatic Ape Theory," in *Outsider Scientists*, ed. Oren Harman and Michael R. Dietrich (Chicago, 2013), 223–47.

a staggering variety of scholarly sources."[98] Tiger fared slightly better at the hands of physical anthropologist Sherwood Washburn (University of California, Berkeley), who nevertheless emphasized the book's "deep problem . . . the nature of the evidence; how can such an interesting point of view become more than a very tentative hypothesis?"[99] Ardrey, unsurprisingly, penned the most positive review, and underscored Tiger's scholarly tone. In his usual metaphor-laden style, he wrote, "footnotes fly by like June bugs in a summer cottage. But let the reader be tolerant. . . . In the academic jungle a footnote can be a man's best friend. Let the reader likewise endure a 'paradigm' or two as scholarly décor. He may read with assurance that no cloistered jargon will muffle the explosion let loose in these pages." In Ardrey's reading of *Men in Groups*, two factions would find their cultural assumptions smashed by Tiger's findings: the Freudians, who held that all male friendship resulted from latent homosexual attraction, and the feminists who sought to challenge the "age-old male bond."[100] All-male groups were not the result of cultural prejudice or psychological deviance, Ardrey concluded, but of our human biological heritage.

Given Tiger and Fox's earlier work, in *The Imperial Animal* they drew evidence from biology, history, and genetics to describe the evolution of human "biogrammar," the political nature of human social interactions, the mother-child bond, and male competition.[101] They devoted the second half of the book to more traditional social scientific topics, illustrating how evolution could have influenced the origins of economic systems, educational practices, and efforts to maintain the health of the social body. In writing *The Imperial Animal*, Tiger and Fox fulfilled what they saw as their scholarly duty of diagnosis, leaving its lessons in the hands of the social scientists and politicians who they hoped would embrace and learn from their biological past.[102] For many readers, the book appeared to build directly on the ethological framework provided by Lorenz, Ardrey, and Morris (see fig. 1). Fox later recalled that the volume attracted reviews by "hostile feminists, turf-defending social scientists, snotty humanists and friendly laymen—and women for that matter."[103] Elizabeth Fisher, for example, questioned the "clear case" for dominance hierarchies in chimpanzees but quipped, "That they exist among academics I have no doubt—one feels that no woman can advance in those circles without making certain ritual submission gestures."[104] Fox also highlighted the infighting among anthropologists, describing their attacks as territorial defense. This allusion was nowhere clearer than in a conversation in which Fox remembers Ardrey cheerfully claiming his antagonists were "being consistent with their own Darwinism: they struggle for existence, for the reproductive success of their own ideas."[105]

As they built their own professional identities, Fox and Tiger sought to include Ardrey in their networks. At a conference, Fox recalls defending Ardrey against at-

---

[98] Christopher Lehmann-Haupt, "The Disturbing Rediscovery of the Obvious," *New York Times*, 18 June 1969, 45.

[99] Sherwood Washburn, "Does Biology Account for the Men's Club?" *New York Times*, 27 July 1969, BR10.

[100] Ardrey, "A Tiger" (cit. n. 76), 12.

[101] Lionel Tiger and Robin Fox, *The Imperial Animal* (1971; repr., New York, 1989). The later reprint includes a foreword by Konrad Lorenz and a new introduction by the authors.

[102] Ibid., 236.

[103] Fox, *Participant Observer* (cit. n. 66), 437.

[104] Elizabeth Fisher, "Nature and the Animal Determinists," *Nation*, 17 January 1972, 89–90.

[105] Fox, *Participant Observer* (cit. n. 66), 395.

Figure 1. *Konrad Lorenz leading Lionel Tiger, Robin Fox, Robert Ardrey, and Desmond Morris down the garden path of human ethology. The original caption read, "The man/ animal bandwagon, undoubtedly on its way to the bank." From Judith Shapiro, "I Went to the Animal Fair . . . the Tiger and Fox Were There,"* Nat. Hist. *80 (1971): 90–8.*

tacks by scientists who lambasted his (and Lorenz's) characterization of aggression as "instinct."[106] According to Tiger, "somebody made some nasty comment about Robert Ardrey, holding him responsible for the Vietnam War and for everything. Fox said, this is just ridiculous, let's just give him an honorary degree and let him be a member of the group and people will stop bothering him—which to my knowledge was the first time anyone had ever said that in public."[107] Tiger suggested that many of Ardrey's professional difficulties stemmed from his lack of academic accreditation. Yet at least according to Fox, Ardrey was happy with life outside the academy. Fox later recalled an illuminating conversation along these lines. After discussing the faults of work by a fellow neo-Darwinian, he remembers Ardrey questioning whether the point was worth a big fight. "Remember . . . you are on the same side in the great struggle: the struggle to get the evolved, biological aspects of behavior recognized and incorporated into a scientific view of human behavior." Try at all times, he encouraged Fox, "to boost and encourage and promote" people on your side.[108]

In fact, both Fox and Tiger saw their work at the HFGF through precisely this lens—it gave them the opportunity to encourage research on the human animal at a time when funds for such work were in short supply. In his memoir, Fox wrote that they "thought of themselves as like the original Royal Society before it became a formal

[106] Segerstråle, *Defenders of the Truth* (cit. n. 10), 90–4.
[107] Tiger, interview by Milam (cit. n. 77).
[108] Fox, *Participant Observer* (cit. n. 66), 395.

institution: an 'Invisible College.' They didn't have an Institute, like [Clifford] Geertz in Princeton, to which people could come. Rather they went out to the people—drew them into the web, put them in touch with each other, gave them support."[109] He additionally submitted that "they were both wary of guruhood; disciples meant trouble and bother," so they self-consciously chose not to create a new journal, hold exclusive meetings, or build "factions, scandal, gossip, just like all the others."[110] Instead, they hoped to spread an evolutionary perspective of human nature in new academic conversations, to "seed the virgin environments." Rather than create a new discipline, Fox and Tiger tried "to return their own discipline," anthropology, to what they saw as its "true mandate." Once anthropology reclaimed its evolutionary roots, they firmly believed, other behavioral sciences would naturally follow suit. "That was the plan."[111]

When Mason W. Gross retired from his position as president of Rutgers University and became president of the HFGF in 1971, he was familiar with Fox and Tiger's work and actively supported their candidacy as research directors. Gross asked them to submit a research plan to the board of directors.[112] When he read it, Henry Moe opposed the idea. He wrote to Gross arguing that Fox and Tiger were "men in a hurry," and as a result they piled "hypothesis upon hypothesis, until, in the end a hypothetical conclusion is stated as fact. This is not science." Moe conceded that he was not an expert in the field, so he sought out reviews of *The Imperial Animal* and disappointingly "found none wholly favorable." He even invoked the memory of the HFGF's illustrious founder, suggesting that during Guggenheim's life, similar proposals had come before him, but he had always rejected the idea of a university-based research center (even though Fox and Tiger had something else in mind).[113] He found himself forced to conclude that the trustees should keep looking. Gross, however, ignored his concerns and on 28 March 1972 recommended to the board of directors that the "Fox-Tiger proposal be approved as a three-year project with a budget of up to $175,000 for the first year."[114] The board agreed, and Fox and Tiger accepted its offer.[115] Shortly thereafter, Tiger reported to Gross two lengthy discussions with Ardrey. "Mr. Ardrey" estimated the Fox-Tiger plan for the HFGF was "both an appropriate and constructive rendition of Mr. Guggenheim's own ideas on the subject."[116] Even after his death, Guggenheim remained a touchstone for members of the organization.

[109] Ibid., 558. Cultural anthropologist Clifford Geertz accepted an appointment at the Institute of Advanced Study in 1970, and Fox and Tiger came to see him as somewhat of a professional nemesis. See especially Tiger's obituary of Geertz, "Fuzz. Fuzz . . . It Was Covered in Fuzz," *Wall Street Journal*, 7 November 2006.

[110] Fox, *Participant Observer* (cit. n. 66), 558.

[111] As recounted by Fox in *Participant Observer* (cit. n. 66), 559.

[112] Robin Fox and Lionel Tiger, "Proposed Program for the Harry F. Guggenheim Foundation," January 1972, in "History of the Harry Frank Guggenheim Foundation" (New York), chap. 4. This photocopied, spiral-bound pamphlet is distributed internally at the foundation (initial date of compilation unknown).

[113] Henry Allen Moe, Memorandum for Dr. Mason Gross, 2 March 1972, Folder: Guggenheim, Harry Frank, Foundation: Tiger-Fox Project #25 and Folder: Tiger and Fox #35, Moe Papers; emphasis in the original.

[114] Folder: Guggenheim, Harry Frank, Foundation #5, Moe Papers.

[115] "Initial terms of agreement, from April 1 1972 to July 1 1973," Folder: Guggenheim, Harry Frank, Foundation: Ardrey, Robert, Moe Papers. According to this contract, Tiger and Fox were each to be paid $1,000/month, plus half of their salary at Rutgers, and any missing benefits because of their half-time work.

[116] Tiger to Mason Gross, 17 April 1972, Folder: Guggenheim, Harry Frank, Foundation: Gross, Mason W., Moe Papers.

Under the guidance of Fox and Tiger's "Invisible College," the HFGF funded research designed to unpack the origins of human aggression and explore the behavioral evolution of humanity more broadly. People applied for, and received, small grants (of approximately $10,000) to support research in a wide variety of fields—psychiatry, zoology, biological and cultural anthropology, psychology, endocrinology, and not least ethology—demonstrating the interdisciplinary nature of human evolutionary studies in the early 1970s.[117] Yet for Moe, it seemed the primary beneficiaries of the foundation's largesse were Fox and Tiger themselves. He wrote to Ardrey, asking him to nominate other leaders in the field who would provide a broader perspective. Ardrey capitulated, listing Nikolaas Tinbergen, Konrad Lorenz, Ernst Mayr, Sherwood Washburn, and René Dubos as important theorists or statesmen of biology—Tinbergen and Lorenz, for example, had shared a Nobel Prize in physiology or medicine earlier that year with physiologist Karl von Frisch.[118] All of these men, however, held academic positions of tremendous power already. Moe also contacted cotrustee Roger Straus, expressing what he hoped was a mutual hesitation at placing "practically all our eggs in the Tiger-Fox basket." Topmost on his mind were the expenses Fox and Tiger were incurring.[119] Moe noted, too, that he had never received a response to his letter to Gross, "either in writing or orally." He felt snubbed. Indeed, at the end of their three-year contract, when Gross proposed giving Fox and Tiger a raise, Straus and Moe attempted to persuade the rest of the trustees to "take a cold hard look at the overall Tiger/Fox relationship philosophically as well as financially."[120] Instead, Fox and Tiger were granted a raise of $3,000 a year.[121]

In 1978, after six years at the helm, the HFGF asked Fox and Tiger to evaluate their work so far as directors. In their responses, Fox and Tiger reclassified most of the grants from previous years, reflecting their evolving interests. In this new scheme, almost half of the grants fell into the categories of ethology (human and animal) and sociobiology, while the bulk of the remaining grantees devoted their research to brains, hormones, and behavior. Paleoanthropology and cross-cultural studies, which had generated two of the most significant forms of data for re-creating a universal human nature in the 1960s, largely disappeared as research categories of interest. The overwhelmingly male grantees included young movers and shakers in biological anthropology and sociobiology, as well as scientists whom Fox and Tiger had read avidly in their early forays into ethology.[122] Although the grantees hailed from a wide array of disciplines, they also knew each other from past symposia, conferences, and edited collections. It was a small but steadily growing community of men and women.

[117] Fox and Tiger stipulated that the grant money (slightly more than $44,000 in 2015 dollars) could be used to pay neither a portion of the primary researcher's salary nor university overhead.

[118] Ardrey to Moe, 30 November 1973, Folder: Guggenheim, Harry Frank, Foundation: Ardrey, Robert, Moe Papers.

[119] Moe to Roger W. Straus, 2 March 1974, Folder: Guggenheim, Harry Frank, Foundation #43, Moe Papers.

[120] Roger W. Straus Jr. to Mr. Peter Lawson-Johnston and Dr. Mason Gross, 4 August 1975, Folder: Guggenheim, Harry Frank, Foundation: Tiger and Fox #23; Moe to Lawson-Johnston and Gross, 8 August 1975, Folder: Guggenheim, Harry Frank, Foundation: Tiger and Fox #23, Moe Papers.

[121] To the Directors of the Harry Frank Guggenheim Foundation, 27 August 1975, Folder: Guggenheim, Harry Frank, Foundation: Tiger and Fox #23, Moe Papers.

[122] See Folder: Guggenheim, Harry Frank, Foundation #18; Minutes of Meeting of Fellowship Committee, Held on 6 April 1973, Guggenheim, Harry Frank, Foundation #7; Minutes of Special Meeting of Board of Directors, Held on 17 May 1973, Folder: Guggenheim, Harry Frank, Foundation #9, Moe Papers.

Crucial to this community-building effort was HFGF support (in part or in whole) for conferences later published as edited collections.[123]

In his self-evaluation, Fox contended, "The Foundation has played a crucial role during the 1970s in facilitating one important development in the thinking of social and biological scientists, namely the greatly increased recognition that evolutionary and genetic factors have an important influence on behavior."[124] Many agencies and groups contributed to this effort, he asserted, but "the Foundation has played a role far out of proportion to the magnitude of its resources."[125] Fox's statement was self-serving but not wrong. For years, the Wenner-Gren Foundation had supported a great deal of primatological and anthropological research, especially through a program called "The Origins of Man," which ran from 1965 to 1972.[126] Because of their increasingly dire economic situation, however, Wenner-Gren dramatically cut back spending on research and conferences in the 1970s. Funding from the National Science Foundation (NSF) failed to make up the difference.[127] In fact, social scientists had been so frustrated with the lack of consideration and attention they received within the hierarchy of the NSF that they strongly considered establishing a separate National Social Science Foundation.[128]

The HFGF thus funded scientific research at the juncture of ethology and biological anthropology at a crucial moment in the crystallization of sociobiology as a discipline.[129] As research directors of the HFGF until 1984, Fox and Tiger successfully created a space where research on the human animal, with all of its gendered tropes, could be supported and sustained. Reacting to a perceived crisis of authority in the social sciences, they sought to incorporate an intellectually rigorous zoological perspective.[130] Earlier writers, like Lorenz, Ardrey, and Morris, had argued for the importance of biological instincts in defining human behavioral patterns, but they could not on

[123] E.g., *Violence and Aggression: Areas of Ignorance* (1972); *Female Hierarchies* (1974); *Biology and Politics, Recent Explorations* (1975); *Conference on the Origins and Evolution of Language and Speech* (1975); *Conference on Brain and Behavior* (1976); *Implications of Sociobiology for the Social Sciences* (1977, in conjunction with the American Academy of Arts and Sciences); *Conference on Kin Selection and Kinship Theory* (1978); and many more cosponsored events. *Reports of the Harry Frank Guggenheim Foundation*, 1929–74 and 1974–8.

[124] Robin Fox, "Overview of Research-Grant Activity of the Harry Frank Guggenheim Foundation," attached to Donald R. Griffin to Peter O. Lawson-Johnson, 9 November 1978, in "History of the Harry Frank Guggenheim Foundation" (cit. n. 112).

[125] Ibid., 6.

[126] E.g., John Buettner-Janusch, ed., *Origins of Man: Physical Anthropology* (New York, 1966); Phyllis C. Jay, ed., *Primates: Studies in Adaptation and Variability* (New York, 1968); Russell Tuttle, ed., *The Functional and Evolutionary Biology of Primates* (Chicago, 1972). The Smithsonian also funded at least one conference, published as J. F. Eisenberg and Wilton S. Dillon, eds., *Man and Beast: Comparative Social Behavior* (Washington, D.C., 1971).

[127] On biological funding within the NSF, see Toby Appel, "Allocating Resources to a Divided Science," in *Shaping Biology: The National Science Foundation and American Biological Research, 1945–1975* (Baltimore, 2002), 207–34.

[128] Mark Solovey, "Senator Fred Harris's National Social Science Foundation Proposal: Reconsidering Federal Science Policy, Natural Science–Social Science Relations, and American Liberalism during the 1960s," *Isis* 103 (2012): 54–82; and Solovey, "Riding Natural Scientists' Coattails onto the Endless Frontier: The SSRC and the Quest for Scientific Literacy," *J. Hist. Behav. Sci.* 40 (2004): 393–422.

[129] The field was given a new name and impetus by the publication of E. O. Wilson's *Sociobiology: A New Synthesis* (Cambridge, Mass., 1975); see Segerstråle, *Defenders of the Truth* (cit. n. 10).

[130] Laura Nader, "Up the Anthropologist: Perspectives Gained from Studying Up," in *Reinventing Anthropology*, ed. Dell Hymes (New York, 1972), 284–311; Naomi Quinn, "The Divergent Case of Cultural Anthropology," in *Primate Encounters: Models of Science, Gender, and Society*, ed. Shirley Strum and Linda Marie Fedigan (Chicago, 2000), 223–42.

their own transform the study of animal and human behavior in the United States. That work had to be done by academic insiders with financial resources at their disposal and a domestic network of influence—men like Fox and Tiger.

CONCLUSION

Whereas Ardrey, Fox, and Tiger saw their theories of human nature reified in their personal experiences with masculine professional networks like the Harry Frank Guggenheim Foundation, feminist scientists and humanists viewed both as illustrating how scientific theories could spring uncritically from the gender norms male scientists took for granted. The masculine stereotypes embodied in these scientific theories were part of a larger reimagining of the roles men played in American social life. Ozzie and Harriet may have represented an ideal life in the 1950s, but the immense popularity of Sloan Wilson's *Man in the Gray Flannel Suit* and David Reisman's *The Lonely Crowd* simultaneously spoke to widespread anxieties about the soullessness of suburban corporate life.[131] By the close of the 1960s, both corporate culture and masculine ideals were rapidly changing and attracting the critical attention of popular writers and scholars alike. Greater numbers of women entered the work force as "career girls," creating office spaces potentially fraught with sexual tension and fluctuating power dynamics.[132] In this context, the importance of reinstantiating homosocial spaces emerged in both social and scientific imaginaries.

Fox has since lamented how wrong he and Tiger turned out to be in predicting the future of anthropology, even as other social sciences have variously embraced their Darwinian message. His faith still resides in the authority of biology, though. Alluding to studies of the human genome, he recently claimed that the "Family of Man is no longer a utopian slogan but a genetic fact."[133] Tiger similarly maintains that after decades of battle over the precepts advanced in the "Zoological Perspective in Social Science," he and Fox have been proven "largely right." "Of course many disagreed then and disagree still," Tiger wrote in 2011. "The low-oxygen post-modern this-and-that fog continues to dull reality . . . with a confidently glad anti-empiricism which neither Fox nor I . . . ever expected to become the viral suffocating force it became."[134] Both men believe their case was won not by social scientists reluctant to adopt their perspective, but by geneticists. Yet as they are all too aware, the field of anthropology continued to change around them.

The (literal and figurative) masculine networks of academic authority exemplified by the HFGF and the sociobiological research it supported did not go unchallenged. Within biological anthropology, conceptions of the human animal quickly extended to encompass the wide variety of strategies female animals and women might utilize to structure the societies in which they lived.[135] Not only were pop-ethologists like Ardrey promulgating the biological origins of "men in groups," so were professionals, and that provided stakes worth fighting for. As scientists confronted cultural changes

---

[131] Sloan Wilson, *The Man in the Gray Flannel Suit* (New York, 1955); David Reisman, *The Lonely Crowd: A Study of the Changing American Character* (New Haven, Conn., 1950).

[132] Helen Gurley Brown, *Sex and the Single Girl* (New York, 1962); Antony Jay, *Corporation Man* (New York, 1971).

[133] Robin Fox, "The Old Adam and the Last Man," *Society* 48 (2011): 462–70.

[134] Lionel Tiger, "Full Circle," *Society* 48 (2011): 500–1.

[135] For a stimulating reflection on this research, see Strum and Fedigan, *Primate Encounters* (cit. n. 130).

at home and in the workplace, their academic debates of the late 1960s and early 1970s laid the groundwork for deep skepticism of biological explanations of human behavior among feminist scientists and humanists in the decades to come.[136]

[136] E.g., Marian Lowe and Ruth Hubbard, "Sociobiology and Biosociology: Can Science Prove the Biological Basis of Sex Differences in Behavior?" in *Genes and Gender II*, ed. Ruth Hubbard and Marian Lowe (New York, 1979), 91–112; Janet Sayers, *Biological Politics: Feminist and Anti-Feminist Perspectives* (New York, 1982); Ruth Bleier, "Sociobiology, Biological Determinism, and Human Behavior," in *Science and Gender: A Critique of Biology and Its Theories on Women* (New York, 1984), 15–48. In the 1990s, feminism and evolutionary theories found more solid common ground; e.g., Patricia Gowaty, ed., *Feminism and Evolutionary Biology: Boundaries, Intersections and Frontiers* (Boston, 1997).

Manliness and Exploration:
The Discovery of the North Pole

by Michael Robinson*

ABSTRACT

Americans crowded newsstands in early 1910 to read Robert Peary's firsthand account of his expedition to the North Pole. As they read "The Discovery of the North Pole," serialized exclusively in *Hampton's Magazine*, few knew that this harrowing, hypermasculine tale was really crafted by New York poet Elsa Barker. Barker's authorship of the North Pole story put her at the center of a large community of explorers, writers, patrons, and fans who were taken with Arctic exploration as much for its national symbolism as for its thrilling tales. The fact that Barker was a woman made her ascent into elite expeditionary circles remarkable. Yet this essay argues that it was also representative: women shaped the ideas and practices of manly exploration at home as well as in the field. Peary's dependence upon women writers, patrons, and audiences came at a time when explorers were breaking away from their traditional base of support: male scientific networks that had promoted their expeditions since the 1850s. Despite the "go-it-alone" ideals of their expedition accounts, explorers adopted masculine roles shaped by the world around them: by the growing influence of women writers, readers, and lecture-goers and, simultaneously, by the declining influence of traditional scientific peers and patrons. Barker and Peary's story, then, reveals a new fault line that opened up between scientists and explorers in the late nineteenth century over the issue of manliness, a fault line still largely uncharted in historical scholarship.

When American explorer Robert Peary took the podium at the Eighth International Geographic Congress in 1904, he addressed an audience well aware of the new competitive spirit of polar exploration. "There is no higher, purer field of rivalry than this Arctic and Antarctic quest," he told the Congress. The rivalry was there for all to see. The audience was filled with polar experts as well as Peary's personal rivals, men who had fielded expeditions to the polar regions over the previous twenty years. By the beginning of the twentieth century, polar exploration had become a new form of geopolitical theater, an expression of what political scientist Joseph S. Nye would later call "soft power" that attracted the attentions of Russia, Austria, Norway, Sweden, Italy, Britain, and the United States. Still, Peary believed that there was a bond connecting the men seated before him, one that transcended the divides of politics and national chauvinism. "If I win . . . be proud because we are of one blood—the man blood."[1]

*Hillyer College, University of Hartford, 200 Bloomfield Avenue, West Hartford, CT 06117; microbins@hartford.edu.

[1] Joseph S. Nye, *Bound to Lead: The Changing Nature of American Power* (New York, 1991). On polar exploration, gender, and national identity, see Lisa Bloom, *Gender on Ice: American Ideologies*

This statement is one that Peary's audience would have understood. Gradually over the course of the nineteenth century, polar exploration had come to represent a place of manly contest rather than scientific investigation. This was especially true of the quest to reach the North Pole. The polar axis was a location made valuable from the efforts of explorers trying to reach it, rather than anything it offered to science or commerce. For thirty years, these efforts had been intense. Between 1870 and 1900, dozens of expeditions had taken up the pursuit of the "Arctic Grail." By the time Peary gave his speech, the sere landscapes of the high Arctic—from the scarps of Ellesmere Island to the estuaries of the Lena Delta in Siberia—were dotted with cairns, campsites, and the wreckage of European and American ships. And graves. As expedition after expedition failed to reach the North Pole, the list of men who failed to return from the high Arctic had grown longer and the value of accomplishing the feat only increased. By 1900, competition to reach the polar axis had become intense. The North Pole developed into the world's most difficult finish line. And like all finish lines, it was a marker for something else, an external benchmark for measuring internal attributes: a way of assessing the mettle of the contestants in the field. This was the true object of Peary's quest and the source of this contest's spectacular popularity among Europeans and North Americans at the turn of the twentieth century. To stand at the North Pole was to achieve an almost impossibly difficult feat, one that, in the doing, might express something rare, perhaps lost, in the industrial age of the Western world: the essential, elemental qualities of manliness itself.[2]

The North Pole had not always carried such meanings. From the 1500s through the early 1800s, it attracted only occasional attention by cartographers who, when forced to render it on a map, adorned it with islands, channels, and magnetic lodestones. Of more serious interest were the lower latitudes of the Arctic regions where Western European countries placed their hopes in finding the Northwest Passage: an ice-free waterway connecting the Atlantic and Pacific Oceans. For England especially, dependent as it was upon maritime trading, the Northwest Passage promised a means of reaching India, China, and the Spice Islands that avoided the long, dangerous, and expensive passage down the coast of Africa and around the Cape of Good Hope. These expeditions did not find the passage, but they did demonstrate one thing. If the Northwest Passage did exist, it would be a mariner's nightmare, a waterway that snaked through thousands of miles of barren islands and was choked with ship-crushing pack ice. As the perils of this region became clear, hope in a commercial route to Asia dimmed. Efforts to find the passage fell off by the late eighteenth century.

Ironically, it was the perils of the Northwest Passage that revived interest in its discovery in the early nineteenth century. Faced with a surplus of officers and ships at the end of the Napoleonic Wars, the British Admiralty decided to put some of them to use

---

of Polar Expeditions (Minneapolis, 1993); Robert G. David, The Arctic in the British Imagination, 1818–1914 (Manchester, 2000); Jen Hill, White Horizon: The Arctic in the Nineteenth-Century British Imagination (Albany, N.Y., 2008); John George Moss, Enduring Dreams: An Exploration of Arctic Landscape (Concord, Ont., 1994); Sarah Moss, Scott's Last Biscuit: The Literature of Polar Exploration (Oxford, 2006); Michael F. Robinson, The Coldest Crucible: Arctic Exploration and American Culture (Chicago, 2006); Francis Spufford, I May Be Some Time: Ice and the English Imagination (New York, 1997); Eric Wilson, The Spiritual History of Ice: Romanticism, Science, and the Imagination (New York, 2003). Robert Peary is quoted in "Eighth International Geographic Congress," National Geographic Magazine 15 (1904): 425.

    [2] Robinson, Coldest Crucible (cit. n. 1), 133–64; John McCannon, A History of the Arctic: Nature, Exploration, and Exploitation (London, 2012).

in an assault upon the "Frost King." There, the British navy could display its mettle by finding the Northwest Passage, advancing the cause of science and geographical discovery at the same time. In this, the Admiralty drew inspiration from Enlightenment expeditions of the 1700s—by Condamine, Cook, Bougainville, Malaspina—that largely avoided the sacking, looting, and slaving that had so animated Europeans in the 1500s and 1600s. The British Arctic expeditions of the early 1800s followed in the track of these Enlightenment expeditions in forgoing overt imperial gains for more pacific ones: filling in maps, gathering specimens, and surveying resources. The looming bergs and desolate wastes of the Canadian Arctic, lit by moonlight and phantasmagoric displays of the Northern Lights, also appealed to popular tastes that now tilted toward the romantic and the gothic. Behind the pressed flowers and specimen jars and the aesthetic appreciation of icebergs, of course, lay other motives. Scientific exploration advanced national interests by gathering intelligence about the wider world. It also gave the world some intelligence of its own. All of the scientific expeditions of the 1700s had operated as majestic acts of display as much as missions of discovery. Pacific Islanders watching naturalists sway up their beaches in the 1700s could not help but notice the leviathan ships from which these strangers disembarked, craft that had crossed two oceans billowing canvas, and, through precise navigation, found them in the middle of the Pacific Ocean. That such ships could navigate their way anywhere in the world—even through oceans of ice—was one of the lessons the Admiralty wanted to impart to the world.[3]

The Admiralty wanted to impart this lesson to its own citizens as well. The search for the Northwest Passage became popular in the British press precisely because it was so difficult. In endeavoring to find it, no longer for commercial advantage but for scientific advancement and national prestige, seamen seemed to express traits of a higher order, of manly—and by extension British—character. Newspapers and journals followed the expeditions closely. Explorers published narratives upon their return. Extending the reach of these stories was the growing power of the British press, which benefited from technological innovations and so was able to offer expedition stories more quickly and cheaply than ever before. An expanding middle class avidly consumed such stories. As American and Western European countries followed the British into the Arctic, they adopted many of their methods and motives for exploration, including their exploitation of exploration stories back home. The polar regions became a means of selling stories and teaching lessons about national character and traits of manliness.[4]

Peary grew up in an age awash in stories of Arctic explorers and their adventures, an ocean of literature that even reached the small, forested towns of southern Maine, where he had grown up in the 1860s. He came of age at a time when the United States had begun to stretch its wings as a nation, sending exploring expeditions into the Pacific Ocean, South America, and the Western Territories. Americans also followed the Arctic

[3] Angela Byrne, *Geographies of the Romantic North: Science, Antiquarianism, and Travel* (New York, 2013), 17–38; Felix Driver, *Geography Militant: Cultures of Exploration and Empire* (Oxford, 2001), 1–23.

[4] Dorinda Outram, "New Spaces in Natural History," in *Cultures of Natural History*, ed. N. Jardine, J. A. Secord, and E. C. Spary (Cambridge, 1996), 249–65; Beau Riffenburgh, *The Myth of the Explorer: The Press, Sensationalism, and Geographical Discovery* (New York, 1994); Michael F. Robinson, "Science and Exploration," in *Reinterpreting Exploration: Reassessing the West's Encounter with the Rest*, ed. Dane Kennedy (New York, 2014), 21–37.

exploration that had become a subject of great interest in America in the early 1800s. Through the last half of the nineteenth century, explorers published hundreds of articles and books about their dangerous adventures in the far north. They lectured to audiences about their travels. Other individuals and institutions joined in the celebration of the Arctic quest. Museums and world's fairs displayed explorers' sledges, dogs, and the Inuit they brought back with them. Prestigious artists such as Frederic Church used expedition accounts as inspiration for their exotic landscapes, while lesser-known artists chronicled expeditions in massive traveling Arctic panoramas. Commercial vendors commemorated Arctic explorers in sheet music, silverware designs, fabric patterns, button molds, and playing cards. The many products of polar culture reinforced the fearsomeness of the Arctic and the fearlessness of the men who explored it.[5]

Beneath their scaffolding of danger and exoticism, these various Arctic productions became vehicles for examining personal character. Narratives became moral tales in which explorers, through their toil and suffering, revealed the highest qualities of personal character. By the late nineteenth century, only the polar regions and central Africa remained "undiscovered." Dozens of expeditions entered the Arctic under the banners of science and commerce. Yet despite their stated goals, explorers generally ignored these elements in their writings. Instead they focused on personal dramas: their battles and victories against the ice, the cold, and starvation. Character had always been an implicit element of Arctic narratives since early narratives had emphasized explorers' endurance, intelligence, and eloquence in describing nature. But by the turn of the century, different elements of character had come to the fore. Arctic narratives increasingly focused on the rugged, rough-and-tumble life that explorers led in the Arctic.[6]

Peary had become enchanted by stories of exploration, particularly the narratives of Elisha Kent Kane, the most celebrated explorer of his age. Kane had led an expedition into the Arctic in 1853 in pursuit of the missing British explorer Sir John Franklin, who had disappeared in 1845 seeking the Northwest Passage. Kane failed to find Franklin but managed to bring his party out of the Arctic alive after their expedition ship, *Rescue*, became trapped in the ice on the northwestern coast of Greenland. Kane's miraculous escape touched off a frenzy of excitement in 1855. His book *Arctic Explorations* sold 150,000 copies when it came to press in 1857. When Peary discovered *Arctic Explorations* in the 1870s as a boy, he wrote, "a chord . . . vibrated intensely within me at the reading of Kane's wonderful book." When Peary died in 1920, he still possessed a clipping about Kane that he had cut out of a Sunday newspaper as a child.[7]

Peary did not discuss which features of Kane's life and work most appealed to him. Yet others did, especially after Kane's early death from rheumatic fever in 1857. These

    [5] Russell Potter, *Arctic Spectacles: The Frozen North in Visual Culture, 1818–1875* (Seattle, 2007), 117–61; Robert Bryce, *Cook and Peary: The Polar Controversy, Resolved* (Mechanicsburg, Pa., 1997), 15–6; Robinson, *Coldest Crucible* (cit. n. 1), 1–14.
    [6] Rebecca Herzig, *Suffering for Science: Reason and Sacrifice in Modern America* (New Brunswick, N.J., 2005), 64–84; Bloom, *Gender on Ice* (cit. n. 1), 15–55; Riffenburgh, *Myth* (cit. n. 4), 5–7, 46–8.
    [7] Mark Metzler Sawin, *Raising Kane: Elisha Kent Kane and the Culture of Fame in Antebellum America* (Philadelphia, 2008); David Chapin, *Exploring Other Worlds: Margaret Fox, Elisha Kent Kane, and the Antebellum Culture of Curiosity* (Amherst, Mass., 2004); George Washington Corner, *Doctor Kane of the Arctic Seas* (Philadelphia, 1972); quotation from Robert Peary, *Northward over the Great Ice* (New York, 1898), xxxiv; Wendell Oswald, *Eskimos and Explorers* (Lincoln, Neb., 1979), 118.

early discussions of Kane prove interesting because they identify features of manliness and character that would contrast with Peary's public persona as an explorer fifty years later. In letters, reviews, and eulogies, Americans of the 1850s admired Kane because of many things, foremost of which was his power of command. In 1854, he had demonstrated his leadership by bringing his crew out of the high Arctic in whaleboats after their ship had become locked in the ice. They also admired his powers of observation, exalting him as a "a genius in science," a man whose "contributions to science laid the whole world under obligation," and by whose death "science has been deprived of an ardent advocate." Yet most of all, they admired Kane's manly self-control, his ability to overcome his own physical limitations, and the refinements of character that enabled him to complete his Arctic mission. Kane's strength of character "raised his feeble frame above bodily weakness," crewman Henry Goodfellow declared, "and enabled him to triumph over cold and hunger." "His palate was delicate," remembered shipmate William Morton, but as conditions in the Arctic worsened, "he accustomed himself to eat puppies and rats." Eulogists seemed willing, even eager, to accentuate the delicacy of Kane's manners and physical features to the point of effeminacy. Belle Burns, a housemaid at Kane's medical school, remembered Kane for "his pretty gentle manners, . . . his sweet young face, and lovely complexion like a girl's." Praising Kane for the care of his sick men, one newspaper observed that he had shown "the gentle qualities of a woman."[8]

Implicit in these eulogies was the assumption that manliness existed as an inner quality rather than as an expression of physical skill or strength. Rather than diminish his status as a man, then, the frail, even effeminate, image of Kane that emerged served to strengthen it. "When something was to be done which required nerve and manhood," Kane's ship's surgeon recalled, "a sleeping power was aroused within him, which sent palpitating heart, puffed cheeks, rheumatic joints, and scurvy limbs hastily to cover." Even Kane's death, coming fifteen months after his return from the Arctic, offered a final demonstration of his manliness: "He literally postponed the fatal crisis," another crew member told audiences, "until his duty was fulfilled." It was in the Arctic, Reverend J. H. Allen told his congregation, that "the seed of death ha[d] been planted."[9]

Peary may not have read these eulogies (which were published as a monograph and included in the first biography of Kane in the late 1850s). If he did, we do not know what he took away from them. Yet it is clear that Peary had a fanatical, almost fetishistic belief in leadership and self-control. "My party might be regarded as an ideal," he wrote, "as loyal and responsive to my will as the fingers of my right hand." Of the Inuit who accompanied him into the Arctic, Peary knew how "to make them useful for my purposes. I had studied their individual characteristics, as any man studies the human tools with which he expects to accomplish results." He led seven expeditions to Greenland and the Arctic Ocean between 1885 and 1909. These missions cost him

<hr/>

[8] Mr. Cuyler, quoted in William Elder, *Biography of Elisha Kent Kane* (Philadelphia, 1858), 273; Amos Bonsall, quoted in ibid., 273; Henry Goodfellow, quoted in ibid., 283; William Morton, *Dr. Kane's Arctic Voyage; Explantory of a Pictoral [sic] Illustration of the Second Grinnell Expedition* (Boston, 1857), 2; Belle Burns, quoted in Corner, *Doctor Kane* (cit. n. 7), 27; *Daily Telegraph* (Harrisburg, Pa.), quoted in David Chapin, "'Science Weeps, Humanity Weeps, the World Weeps': American Mourns Elisha Kent Kane," *Penn. Mag. Hist. Biogr.* 123 (1999): 275–301. Kane's eulogies are excerpted from Robinson, *Coldest Crucible* (cit. n. 1), 50–1.

[9] Isaac I. Hayes, in Elder, *Biography*, 271; Morton, *Dr. Kane's Arctic Voyage* (both cit. n. 8), 2; J. H. Allen, *Elisha Kent Kane* (1857), 2.

eight toes and nearly broke his health. Even during his time at home, he poured his energies into lecturing, courting wealthy patrons, paying debts, and preparing for the next expedition. Peary's interest in Arctic exploration, and the meanings he attached to it, recapitulated those of nineteenth-century America. It was an occupation that gained focus over time, shifting from the Northwest Passage to the latitudes of the high Arctic, and finally the North Pole itself. The North Pole gradually became the geographical axis of his life, the point around which all other aspects—family, military career, friends, and hobbies—circled at a distance. Yet Peary's persona as an explorer differed significantly from Kane's. His narratives rippled with the muscular imagery that had become hallmarks of a new masculine sensibility of the late 1800s.[10]

SCIENCE, EXPLORATION, AND THE NEW MASCULINE IDEAL

The increasing muscularity of the Arctic quest grew out of changes back home rather than the nature of Arctic exploration itself. By the 1890s many cultural critics warned that white Americans were becoming soft. This anxiety about overcivilization extended beyond the United States to affect other regions of the industrial world, particularly Britain and Germany. As critics saw it, mechanization, and the increasing importance of urban culture, had divorced people from nature and made them weak. Overtaxed by the mental demands of civilized life, people increasingly appeared to suffer from nervous disorders such as neurasthenia. Men, critics warned, were especially vulnerable to overcivilization. In the United States, Theodore Roosevelt became a celebrity advocate of physical improvement. Roosevelt had grown up as a sickly child in New York City, and the experience had led him to believe that urban ills and lack of physical activity threatened to make modern men effeminate. Along with psychologist G. Stanley Hall and Boy Scouts founder Ernest Seton, many cultural critics advocated a regimen of physical culture for men and boys to stave off civilization's emasculating influence. Sporting clubs and mountaineering societies took special interest in the improvement of the male body. Within this context, Arctic explorers became new icons of "the strenuous life." While explorers still embodied American character, their toil in the Arctic kept them from succumbing to the emasculating effects of modern American culture. In their narratives, explorers increasingly adopted this new masculine ideal, emphasizing their battles against nature and their return to a primitive lifestyle in the Arctic.[11]

Adopting this image posed something of a dilemma for Peary. He carried the flag of

[10] Peary, *The North Pole: Its Discovery in 1909 Under the Auspices of the Peary Arctic Club* (New York, 1910), 270–1 (quote about "right hand"), 44 (quote about "human tools"); Bryce, *Cook and Peary* (cit. n. 5). Robert Peary, *Nearest the Pole: A Narrative of the Polar Expedition of the Peary Arctic Club in the S. S. Roosevelt, 1905–1906* (New York, 1907), 44.

[11] The "crisis of masculinity" in turn-of-the-century America has been chronicled by many scholars, most thoroughly by Gail Bederman, *Manliness and Civilization: A Cultural History of Gender and Race in the United States, 1880–1917* (Chicago, 1995), and Michael Kimmel, *Manhood in America: A Cultural History* (New York, 1996), 81–190. Yet the causes and scope of this crisis are still debated, largely because ideals of manliness were shifting elsewhere as well, sometimes in parallel fashion, particularly in northern and western Europe. See Michael S. Reidy's essay, "Mountaineering, Masculinity, and the Male Body in Mid-Victorian Britain," in this volume; James Eli Adams, *Dandies and Desert Saints: Styles of Victorian Masculinity* (Ithaca, N.Y., 1995), 149–82; George L. Mosse, *The Image of Man: The Creation of Modern Masculinity* (Oxford, 1996), 77–106; Matti Goksøyr, "Taking Ski Tracks to the North: The Invention and Reinvention of Norwegian Polar Skiing: Sportisation, Manliness, and National Identities," *Int. J. Hist. Sport* 30 (2013): 563–80.

American interests in the Arctic, both in symbol and in practice. As such, he became an icon of American civilization and progress. Yet Peary's quest for the North Pole was also built upon a critique of modern life and its seemingly emasculating ways. If the North Pole itself represented a contest among civilized nations, it also represented a place of atavistic escape, a realm in which explorers abandoned civilized life in order to find the qualities of manliness that had been impeded and restrained by the industrial, urban world. As the mountaineer Robert Dunn wrote after attempting a first ascent of Mount McKinley with Frederick Cook in 1906, modern exploration was important because it resulted in "men with the masks of civilization torn off . . . human beings tamed for centuries, then cast out to shift for themselves like the first victims of existence." Exploration had the power to advance the fields of science, geography, and commerce and, in so doing, help humanity climb the ladder of social progress. Yet "the best field of all," argued Dunn, "[is] to help this knowledge of ourselves," a stripping away of civilized practices that would liberate the overcivilized man trapped beneath. Reconciling these incommensurable aims of exploration—of advancing civilization even as one was trying to escape from it—was the task in front of Peary as he labored to launch an expedition that would place him at the North Pole.[12]

These conflicting ideals of exploration affected Peary's expeditions in many ways. First, they complicated his use of modern technology in the Arctic. After failed attempts to reach the North Pole from 1898 to 1902, Peary decided that he needed a more powerful ship, one equipped with both steam and sail, strong enough to pass through the ice-choked channel of Smith Sound, the gateway to the Polar Sea. For fifty years, the pack ice of Smith Sound had dashed the hopes of Arctic explorers by imprisoning their ships in ice and crushing them to pieces. Few vessels survived this gauntlet. As a result, most expedition parties were forced to make camp on the southern shores of the sound and then travel overland using dogs and sledges. By the time expedition parties reached the northern shores of Greenland and Ellesmere Island, they were exhausted, their condition too poor to attempt the second, and more difficult, part of the journey: a 500-mile trek across the Polar Sea, a landscape of shifting pack ice, yawning fissures of open water, towering hummocks, and icy rubble.

Peary was well suited to the task of designing an Arctic expedition ship. In addition to his years of experience sailing through Smith Sound's treacherous waters, he had studied civil engineering at Bowdoin College and served as a US naval engineer in Nicaragua. He became a member of the American Association for the Advancement of Science after giving a lecture on engineering features of the Nicaragua Canal. Engineering was not merely a means of professional advancement. He had an engineer's eye. His Arctic field journals were filled with designs for new stoves, snowshoes, sledges, and other expeditionary equipment that he often put to use on later expeditions. Working with the shipbuilding firm of McKay and Dix in Bucksport, Maine, Peary developed plans for a powerful new ship. He announced the building of the new ship in his remarks before the International Geographical Congress in 1904, the same assembly to which Peary had expressed pride in the "man blood" of explorers engaged in the North Pole quest.

Yet introducing a modern ship posed its own jeopardies. The attainment of the North Pole had become valuable because it was so difficult to reach. If Peary succeeded in reaching the North Pole because his new state-of-the-art expedition ship

[12] Robert Dunn, *The Shameless Diary of an Explorer* (New York, 1907), 3–4.

made it easier, he risked diluting the value of the accomplishment and, as a result, diminishing his image as an antimodern, atavistic, go-it-alone explorer. Peary had already written disparagingly of those who believed that the North Pole quest would be solved by the use of motor-powered balloons, rockets, and other transportation devices. "Man and the Eskimo dog," he told members of the National Geographic Society, "are the only two mechanisms capable of meeting all the varying contingencies of Arctic work."[13]

When Peary described the new ship to the members of the International Geographical Congress, therefore, he emphasized the rugged qualities of his expedition vessel rather than its state-of-the-art features. "She will possess such strength of construction as will permit her to stand this [ice] pressure, without injury. She will possess such features of bow as will enable her to smash ice in her path, and will contain such engine power as will enable her to force her way through the ice."[14] In keeping with tradition, Peary gendered his new ship female, but he made clear to his audience that it was a craft inhabited with a manly spirit. He named the new vessel the *Roosevelt* in honor of Theodore Roosevelt, who had become, by the early 1900s, the patron saint of manly physical culture (fig. 1). Moreover, he wrote about the ship as if it, too, personified a premodern warrior rather than a twentieth-century marvel of technology. Later describing his voyage through Smith Sound in 1906, he wrote, "The *Roosevelt* fought like a gladiator, turning, twisting, straining with all her force, smashing her full weight against the heavy floes whenever we could get room for a rush. . . . Ah, the thrill and tension of it, the lust of battle, which crowded days of ordinary life into one." In so doing, Peary framed his expedition ship as an extension of manly character rather than a substitute for it.

Other difficulties were harder to resolve in this way. In particular, Peary's manly, antimodern persona began to put him at odds with scientists and geographers. For almost 200 years, the association of scientists and explorers had been mutually beneficial. Enlightenment voyages gained prestige from their official associations with science organizations such as the Royal Society (UK), the Académie des sciences (France), and the Smithsonian Institution (USA), which gave legitimacy to their efforts to survey the world and its natural treasures. For their part, naturalists were generally happy for the opportunity to gather specimens and observations aboard ship or, when this was impossible, to inspect them upon the expedition's return. Even leaders of smaller, privately led expeditions benefited from their association with the scientific community. For example, in the 1850s and 1860s, Arctic explorers seeking the missing explorer Sir John Franklin pledged to conduct scientific work and, as a result, received scientific instruments as well as letters of support from prominent scientists. While scientific societies rarely had the funds to bankroll expeditions themselves, their prestige made it easier for explorers to raise money from the federal government and among private donors.[15]

Yet by the 1880s, this tacit quid pro quo between scientists and Arctic explorers had begun to break down, particularly in the United States. Scientists had grown skeptical of explorers who pledged their commitment to conducting scientific work in the polar

---

[13] Robert Peary, "Honors to Peary," *National Geographic Magazine* 18 (1907): 57.

[14] Peary, "Eighth International Geographic Congress," *National Geographic Magazine* 15 (1904): 425.

[15] On the dynamics of exchange between scientists and explorers, see the essays in Roy MacLeod, ed., *Nature and Empire: Science and the Colonial Enterprise*, vol. 15 of *Osiris* (2000); and James Delbourgo and Nicholas Dew, eds., *Science and Empire in the Atlantic World* (New York, 2008).

Figure 1. *Robert Peary aboard the* Roosevelt. *Photo credit: Library of Congress.*

regions only to return, time after time, empty-handed. This skepticism increased as
explorers directed their energies toward the North Pole. Reaching the earth's northern
axis was so difficult because the pack ice of the Polar Sea offered almost nothing in the
way of food, equipment, or shelter, all of which had to be ported from locations farther
south. Once an expedition party left the northern shores of Greenland or Ellesmere
Island, the consumption of provisions became the clock by which explorers measured
their progress. As a result, the race for the North Pole became exactly that: a dash to
reach 90°N and return before provisions ran out. It was, therefore, poorly suited for
conducting science. There was little time for making careful observations apart from

those determining positions of latitude and longitude. There was little space on the sledges for collecting specimens. Was it proper to even call these efforts scientific? So concerned was the American Geographical Society with the state of Arctic science that it sent a questionnaire to a dozen prominent scientists asking them "whether any problems in your department of science would be . . . promoted by further exploration toward either pole." Some scientists offered their support, but others were pessimistic. "I have no doubt that the value of the example of courage, hardiness, and fearlessness in the face of great peril is considerable—especially to an ease-loving race," wrote W. T. Sedgewick. "But I do *not* believe that *science* demands these periodic battles with hostile nature."[16]

For their part, Arctic explorers had become less dependent upon scientists to enhance their prestige and open federal coffers. By 1900, almost all polar expeditions were financed privately, by wealthy patrons or commercial enterprises—usually large newspapers—that could capitalize on stories of manly exploration. The money of wealthy patrons and lucrative newspapers worked as well in outfitting ships, and it did not come with the same strings: explorers did not have to receive scientific training, petition scientific societies, or conduct precise measurements of wind speed, tides, magnetic variation, and barometric pressure in the field.

Yet there were other reasons why Peary and others were more than willing to relinquish their former scientific allies: science clashed with their image as explorers. This may seem counterintuitive. At the turn of the twentieth century, American science remained a male preserve, notwithstanding the extensive participation of women at many levels of scientific work. Analytical and mathematical reasoning were aligned— doctors and educators argued—with male, rather than female, natures. From a practical point of view, men dominated the membership of scientific societies, appointments in academic departments, and the authorship of scientific papers. In both principle and practice, then, science retained a male identity.[17]

Yet to say that Americans thought of science as the work of men in turn-of-the-century America is not to say that it was seen as manly. While the authority and respect attached to science extended to scientists as well, these were men who represented the acme of civilization. Among the growing ranks of Americans who worried about civilization's effects upon the physical and mental health of young men, science became a suspect influence. Social critics who worried about boys losing their way by becoming too bookish only had to look at budding science majors to see their fears realized. More and more science work was moving indoors by 1900, and scientific training increasingly took place within the laboratory rather than in the field. So worrisome was the rise of the overly bookish scientist that William Harmon Norton, professor of geology at Princeton and president of the National Education Association (NEA) Science Division, made it the subject of his NEA address in 1902. "A college course may easily dim the eye and leave one a duller observer of things than at entrance. I once asked [Vanderbilt University Professor of Geology] Dr. Alexander Winchell if he found his seniors good observers. 'Yes,' he replied, 'unless they have had too much Latin and

[16] Sedgewick to Holt, undated, American Geographical Society Archives, New York; emphasis in the original.

[17] Miriam R. Levin, *Defining Women's Scientific Enterprise: Mount Holyoke Faculty and the Rise of American Science* (Lebanon, N.H., 2005), 101–51; Margaret W. Rossiter, "'Women's Work' in Science, 1880–1910," *Isis* 71 (1980): 381–98; Toby A. Appel, "Physiology in American Women's Colleges: The Rise and Decline of a Female Subculture," *Isis* 85 (1994): 26–56.

Greek.'"[18] The Reverend Andrew Morrissey, president of Notre Dame, was even more blunt in his address before the students: "The world does not want mere scholars or scientists. It needs manly men."[19] When Commanding General of the Army William T. Sherman testified about the scientific work of the Army Signal Corps—then responsible for telegraph systems and weather reporting—he responded that they were "no more soldiers than the men of the Smithsonian Institution."[20] In part to defend themselves against the charge of being men of science rather than soldiers, the Signal Corps began organizing the Lady Franklin Bay Expedition to northern Ellesmere Island.

These ideas would come to a head in the North Pole controversy of 1909 when Robert Peary and rival explorer Frederick Cook both claimed priority of discovery. When a number of scientists weighed in on the claim, they were quickly rebuked:

> It seems strange that a reputable man's claim to have accomplished such a purely sporting feat as the discovery of the North Pole should be questioned in angry language, showing all the traces of envy and malice. It seems not without significance that the carping comes from men who have failed in similar enterprises and from armchair and laboratory scientists.[21]

This is not to say that the expression of a rugged, physical manliness was incompatible with science. As historians Bruce Hevley, Rebecca Herzig, and Michael S. Reidy have shown, many Victorian scientists took up research projects in hazardous locations—mountains, glaciers, and the high seas—and then proudly recounted their adventures in their books and articles.[22] Yet the fact that scientists drew attention to their acts of arduous fact gathering is telling. Men of science wrote about manly adventure not because they felt it was relevant to the work of science, but precisely the opposite. Despite the fact that most scientists were male, manliness and science were concepts that, by the late nineteenth century, belonged to separate categories. They required effort to connect.

Indeed, the concepts of "manly science" and "muscular science," so frequently used by historians, are largely modern inventions. Both phrases existed at the turn of the century but carried different connotations. In turn-of-the-century America, one found practitioners of manly science in the boxing ring, not the laboratory. Boxers, wrestlers, and weightlifters were routinely described in daily newspapers as being either rugged or scientific, using strength or wits in their battles with opponents. When featherweight champion Abe Attell defeated Ed Kelly in the Dreamland Ring of San Francisco, for example, an Associated Press writer gushed, "It was a case of the polished and scientific boxer against the more rugged slugger and science won."[23] (It is interesting to note that, even within the context of sport, science was defined as a cerebral skill that distinguished it from ruggedness.)

For Peary, then, establishing his relationship with science required care. As

---

[18] "Teachers of Science," *Minneapolis Journal*, 9 July 1902, 3.

[19] Sermon by Dr. Morrissey, CSC, of Notre Dame, in "New Scholastic Year at Notre Dame," *Intermountain Catholic* (Salt Lake City, Utah), 3 Oct 1903, 7.

[20] Sherman, quoted in Rebecca Robbins Raines, *Getting the Message Through: A Branch History of the U.S. Army Signal Corps* (Washington, D.C., 1996), 61.

[21] Defense of Cook's claim from "Discovery of the North Pole," *Bismarck Daily Tribune*, 5 September 1909, 3 (originally printed in *Chicago InterOcean*).

[22] Bruce Hevly, "The Heroic Science of Glacier Motion," *Osiris* 11 (1996): 66–86; Herzig, *Suffering* (cit. n. 6), 1–16.

[23] "Science Won," *Arizona Republican*, 29 February 1908, 5.

president of the International Geographic Society, he was aware of the historical links between science and geographical discovery. Certainly he was aware of Kane—who had been the darling of the scientific community in the 1850s—and others who had used their alliances with scientific institutions to help launch their expeditions. Yet the world had changed. In 1900, Peary did not require scientific connections for the purposes of raising money, nor did he need a scientific persona for the purposes of enthralling audiences. At the turn of the twentieth century, science seemed to have less to do with manliness than it did with civilization, and civilization—with its luxuries and excesses—was precisely what Peary was trying to escape. Thus, while Kane had worn the mantle of "man of science" proudly, writing with delight about Arctic flora and fauna, icebergs and weather, Peary seemed uninterested in such matters. The giant Arctic walrus that imperiled the crews of many Arctic expeditions had become so commonplace to Peary that he left the account of it to his assistant. "I have seen so much of it in the past," he admitted, "that my first vivid impression is somewhat blunted."[24] Peary's tedium even extended to Arctic hunting, despite its hallowed position within Arctic narratives for over a hundred years. "Twenty odd years of arctic experience had dulled for me the excitement of the polar-bear chase." Evocative passages were exceptional in Peary's writing, most of which moved tersely from one scene to another. "The reader will find no padding," he warned in one narrative; "my constant aim has been condensation." Peary's writing style framed him as a kind of Arctic McTeague, whose bland language served to highlight his physical strength and determination. Peary's spartan prose was consonant with the new masculine ideal, but it threatened to rob his narratives of the color and suspense that appealed to readers. After hearing Peary's lectures, one woman advised him to improve his storytelling. "I think in your book," she wrote, "you may well dwell longer on that soul-stirring incident crossing the young ice on snow-shoes, and may appeal to your readers by sentences more intense than—'I was more than glad when we stepped on firm ice.'"[25]

ELSA BARKER AND "PEARY'S OWN STORY"

In 1909, events made it imperative that Peary tell a convincing and compelling story. When the *Roosevelt* steamed into the tiny Labrador port of Indian Harbor after a year in the Arctic, he telegraphed his wife Josephine: "Have made good at last. I have the Pole. Am well. Love Bert." The news ricocheted through the popular press. By the time the *Roosevelt* dropped anchor at the fishing station of Battle Harbor farther down the coast, two Associated Press reporters were waiting for him. Peary's announcement generated spectacular interest in the United States. It was a story that seemed designed for the front page of American dailies. After twenty-three years of Arctic hardship, an American naval officer had planted the Stars and Stripes on the polar axis, one of the world's last remaining terrae incognitae. But what made the story irresistible to newspaper publishers was that Peary's news came on the heels of another announcement by another American explorer, Frederick Cook, who claimed to have reached the North Pole first.

A year before Peary steamed north for Greenland in the *Roosevelt*, Cook had quietly attached himself to a private hunting expedition with the wealthy big game hunter

---

[24] Peary, *North Pole* (cit. n. 10), 79.
[25] Alice E. Johnson to Robert Peary, 4 February 1907, Box 32, Folder J, Robert E. Peary Papers, National Archives.

John Bradley. When Bradley's ship sailed into Smith Sound in August 1907, it deposited Cook, along with crewman Rudolph Franke, on the eastern shore of the sound. In the spring of 1908, Cook, Franke, and a party of Smith Sound Inuit reached the Polar Sea, on the west coast of Axel Heiberg Land. There, he sent all of the party back except for two Inuit men, Etukishuk and Ahwelah. When Cook reappeared at Annoatok a year later, in the spring of 1909, he reported reaching the North Pole on 22 April 1908, after a long trek over the polar ice cap. He had to wait until the fall of 1909 to leave Greenland, on a Danish ship headed for Copenhagen. While stopped in the Shetland Islands, Cook wired the news of his discovery to the *New York Herald* on 1 September 1909, and the story ran as a front-page headline the following day.[26]

Learning of Cook's story four days later, Peary dismissed his rival's claim, stating that "Cook's story should not be taken too seriously," stoking the flames of controversy in the popular press. Supporters of Cook countered with claims that Peary had fabricated elements of his own North Pole quest. Unlike other regions of the earth, the North Pole had no natural features that could be used to confirm or deny discovery. Floating on a sea of shifting pack ice, the polar axis could only be verified by mathematical calculations based upon astronomical measurements. All of these could be easily fabricated. Even photographs of the North Pole were useless for the purposes of proof. As a result, the question of proof became a matter of trust. Which man exhibited more honesty of character? Whose story of discovery seemed more believable? As newspaper, magazine, and book publishers barraged both men with offers for their stories, Peary quietly sought out a ghostwriter for this most important piece of writing in his career.[27]

Peary chose Elsa Barker, a New York writer and poet, to write his narrative for *Hampton's Magazine*. At first glance, Barker seemed an unlikely candidate to ghost Peary's narrative. Whereas Peary promoted himself as a rugged, Rooseveltian individualist, Barker gave her allegiance to the Socialist Party. When he had been sledging over the Arctic ice cap, she had been writing labor plays and founding a socialist theater. Barker's eclectic interests also included comparative mythology, which had eventually drawn her attention to the symbolism of Arctic exploration. In 1908, Barker had gained much publicity when she published a poem about Peary's upcoming expedition in the *New York Times*. "The Frozen Grail" cast Peary as an Arctic Galahad who, in seeking the North Pole, was reliving the ancient quest of the Holy Grail. In a number of ways the poem embodied the new masculine ethos. At the turn of the century, knight-errantry had come into vogue among many cultural critics as the symbol of a premodern age more honorable and more muscular than their own. The poem also set up sharp divisions of masculine and feminine ideals, by coupling Peary with "the white immaculate Virgin of the North" who represented nature and guarded the Grail. Finally, it highlighted the importance of physical strength in Peary's quest. The citadel of the North Pole, "like the heavenly kingdom," she wrote, "must be taken only by violence."[28]

With "The Frozen Grail," Barker had proven her ability to write within the new

---

[26] Excerpted from Robinson, *Coldest Crucible* (cit. n. 1), 142. On Cook's expedition, see Bryce, *Cook and Peary* (cit. n. 5), 294–348.

[27] "Peary's Statement," *New York Tribune*, 9 September 1909, 1; Bryce, *Cook and Peary* (cit. n. 5), 294–5, 349.

[28] Elsa Barker, "The Frozen Grail," *New York Times*, 7 July 1908. On the turn-of-the-century fascination with knight-errantry, see T. J. Jackson Lears, *No Place of Grace: Antimodernism and the Transformation of American Culture, 1880–1920* (Chicago, 1994), 101–21.

masculine genre that pervaded Arctic narratives at the turn of the century. After publication in the *New York Times*, it was reprinted in *Literary Digest* and later in an
anthology of Barker's poetry. It was reported that Peary took the poem to the North
Pole. Barker used the poem's popular success to elevate her status as a writer. In early
1909 she secured a position as a writer at *Hampton's Magazine*, a popular middle-
brow magazine with a circulation of 350,000. When Peary returned from the Arctic
later that year, Barker approached him as Hampton's agent, seeking to secure Peary's
narrative for the magazine. Hampton offered Peary an enormous sum—$40,000—
for rights to the narrative. Under their agreement, Peary would supply Barker with
his notes and recollections of the expedition from which she would write the narrative. For the editors of *Hampton's*, Barker's passionate prose offered a means of
compensating for Peary's stilted language. While Peary may have cultivated his terse
prose as an expression of manly reserve, it increasingly struck the public as evasive,
especially as the controversy with Cook grew. In a confidential memo, Hampton's
editorial department wrote Barker to emphasize the importance of making Peary's
narrative more exciting and convincing: "It has been said of his lectures that they
lack detail in the events of the last few days of the dash to the Pole, and that he describes too quickly the return journey. I am convinced that careful attention to detail
and putting into popular form the commander's scientific observations are the chief
essentials."[29]

With her new position, and the promise of Peary's narrative, Barker seemed to
transcend the gender barriers traditionally placed on women journalists. Yet Barker's
success depended in part upon her status as a woman writer. Her ability to find prominent publishers for her work (*New York Times*, *Literary Digest*, *Bookman*, *Hampton's
Magazine*) reflected not only the popularity of Arctic exploration but also the new
opportunities available to women writers at the turn of the century. Since the 1880s,
women reporters had become increasingly numerous in American newspapers. *Ladies' Home Journal* had become the most popular magazine in America, and publishers had become convinced of the importance of their female readers. Women not only
read more than men, they concluded, but controlled family purse strings as well. In
hopes of captivating this audience, editors hired more women to write for women's
and family magazines.[30]

Clearly the editors of *Hampton's* saw in Barker's poetry the detailed and passionate
language that Peary had been unable, or unwilling, to develop in his narratives, qualities that were consistent with the view that women were more evocative writers than
men. Because Arctic exploration stories appealed to both men and women, Barker's
status as a woman writer made good business sense. Gender also provided Barker
with certain advantages in her dealings with Peary. As male writers vied with Barker
for the opportunity to publish Peary's narrative, she found ways of gaining access to
Peary that were not available to them. Barker struck up a correspondence with Peary's
wife and closest confidante Josephine. Through this correspondence, Barker assured

[29] After publishing "The Frozen Grail" in 1908, the *New York Times* republished the poem after
Peary's return on 7 September 1909, 8, and again on 22 December 1909, 10. It was also printed in the
Springfield (Mass.) Republican, 25 October 1908, 21, and *Washington (D.C.) Times*, 19 July 1908, 10.
A few months later, Barker published the poem as part of an anthology, also titled *The Frozen Grail*
(New York, 1910).

[30] Helen Damon-Moore, *Magazines for the Millions: Gender and Commerce in the Ladies' Home
Journal* (New York, 1994), 29–80.

Josephine that she would adhere to her husband's vision of the narrative. "I have told [Hampton] straight to his face," Barker wrote to her, "that my first loyalty is to Peary and my second to him."[31]

Barker's strategy worked. When Josephine and her two children sailed to Canada to meet Peary on his way back from the Arctic, Barker traveled with them. There she offered Peary "a pile of money for the . . . North Pole story." Peary accepted Barker's offer, and she began to write the narrative a few months later. But she soon encountered new problems that required her to use her gender in her dealings with Peary. Despite his arrangement with Barker, Peary proved reluctant to give her important details about his North Pole journey. As deadlines loomed for publication, Barker's position grew more and more difficult, not only as a writer, but as someone who described herself as a woman of delicate constitution. When Peary's silence endangered her deadlines, she appealed to his respect for womanhood's weaker nature. "While I may have a man's ability for work and a man's will," she wrote to him, "I have only a woman's nerves." Peary consented to her requests, and the installments of the narrative were published on time.[32]

Yet gender was a two-edged sword for Barker. Because she was the ghostwriter for "The Discovery of the North Pole," Barker's authorship remained hidden. The sales of the magazine were lower than expected, and this did not help her status with Peary or Benjamin Hampton. Barker's writing was not to blame. Her narrative was more colorful than Peary's but still hewed closely to the brawny, vigorous style of his earlier works, even in her descriptions of the mighty *Roosevelt*:

> As the steel-shod stem of the *Roosevelt* split a floe squarely in two, it would emit a savage snarl that seemed to have behind it all the rage of the invaded immemorial Arctic, struggling with the self-willed usurper—Man.[33]

Yet, the controversy between Peary and Cook had died down by 1910, and public interest in Peary's narrative had cooled with it. As Peary set out to publish his narrative as a book, he started looking for a new ghostwriter with whom he had more in common: "Someone who has the big, masculine literary instinct," he wrote to his publisher.[34] Although Barker's served as the foundation for this new narrative, *The North Pole*, she was not invited to rewrite it. "The Discovery of the North Pole" marked the end of Barker's foray into exploration writing. Even at the height of her success at *Hampton's*, she had realized the fleeting quality of the achievement. "I wish that I were not a woman," she wrote in September 1909, "that I might go to every newspaper office in New York, Washington, and Chicago and do what I have done in the office of Benjamin Hampton. But as it is, I can only fight in my own way." In the years to come, Barker returned to spiritualism and mythology in her novels and articles. But her work occasionally offered social commentary, as in her 1913 play "Freedom and Ida Fleming," in which we get a glimpse of Barker's own feelings about her career. In it two men examine the issue of women's work:

[31] Barker to Josephine Peary, 27 September 1909, Elsa Barker Correspondence, Robert E. Peary Family Collection, National Archives, College Park, Md.

[32] Bryce, *Cook and Peary* (cit. n. 5), 478.

[33] "The Discovery of the North Pole," *Hampton's Magazine* 24 (March 1910): 340.

[34] Quoted in Bryce, *Cook and Peary* (cit. n. 5), 482.

Crawford: I pity the women who work.

Burke: I think they are splendid, the women who work. They are free.

Crawford: Do you call that freedom? I call it drudgery.

Burke: That depends on how they take it.

Crawford: Most of them take it hard.

Burke: That's because they have none of the privileges which ought to go with economic independence.

Crawford: Privileges such as?

Burke: Oh! Living their lives in their own way.[35]

It would be easy to think of Barker's foray into exploration writing as something exceptional: a lone woman writer who established a niche for herself briefly at the pinnacle of masculine writing. Yet Barker's participation is more representative than this. Women were involved at every level of the Arctic enterprise: as explorers and expeditionary assistants, and as agents, patrons, readers, and lecture-goers at home. What is more striking still: women reached the height of their involvement in the Arctic enterprise precisely when this "hypermasculine" phase of expeditionary discourse was at its zenith. During the period when Peary presented himself and his Arctic work in the most exclusionary gendered terms, women became most crucial to his campaign, both to reach the North Pole and to capitalize on its attainment back home.

Women had been writing about Arctic exploration since the early 1800s in a variety of formats, including narratives, biographies, poetry, and exposés. The themes of these writings were broad. Some writers, such as Barker, Euphemia Blake, and Emma DeLong, took on a masculine voice, disguising or downplaying their roles. Others, such as Mina Hubbard and Margaret Fox (known for her love affair with Elisha Kane), used their identities as women to appeal to their readers. Still others, including Mary Shelley, Emily Dickinson, and Charlotte Brontë, used the Arctic and its hardships as material for literary inspiration.

This body of literature expressed a range of attitudes about exploration as an activity, from those who encouraged Arctic discovery as an act of manliness, to those who critiqued these expeditions or described the Arctic as a place of female agency.[36]

[35] Elsa Barker, "Freedom and Ida Fleming," Box 2, Folder 6, Elsa Barker Papers, Brown University Library, Providence, R.I.

[36] Barker, "Frozen Grail" (cit. n. 28), and "Discovery of the North Pole" (cit. n. 33); Euphemia Vale Blake, *Arctic Experiences: Containing Capt. George E. Tyson's Wonderful Drift on the Ice-floe, a History of the Polaris Expedition, the Cruise of the Tigress, and Rescue of the Polaris Survivors* (New York, 1874); Emma DeLong, *The Voyage of the Jeannette: The Ship and Ice Journals of George W. De Long, Lieutenant-Commander U.S.N. and Commander of the Polar Expedition of 1879–1881* (Boston, 1884); Mina Hubbard, *A Woman's Way through Unknown Labrador: An Account of the Exploration of the Nascaupee and George Rivers* (New York, 1908); Margaret Fox, *The Love Life of Dr. Kane: Containing the Correspondence, and a History of the Acquaintance, Engagement, and Secret Marriage between Elisha K. Kane and Margaret Fox* (New York, 1866); Mary Wollstonecraft Shelley begins her story of *Frankenstein, or the Modern Prometheus* (London, 1823) in the high Arctic, where Robert Walton, commander of a British Arctic expedition, encounters Dr. Frankenstein on the pack ice chasing his monster (1–38); Brontë opens *Jane Eyre* (London, 1847) with her ten-year-old protagonist contemplating the Arctic regions, an allusion to the chilly, turbulent life she experiences with the Reeds: "Of these death-white realms I formed an idea of my own: shadowy, like all the half-comprehended notions that float dim through children's brains, but strangely impressive." As Timothy Morris explains in "Dickinson's Arctic" (*Emily Dickinson J.* 6 [Spring 1997]: 90–108), Emily Dickinson was fascinated with Arctic exploration, particularly the search for missing British explorer Sir John Franklin. The Arctic also found its way into a number of her poems. On women explorers and travel writers, see David Chapin, *Exploring Other Worlds: Margaret Fox, Elisha Kent Kane, and the*

While manliness became a dominant theme in late nineteenth-century polar literature, it was also a theme that circumscribed the tone and style of the men writing it. Gone was the exuberant tone of Elisha Kane's narratives, with its discourses on every aspect of Arctic life. These stories were out of place in turn-of-the-century polar literature, which required "the big, masculine literary instinct." As Peary's writings implied, there were only so many ways that one could experience—and describe—a polar bear chase. Within this line of thinking, the toughness required for summiting 20,000-foot peaks or reaching the polar axis were at odds with the sensibility of writing itself. As Robert Dunn admitted, "One requisite of the explorer . . . is insensitiveness. I understand why their stories are so dry. They can't see, they can't feel; they couldn't do these stunts if they did." Those seeking themes beyond the hardships of men and the race for the poles had better luck turning to women's narratives, poetry, or literature. Women's sentimental literature, as it was called, often expressed a broader range of ideas than those found in journals targeted to men. While sentimental writing and soft reporting are often dismissed today as fluff, they provided an important vehicle for expressing contrarian views absent from genres of male-oriented literature. Thus, while Barker and other women faced their own restrictions—as travelers as well as professional writers—they were often freer to articulate a broader range of ideas about the Arctic, its inhabitants, and its explorers.[37]

Beyond contributing voices to the project of polar exploration, women also gave it muscle. When the Admiralty reignited the search for the Northwest Passage in the 1820s, it was an all-male project, organized by the British Navy and funded by the British government. The disappearance of Sir John Franklin's 1845 Arctic expedition, however, triggered a number of private expeditions as well as some from other countries. Central to both were the actions of Jane, Lady Franklin, wife of the missing explorer. From 1848 until the mid-1860s, she became a tireless advocate of new exploration, using her moral authority as Franklin's widow to push for rescue campaigns that would also contribute to geographical discovery. It was at her behest that the United States fielded the Grinnell Expedition into the Arctic in 1850 (where Elisha Kane served as medical officer and wrote the expedition's official narrative). Wives and widows became powerful figures in the polar enterprise, including Robert Falcon Scott's wife, the artist Kathleen Scott (née Bruce), who commemorated her husband in sculpture, and Madame Leonie Osterrieth, who was given the title "Mother Antarctica" for her role in raising funds for the Belgica Expedition to Antarctica. Through their association with explorer husbands, a number of women became important advocates and patrons of polar exploration.[38]

Antebellum Culture of Curiosity (Amherst, Mass., 2004); Spufford, *I May Be Some Time* (cit. n. 1), 15–6, 30–5; Heidi Hansson, "Feminine Poles: Josephine Diebitsch-Peary's and Jennie Darlington's Polar Narratives," in *Cold Matters: Cultural Perceptions of Snow, Ice, and Cold*, ed. Heidi Hansson and Catherine Norberg (Umeå, 2009), 105–24. On Josephine Peary's complicated relationship with her husband, the Arctic, and its peoples, see Patricia Pierce Erikson's essay, "Homemaking, Snowbabies, and the Search for the North Pole: Josephine Diebitsch Peary and the Making of National History," in *North by Degree: New Perspectives on Arctic Exploration*, ed. Susan A. Kaplan and Robert McCracken Peck (Philadelphia, 2013), 257–88.

[37] Robert Dunn, "Highest on Mt. McKinley," *Outing* 44 (1904): 31. As Eugenia Lean points out in "Recipes for Men: Manufacturing Makeup and the Politics of Production in the 1910s" (in this volume), this function of women's sentimental literature was not limited to American print culture.

[38] Kenneth McGoogan, *The Ambitions of Jane Franklin* (Crows Nest, New South Wales, 2013); Janice Cavell, "Manliness in the Life and Posthumous Reputation of Robert Falcon Scott," *Can. J. Hist.* 45 (2010): 544–5; Hugh Robert Mill, *The Siege of the South Pole* (New York, 1905), 386.

Yet women also became important in their own right. As government funding for Arctic exploration waned in the late nineteenth century, explorers increasingly looked to wealthy patrons and businesses to bankroll their expeditions. For Peary, funding for large-scale items, such as the construction of the *Roosevelt*, came from his wealthy men's organization, the Peary Arctic Club. He looked to other funding streams such as journal and book writing to raise money for his North Pole expeditions. In both of these arenas, the influence of women became vital to his campaigns. While Peary promoted his expeditions by writing for magazines that catered to upper-class male readers such as *Outing*, he also published his stories in *Ladies' Home Journal*.[39] This was also true for Adolphus Greely and other male explorers who, at the end of century, actively sought to increase their appeal to women. The desire to write for female audiences was not new in the late 1800s, but it received a boost of enthusiasm from magazine publishers who increasingly believed that women had become the primary consumers of the American family. This growing focus on the female reader created demand for women writers and reporters.[40]

In 1870, only a handful of women worked for New York papers; by 1886 the number was closer to two hundred. Five hundred women were employed by newspapers nationwide. Like magazines, newspapers increasingly found advertisers who pitched their wares at women, led especially strongly by department store advertisers. In turn, newspapers devoted more copy to stories that would interest women, especially in evening editions and Sunday sections. In 1903, the *Ladies' Home Journal* became the first magazine to have one million subscribers and by 1910 boasted that one of every five women in the country read their magazine. As explorers turned to the popular press as a market for their manly stories, they found it increasingly focused upon women. Not surprisingly, then, Peary's interactions with editors and publishers were increasingly mediated by women. Long before he started working with Elsa Barker, Peary was in continual correspondence with Miss Galbraith Welch, American Manager of Curtis Brown & Massie, who promoted his international rights of publication.[41]

Yet the importance of women to Peary's missions was most evident in his lecture campaign, where he relied upon women organizers and ladies clubs for large sums of money. Even before his celebrated and controversial expedition of 1908–9, Peary was approached by women's organizations that wanted him to speak to their memberships. Over the course of 1907, when Peary was actively raising money for his next campaign, 80–90 percent of his lecture requests came from women's organizations across the country. This number does not include the women who wrote to Peary hoping to secure a speaking engagement to mixed audiences or children's organizations. Peary accepted most of these requests, for sums ranging from $100 to $300. He made a point to publically thank wealthy male donors, such as Morris K. Jesup, who contributed

[39] Robert Peary, "What I Expect to Find at the North Pole," *Ladies' Home Journal* 22 (March 1905): 5, 58. Peary's article was also listed by the *Women's Missionary Friend* (37–8 [1905]: 127) in their "Worth Reading" column.

[40] Explorers who targeted their writing for women readers include Adolphus Greely, "What There Is at the South Pole," *Ladies' Home Journal* 2 (1897): 2; A Midshipman, "Lost in the Frozen North," *Godey's Lady's Book* 109 (1884): 384–93. On the rise of magazines targeting women audiences, see Damon-Moore, *Magazines* (cit. n. 30), 2.

[41] Frank Luther Mott, *American Journalism: A History, 1690–1960* (New York, 1962), 406, 489, 598–9; Damon-Moore, *Magazines* (cit. n. 30), 1. On Galbraith Welch, see Folders "Curtis Brown and Massie," Letters Received 1909, Box 36, and "Miss Galbraith Welch," Letters Sent 1910, Box 14, Robert E. Peary Papers, National Archives, College Park, Md.

large sums to his North Pole campaigns. He also honored these men by naming dozens of Arctic bays, capes, and mountains after them. Yet for his grassroots supporters, the thousands of women who bought lecture tickets or paid membership dues, thanks were offered more quietly. The Peary archives are full of letters from women lecture organizers writing to Peary thanking him for sending them blue fox skin pelts.[42]

Why did women attend Peary's lectures in droves? Since the 1850s, women had expressed their interest in polar exploration, buying the books and attending the lectures of famous Arctic explorers. While it is difficult to establish with certainty what elements most attracted these women to stories of Arctic exploration, evidence suggests that they thrilled at the same themes that appealed to men: heroism and manly adventure. At the same time, their interests had also shifted to reflect the attitudes of their age. When Arctic explorers in the early 1800s had reveled in the Romantic Zeitgeist, women had praised their dreamy, gothic sensibility. "I am with the party in all their weary journeys," one woman wrote in 1857 after reading Kane's narrative, "and when I turn to gaze on the dark magnificent landscape, I can almost realize the solemn, the dreadful stillness of the Arctic night." Yet as explorers' accounts became more masculine in the 1890s, women's views changed to reflect—and often embrace—the new muscular ethos. Among articles on woolen hats and wrinkle treatments, *Harper's Bazaar* published stories about sledge parties, advertised polar narratives, and instructed readers on the cultivation of endurance (its advice: look to "the hardy and intrepid searchers for the North Pole"). If manliness seemed to be in crisis at the turn of the century, it was a crisis that vexed men and women alike. "The glory of manhood seems to have departed," a woman wrote Peary in 1907. "But you, and your ideals, justify it to my mind—and the response from the people, the men and growing boys, as their spirits still ring true to the appeal of noble adventure, is so encouraging that we need not yet doubt the future of America."[43]

CONCLUSION

The career of Robert Peary, America's most celebrated turn-of-the-century explorer, reveals the powerful, if fraught, relationships that had developed between the concepts of manliness, science, and exploration. Science and technology, concepts so closely associated with exploration during the eighteenth and nineteenth centuries, had taken on meanings that were not always welcomed by a new generation of "extreme" explorers, particularly those who, like Peary, had adopted an antimodern persona in their pursuit of geographical discovery. That a number of explorers felt free to shed the mantle of science during the height of the Progressive Era, when—scholarship

[42] Over the course of one year, 1907, he received requests from the Women's Club of Port Chester (N.Y.), the Vassar Student Aid Society, Ladies Aid Society of Washington Heights Baptist Church (D.C.), Women's Club of Ausmire (Conn.), Women's Hospital Board, Grand Rapids (Mich.), Civic League of Englewood, N.J., College Women's Club of Rochester, N.Y., Medford Women's League (Mass.), Salem Women's Club (Mass.), Laconia Women's Club (N.H.), Collegiate Equal Suffrage League (N.Y.), Veltin School for Girls, and the Daughters of the American Revolution. Letters from Boxes 30–2, Robert E. Peary Collection, "Letters Received 1907."

[43] Quotation from Emma Lou Sprigman to Jane Duval Leiper Kane, 20 April 1858, Elisha Kent Kane Papers, American Philosophical Society, Philadelphia, Pa.; "A Sledge Party from the 'Alert,'" *Harper's Bazaar* 9 (9 December 1876): 792; "Nansen's Great Book," *Harper's Bazaar* 30 (27 March 1897): 263; Grace Peckham Murray, "The Cultivation of Endurance," *Harper's Bazaar*, 31 (3 December 1898): 1035; quotation from Constance G. Dubois to Robert Peary, 15 April 1907, Box 30, Folder G, Robert E. Peary Family Collection.

Figure 2. *The Eighth International Geographic Congress, Washington D.C., 1904. Photo credit:* National Geographic Magazine, *1904.*

suggests—the authority of science was so high, requires us to look more carefully at the role science played in other extreme-environment activities such as mountain climbing, ballooning, and aviation and submersible testing, all of which became increasingly important in the twentieth century.

Peary's career indicates that an important distinction needs to be made between the way explorers used ideas about manliness for rhetorical purposes and the roles men actually played in expeditionary campaigns. Peary's commitment to a new muscular ethos of exploration, so exclusionary of women as a matter of course, remained highly dependent upon women as a matter of practice. Because discussions of women and exploration have, for many years, focused on the practices of women explorers, little attention has been spent examining the role of women in constructing masculine narratives themselves, directly as ghostwriters and editors, and indirectly as fans, readers, and lecture-goers.[44]

As a result, when Peary promoted polar exploration as the world's most exclusive men's club, scholars have taken him at his word. Yet even within the Eighth International Geographical Congress—that fellowship of geographers and explorers that Peary praised for being "of one blood—the man blood"—about 10 percent of his audience was women. It is unlikely that Peary—a man who worked closely with women in every aspect of his expeditionary work—failed to notice the women participants in the congress, listening to his lecture in their high collars and broad-brimmed hats (fig. 2). Rather, he saw no contradiction in describing this mixed party, so committed to the goal of manly exploration, as blood brothers even if, to be accurate, some of them were sisters.

---

[44] As Mary Terrall points out in "Masculine Knowledge, the Public Good, and the Scientific Household of Réaumur" (in this volume), so-called male preserves were often built upon a foundation of collaborative work by men and women.

Certainly he saw no contradiction in hiring Barker to craft his story of polar adventure. This is because he knew that she was—despite her cosmopolitan life, left-leaning politics, and "women's nerves"—of the same mind when it came to polar exploration. The Arctic represented many things to both of them: a stage for national drama, a field for personal fame, and a place where, through manly toil, the poles of male and female nature might resolve themselves. Yet the very fact of their strange partnership, and other mixed-sex collaborations that became integral to the Arctic project, suggests a different story. The poles of gender were no easier to secure than their geographical counterparts. Like the ends of the earth, these norms proved difficult to attain and impossible to verify. They became real only as the axis of an argument, forever unresolved, always out of reach.

SCIENTIFIC LABOR AND GENDERED SPACES

Prosthetic Manhood in the Soviet Union at the End of World War II

by Frances Bernstein*

ABSTRACT

Millions of Soviet soldiers were disabled as a direct consequence of their service in the Second World War. Yet despite its expressions of gratitude for their sacrifices, the state evinced a great deal of discomfort regarding their damaged bodies. The countless armless and legless veterans were a constant reminder of the destruction suffered by the country as a whole, an association increasingly incompatible with the postwar agenda of wholesale reconstruction. This article focuses on a key strategy for erasing the scars of war, one with ostensibly unambiguous benefit for the disabled themselves: the development of prostheses. In addition to fostering independence from others and ultimately from the state, artificial limbs would facilitate the veterans' return to the kinds of socially useful labor by which the country defined itself. In so doing, this strategy engendered the establishment of a new model of masculinity: a prosthetic manhood.

On 24 June 1945, roughly seven weeks after Germany's surrender,[1] an immense military parade was held on Moscow's Red Square to celebrate the Soviet victory in what was known as the Great Patriotic War. With its lengthy columns of soldiers marching in tight formation and its display of the latest military hardware, the meticulously choreographed spectacle projected an image of vigor and confidence, leaving no doubt of the country's readiness in the event of any future threat to its land or values (see fig. 1).[2] Missing from this celebration, however, were representatives of two groups who also had served and made important contributions to Germany's defeat. According to the parade order drawn up by Army General Aleksei Antonov, chief of the

*Department of History, Drew University, Madison, NJ 07940; fbernste@drew.edu.

My thanks to Eliot Borenstein, Greta Bucher, Christopher Burton, Erika Fraser, Dan Healey, Lisa Kirschenbaum, Claire McCallum, Misha Poddubnyi, and the editors for their input. Support for the research and writing of this article was provided by a grant from the National Council for Eurasian and East European Research.

[1] Germany's military surrender to the Soviet Union took place on 9 May, one day after it had surrendered to the Allies (VE Day).

[2] Nina Tumarkin, *The Living and the Dead: The Rise and Fall of the Cult of World War II in Russia* (New York, 1994), 92–4. Appearance made a difference—David Abramovich Dragunskii and two other senior officers were to be excluded because of their short stature despite their status as war heroes. After Marshal Ivan Konev intervened on their behalf, they were permitted to march. See Albert Axell, *Russia's Heroes, 1941–1945* (London, 2001), 161–2. For a discussion of earlier Russian military masculinity, see Karen Petrone, "Masculinity and Heroism in Imperial and Soviet Military-Patriotic Cultures," in *Russian Masculinities in History and Culture*, ed. Barbara Evans Clements, Rebecca Friedman, and Dan Healey (New York, 2002), 172–93.

Figure 1. Column of sailor-paratroopers and submariners in Red Square Victory Parade, Moscow. Stalin set the date for the parade right after Germany's surrender, giving soldiers abroad the chance to return home and participants the same amount of time to perfect their performance, for which they were drilled and rehearsed continually. Photographer: Ivan Mikhailovich Shagin, taken on 24 June 1945. Source: Russian State Documentary Film and Photo Archive, ed. khr. 0-213786, http://victory.rusarchives.ru/index.php?p=31&photo _id=410 (accessed 13 September 2013).

general staff of the Soviet Armed Forces, participant regiments would be comprised of "male, active duty personnel." Explicitly excluded, therefore, were women, a significant number of whom had fought at the front, and those (male) soldiers whose battlefield wounds barred them from further service.[3]

As a result of injury, frostbite, gangrene, or improper medical care, millions of Soviet soldiers were disabled as a direct consequence of their wartime experience.[4] Conservative estimates place this number at 2.75 million, but it is most likely much higher given the bureaucratic and material obstacles to receiving special disability status (and hence the modest state benefits to which a disabled veteran would be entitled).[5] Considering the sheer numbers involved, one would expect to find the presence of so

[3] Beginning in the summer of 1942, women were actively recruited into the armed forces, including for active frontline service. More than 800,000 women served over the course of the war. See Anna Krylova, *Soviet Women in Combat: A History of Violence on the Eastern Front* (New York, 2010); Roger D. Markwick and Euridice Charon Cardona, *Soviet Women on the Frontline in the Second World War* (New York, 2012); Svetlana Alexiyevich, *War's Unwomanly Face* (Moscow, 1988).

[4] G. A. Khorokhorina, *Politika gosudarstva v oblasti sotsial'nogo obespecheniia i reabilitatsii invalidov voiny i truda v period 1941–45 (na materialakh rsfsr)* (Moscow, 2009), esp. sec. 2; Amnon Sella, *The Value of Human Life in Soviet Warfare* (New York, 1992), 67; Russian State Archive of Socio-Political History (hereafter cited as RGASPI), f. 17, op. 117, d. 511, ll. 107–10.

[5] This constituted roughly 8 percent of the entire armed forces. See G. F. Krivosheev, *Soviet Casualties and Combat Losses in the Twentieth Century* (London, 1997), 92. See also Mark Edele, *Soviet Veterans of the Second World War: A Popular Movement in an Authoritarian Society, 1941–1991* (Oxford, 2008); Beate Fieseler, "The Bitter Legacy of the 'Great Patriotic War': Red Army Disabled Soldiers

many permanently injured veterans to resonate widely in the postwar experience.[6] But the state's attitude toward the "Invalids of the Patriotic War," as they were called, was decidedly complex and conflicted. Despite the expressions of gratitude for the soldiers' sacrifices and pledges of material support encountered in newspaper editorials and public addresses, the state evinced a great deal of discomfort regarding their damaged bodies.[7] In the context of a culture long hostile to physical impairment, such substantial numbers of armless and legless men constituted a potent threat to the myth of Soviet invincibility already being manufactured as the Cold War heated up.[8]

If economic and political motives dictated that veterans as a special interest group be disbanded as soon as possible,[9] this was doubly true for the disabled among them. For one thing, they were a constant reminder of the destruction suffered by the country as a whole, an association increasingly incompatible with the postwar agenda of wholesale reconstruction. For another, the danger posed by injured male bodies was perceived in specifically gendered terms, with disability threatening feminization. The postwar vision of a remasculinized Soviet Union necessitated a return to the traditional male and female norms that had been so disrupted by the war years.[10] Because disability was perceived as a problem of men, it required a specifically masculine solution.[11]

Of the many forces working to "unmake" veterans, three policy agendas in particular targeted those with disabilities and facilitated their disappearance. The first involved simply excluding them from any official commemoration, representation, or

---

under Late Stalinism," in *Late Stalinist Russia: Society between Reconstruction and Reinvention*, ed. Juliane Furst (London, 2006), 46–61, on 47.

[6] My focus on visible signs of impairment and specifically on soldiers with amputations is not to suggest the absence of "invisible" disabilities (sensory and psychiatric) among Red Army soldiers: on the contrary, both types of battlefield injuries and conditions were pervasive.

[7] According to Beate Fieseler, Stalin himself never publicly referred to the war disabled, either during or after the war; Fieseler, "'Nishchie pobediteli': invalidy Velikoi Otechestvennoi voiny v Sovetskom Soiuze," *Neprikosnovennyi zapas'* 2–3 (40–41) (2005), http://magazines.russ.ru/nz/2005/2/ (accessed 12 September 2011).

[8] On the absence of disabled soldiers from monuments and other forms of commemoration, see Robert Dale, "Re-Adjusting to Life after War: The Demobilization of Red Army Veterans in Leningrad and the Leningrad Region, 1944–1950" (PhD diss., Univ. of London, 2010), 20. On the emergence of a Cold War masculinity, see Erica L. Fraser, "Masculinities in the Motherland: Gender and Authority in the Soviet Union during the Cold War, 1945–1968" (PhD diss., Univ. of Illinois at Urbana-Champaign, 2009). As an interesting point of comparison, see Gregory Weeks's description of how the Cold War affected disability policy in occupied Austria: "Fifty Years of Pain: The History of Austrian Disabled Veterans after 1945," in *Disabled Veterans in History*, ed. David A. Gerber (Ann Arbor, Mich., 2000), 229–50, on 241–2.

[9] On veterans as a political threat, see Edele, *Soviet Veterans* (cit. n. 5).

[10] Because of space constraints, I am unable to address in any detail the demands made upon Soviet women during these years. Of course they were obligated to make bodily sacrifices of their own in the name of postwar reconstruction and stability. In addition to caring for their disabled husbands and sons, women were expected to erase the memory of their wartime experience, relinquish their newly acquired independence, leave higher-paying and skilled jobs to make way for demobilized men, and produce as many children as possible (a particularly daunting challenge, given the highly skewed female-to-male ratio). See Greta Bucher, *Women, the Bureaucracy and Daily Life in Postwar Moscow, 1945–1953* (New York, 2006); Mie Nakachi, "Replacing the Dead: The Politics of Reproduction in the Postwar Soviet Union, 1945–1955" (PhD diss., Univ. of Chicago, 2008); Anna Krylova, "'Healers of Wounded Souls': The Crisis of Private Life in Soviet Literature, 1944–1946," *J. Mod. Hist.* 73 (2001): 307–31, esp. 324–7.

[11] The term "remasculinization" comes from Susan Jeffords, *The Remasculinization of America: Gender and the Vietnam War* (Bloomington, Ind., 1989). For more on the Soviet remasculinization narrative, see my "Rehabilitation Staged: How Soviet Doctors 'Cured' Disability in the Second World War," in *Disability Histories*, ed. Susan Burch and Michael Rembis (Urbana-Champaign, Ill., 2014), 218–46.

association with the Red Army, as they had been from the victory parade.[12] A second effort concerned the state's ongoing practice of redefining disability classifications to reduce its substantial support obligations to veterans with impairments. That so many previously identified as disabled were now categorized as fit to work greatly boosted the state's triumphant claims about its success in treating and rehabilitating this population.[13]

This article focuses on a third strategy for erasing the scars of war, one with more ostensibly unambiguous benefit for the disabled themselves: the development of prostheses. Offering the possibility of elevating the wounded from the ranks of the child-like invalids to which they would otherwise be consigned (rhetorically and actually), these devices served a variety of objectives, at once aesthetic, political, and cultural. In addition to fostering independence from others and ultimately from the state, artificial limbs would facilitate the return of the war disabled to the kinds of socially useful labor by which the country defined itself. Henceforth, the disabled male body would be configured not in terms of absence (the missing limb), but presence (the mechanical replacement). Implicitly, the agenda for disabled veterans required the establishment of a new kind of masculinity: a prosthetic manhood.

By design, execution, and representation, Soviet postwar prosthetics were masculine objects and assumed a male recipient.[14] Moreover, the function of prostheses as a technology of masculinity extended beyond the objects themselves to those who designed them. During the war and especially in the period that followed, a significant amount of publicity was given to so-called invalid-inventors [*invalidy-izobretateli*]: men, disabled themselves, who made devices that were then put into production for other war amputees.[15] These inventions and inventors conveniently compensated for the shortcomings of professional prosthetic design, a point to which I will return below.[16] Through their creations, inventors made possible a return to manhood that was, in principle, accessible to any disabled veteran. By employing the same modern technology that helped win the war to engineer masculinity, the bodies of soldiers, like the country itself, could be made whole again. In the attention shown to these inventors, as well as to other prosthesis success stories, a model of behavior—an exemplary invalidism—was articulated to which others with the same impairments could aspire. Yet, as this article demonstrates, the prototype of prosthetic manhood would remain confined to the blueprint stage, ultimately unattainable by those expected to adopt it.

---

[12] By removing them from public view, the relegation of the most severely disabled to special care facilities served a similar function.

[13] Central State Archive of St. Petersburg (hereafter cited as TsGASPb), f. 2554, op. 2, d. 471, ll. 1–2; d. 502, ll. 2–9; d. 533, ll. 48–56; Fieseler, "'Nishchie pobediteli'" (cit. n. 7).

[14] On the normative and neutral status of the male image in visual propaganda, see Elizabeth Waters, "The Female Form in Soviet Political Iconography, 1917–1932," in *Russia's Women: Accommodation, Resistance, Transformation*, ed. Barbara Evans Clements, Barbara Alpern Engel, and Christine D. Worobec (Berkeley and Los Angeles, 1991), 225–42, on 227, 232.

[15] State Archive of the Russian Federation (hereafter cited as GARF), f. 438, op. 1, d. 550, ll. 109–10. For other examples, see N. Shenk, "Izobretateli-invalidy," *Sotsial'noe obespechenie* 2 (1941): 16–8; "Konstruktor-izobretatel'," *Sotsial'noe obespechenie* 5 (1941): 12–3; D. S. Reshchikov, *S pomoshch'iu rabochego proteza* (Moscow, 1958); I. Treskov, "O protezakh," *Pravda*, 30 July 1945.

[16] A. Miftakhov, "Bol'she vnimaniia proteznomu delu," *Sotsial'noe obespechenie* 1 (1956): 29–30; S. Pozniakov, "Pochemu invalidy vozrashchaiut protezy?" *Sotsial'noe obespechenie* 10 (1956): 43–6. The Ministry of Social Welfare eventually held republic- and countrywide contests to encourage non-professionals to submit designs for new models. See Ia. A. Rants, "O konkurse na usovershenstvovannye protezy," *Vrachebnoe delo* 1 (1949): 77–8.

BODIES AND THEIR VALUE IN THE SOVIET UNION

A great deal can be learned about the nature of a state by considering the values it attaches to the bodies of its citizens. Of all Western nations in the twentieth century, the Soviet Union has arguably been the most permissive—even encouraging—of damage inflicted upon its own population and justified in the name of the national interest: from the moment of its founding the Soviet Union had endured world war, revolution and civil war, dekulakization and collectivization of the peasantry, the rapid and unforgiving pace of industrialization, the GULAG system of forced labor, mass deportations, resettlements, and famines. We need only recall Stalin's notorious 1931 speech to industrial managers, justifying the physical sacrifices required for industrialization, to appreciate how bodily harm at the hands of the country's enemies also could be mobilized for political purposes:

> To slacken the tempo would mean falling behind. And those who fall behind get beaten. But we do not want to be beaten. No, we refuse to be beaten! One feature of the history of old Russia was the continual beatings she suffered because of her backwardness. She was beaten by the Mongol khans. She was beaten by the Turkish Beys. She was beaten by the Swedish feudal lords. She was beaten by the Polish and Lithuanian gentry. She was beaten by the British and French capitalists. She was beaten by the Japanese barons. All beat her—because of her backwardness, because of her military backwardness, cultural backwardness, political backwardness, industrial backwardness, agricultural backwardness. They beat her because it was profitable and could be done with impunity. . . . They beat her, saying: "You are abundant," so one can enrich oneself at your expense. They beat her, saying: "You are poor and impotent," so you can be beaten and plundered with impunity. Such is the law of the exploiters—to beat the backward and the weak.[17]

In similar fashion, Soviet propaganda following the German invasion of June 1941 portrayed Russia as the victim of abject violence perpetrated by the Nazis. Countless exhortations called on its citizens to avenge the country's despoiled villages, violated women, and massacred children, sacrificing all in defense of the motherland. If Stalin was portrayed as the great father of the people, the country itself was personified as a mother (literally mother-motherland, *mat'-rodina*), a symbol of endurance, vulnerability, and devotion, in whose name her children were called upon to lay down their lives.[18] With Russia at its most powerless coded as feminine, (male) Red Army soldiers with disabilities thus posed a particular representational problem.[19] Such visual reminders of the violence of war were at odds with the image of military masculinity the state sought to project, especially following the decisive Soviet victory at Stalingrad (February 1943), which marked the turning point of the war.

[17] Joseph Stalin, "The Tasks of Business Executives: Speech Delivered at the First All-Union Conference of Leading Personnel of Socialist Industry, February 4, 1931," Marxist Internet Archive, http://www.marxists.org/reference/archive/stalin/works/1931/02/04.htm (accessed 1 July 2013).

[18] Lynne Attwood, *Creating the New Soviet Woman: Women's Magazines as Engineers of Female Identity, 1922–1953* (London, 1999), 136, 138; Elena Baraban, "The Return of Mother Russia: Representations of Women in Soviet Wartime Cinema," *Aspasia* 4 (2010): 121–38; Susan Corbesero, "Femininity (Con)scripted: Female Images in Soviet War Time Propaganda Posters, 1941–1945," *Aspasia* 4 (2010): 103–20; Linda Edmondson, "Putting Mother Russia in a European Context," in *Art, Nation and Gender: Ethnic Landscapes, Myths and Mother-figures*, ed. Tricia Cusack and Sighle Bhreathnach-Lynch (Burlington, Vt., 2003), 53–66.

[19] For an early analysis of representations of disabled soldiers, see Vera Dunham, "Images of the Disabled, Especially the War Wounded, in Soviet Literature," in *The Disabled in the Soviet Union*, ed. William O. McCagg and Lewis Siegelbaum (Pittsburgh, 1989), 151–64.

That this was not the only possible reaction of a country to its disabled soldiers can be seen in the responses of the USSR's most important wartime adversary and ally, both of whom recognized the value of mobilizing visual images. In Nazi Germany, representations of invalided veterans served to distinguish between the worthy disabled—who sacrificed their bodies in the service of their country—and the unworthy, who would be targeted for the country's coercive eugenic measures. In the United States, in contrast, representations of amputee soldiers underscored the sacrifices required of all citizens during wartime. This was a reversal of an earlier War Department policy prohibiting media depictions of wounded or dead soldiers, an approach much closer to that of the Soviets.[20]

Visual acknowledgment of the disabling consequences of war at the front could be found elsewhere, in particular in the pervasive depictions of mutilation inflicted by the Soviet Union upon its enemies, onto whom its own staggering wounds could be projected.[21] Cartoons, illustrations, and posters showed Germans decapitated or blown apart, arms and legs raining down from the sky, stumps prominently displayed (fig. 2). Through this transposition, the physical horrors of war that befell the Red Army could be discursively managed. By ascribing disabling injuries solely to the country's opponents, however, disability itself became politically suspect at home: something to which only its enemies were susceptible. Indeed, it was far less problematic to show deceased soldiers—marked by an empty helmet or an eternally waiting mother—whose loyalty and commitment were unimpeachable.

The wariness with which disabilities were treated in the Soviet context stemmed in part from ideological opposition to charity (as a hypocritical practice of the bourgeoisie) and begging (in which stumps and scars served as tools of the trade).[22] Relegated safely to Russia's tsarist past, both endeavors relied on pity, a concept deemed inimical to state ideology and also unnecessary, given the social welfare system established after the revolution. In the postrevolutionary context, disabilities were likely associated instead with personal fallibility, most frequently by way of drunkenness, or with poor work performance, as there could be no public acknowledgment of the countless workplace accidents owing to unsafe conditions, shoddy equipment, or unrealistic production targets.

The Soviet Union's experience in the Second World War triggered the reappearance of veterans with disabilities crowding train stations and streets begging for alms, some with empty sleeves pinned up, others propelling their legless bodies on makeshift rolling platforms. Eventually, the former servicemen departed these spaces, some voluntarily and others as a result of decrees cracking down on "antisocial and parasitical elements" issued in the early 1950s.[23] Other, more sinister motivations were found to

[20] On Germany, see Carol Poore, *Disability in Twentieth Century German Culture* (Ann Arbor, Mich., 2007), chap. 2. On the United States, see David Serlin, *Replaceable You: Engineering the Body in Postwar America* (Chicago, 2004), chap. 1; and George Roeder Jr., *The Censored War: American Visual Experience during World War Two* (New Haven, Conn., 1993).

[21] Similar illustrations appeared regularly in popular publications such as *Ogonek* 17 (1943): 11–2; and *Krokodil* 4 (1942): 4; 28 (1942): 4. Identifying this same phenomenon, Claire E. McCallum argues alternately that wartime depictions of severely injured enemies served to express the Soviet Union's military might and the health of the body politic; McCallum, "The Fate of the New Man: Reconstructing and Representing Masculinity in Soviet Visual Culture, 1945–65" (PhD diss., Univ. of Sheffield, 2011), 110–2. For a more general discussion of these conventions, see Victoria E. Bonnell, *Iconography of Power: Soviet Political Posters under Lenin and Stalin* (Berkeley and Los Angeles, 1997), 260.

[22] N. P. Priorov, *Kak pol'zovat'sia protezom* (Moscow, 1943), 3.

[23] On the disappearance of disabled war veterans from public view see Edele, *Soviet Veterans* (cit. n. 5), esp. 93–5; Beate Fieseler, "La protection sociale totale: les hospices pour grandes mutilés de guerre

Figure 2. *Petr Aliakrinskii, "The Work Is as Good as Its Master," poster from 1941, Moscow, Iskusstvo. Drawing an equivalency between labor front and battlefront, the poster stresses the worker's expertise and potency as a vital weapon in the country's arsenal. Source: Hoover Institute Archives Poster Collection, RU/SU 2191.*

account for at least some of those invalid-veterans soliciting handouts as well as other kinds of assistance. As P. P. Verzhbilovskii, a social worker and the author of several publications on the employment of the war disabled, cautioned his readers:

> The enemy could exploit the advantages and priorities the Soviet Union awarded its wounded fighters. There are notorious cases of infiltrators and saboteurs being sent into our country to spy disguised as an invalid. There are cases when the Hitlerites intentionally maimed those being sent here as spies and saboteurs, since masquerading as an invalid made it easier for them to conduct their vile espionage.

Verzhbilovskii concluded with the following note of caution: "This is why in questions concerning our service to invalids we must not forget about vigilance."[24]

Whatever the values—whether discomfort, gratitude, or suspicion—associated

---

dans l'Union soviétique des années 1940," *Cah. Monde Russe* 49 (2008): 419–40; Elena Zubkova, "S protianutoi rukoi: nishchie i nishchestvo v poslevoennom SSSR," *Cah. Monde Russe* 49 (2008): 441–74; Dale, "Re-Adjusting" (cit. n. 8), 146–7.

[24] P. P. Verzhbilovskii, *Trudovoe ustroistvo invalidov otechestvennoi voiny* (Moscow, 1948), 29.

with the disabled soldier during the war, with the transition to peacetime new demands were made upon all bodies, impaired or not. Becoming injured in defense of one's country was no longer sufficient evidence of dedication to the state; additional sacrifices were called for in the name of postwar reconstruction.[25] To rebuild the country's devastated economy and infrastructure, those with disabilities, like everyone else, were expected to work to the best of their abilities.[26] Beyond the general ideological value of labor, the ability to work was deemed to be central to a wounded soldier's recovery, and labor therapy was employed in all treatment and rehabilitation centers.[27] Moreover, disability benefits were too low to live on without supplemental income from working.[28]

In her analysis of representations of disabled Soviet soldiers, historian Claire McCallum draws a sharp distinction between images produced during and after the war. She finds that wartime depictions were confined to scenes of battle, with soldiers' wounds emphasizing their heroism and willingness to sacrifice themselves in the cause of victory. After 1945, visual portrayals of disabled veterans all but disappeared; other genres (such as literature and official discourse) emphasized the veterans' ability to "overcome" their conditions and successfully reintegrate into peacetime society.[29]

McCallum highlights soldiers surmounting their disabilities, but according to literary scholar Lilya Kaganovsky, socialist realism (the artistic approach that became the official—and obligatory—state cultural style in 1934) dwelled upon and required such impairments.[30] Kaganovsky identifies two contradictory models of "exemplary masculinity" in Stalinist-era[31] representations: juxtaposed against the "virile male body" of the iconic steelworker was the image of "the wounded, long-suffering invalid." Through the sacrifice of his "traitorous" body, the "heroic invalid" of socialist realism offered an easily deciphered symbol of ideological dedication to the state. At the same time, these texts established "real" men's distance from political power since they posed no threat to Stalin's own perfection.[32] This, she argues, was the "real goal of Stalinist masculinity."[33]

[25] On reconstruction and the state of the economy, see Sheila Fitzpatrick, "Postwar Soviet Society," in *The Impact of World War II on the Soviet Union*, ed. Susan J. Linz (Totowa, N.J., 1985), 129–56; Mark Harrison, "The Soviet Union: The Defeated Victor," in *The Economics of WWII*, ed. Mark Harrison (New York, 1998), 268–301; Jeffrey Jones, *Everyday Life and the Reconstruction of Soviet Russia* (Bloomington, Ind., 2008).

[26] The only exceptions to this obligation were those classified as category 1 disabled, whose severe disabilities not only prevented them from working but necessitated full-time care.

[27] There is an extensive literature dealing with the use of labor therapy to treat soldiers with disabilities. See, e.g., V. P. Makridin, "Voprosy trudoterapii," *Gospital'noe delo* 6 (1944): 7–9. On the contribution of psychologists to the development of labor therapy during the war, see Albert R. Gilgen, Carol K. Gilgen, Vera A. Koltsova, and Yuri N. Oleinik, *Soviet and American Psychology during World War II* (Westport, Conn., 1997), chap. 3.

[28] Mark Edele argues that the provision of welfare services and pensions were intentionally inadequate (rather than accidentally or circumstantially so) to force those with disabilities back to work; Edele, *Soviet Veterans* (cit. n. 5), 90.

[29] McCallum, "Fate of the New Man" (cit. n. 21), 123–4.

[30] Socialist realism is a style of art whose purpose was to advance the goals of communism and socialism. See Katerina Clark, *The Soviet Novel: History as Ritual* (Chicago, 1981).

[31] For the purposes of this article, "Stalinism/Stalinist" refers both to the chronological era during which Stalin was in power and to the repressive and authoritarian form of governance characterizing his regime.

[32] Lilya Kaganovsky, *How the Soviet Man Was Unmade: Cultural Fantasy and Male Subjectivity under Stalin* (Pittsburgh, 2008), 4, 22, 120.

[33] Ibid., 146.

This essay seeks to draw from these seemingly contradictory interpretations. It builds upon McCallum's analysis by suggesting that the temporal distinction she makes between the wartime and postwar functions of these images extends to the injuries themselves. With few exceptions, battlefield injuries remained invisible in visual media, marked instead by a modest bandage wrapped around the forehead. Disabled soldiers were likewise missing from the war monuments erected after the Soviet victory, with the first appearing only in the mid-1950s.[34] Moreover, the shift from wound, however grievous, to disability was likewise one of location. As long as the soldiers were at the front, in the heat of battle, the wounds remained wounds, that is, temporary. Disability—both as permanent status and sign—was a product of the rear.

Furthermore, the model of prosthetic manhood discussed here mediates between Kaganovsky's opposing figures: through the use of artificial arms and legs, the heroic invalid could be engineered into the iconic steelworker. Regardless of the authenticity of their mechanical limbs, what made disabled veterans exemplary was the awareness that they were still disabled underneath. Writing about American advertisements for prosthetics at the turn of the twentieth century, Marquard Smith identifies a central dilemma encountered by commercial manufacturers of artificial limbs: how to market a device that, given the optimal outcome, would be indistinguishable from a real limb.[35] A mechanical limb needed to be inconspicuous enough to remain unnoticed but visible enough to be identified as artificial and therefore consumable: simultaneously erased and emphasized.

Even though Soviet prostheses were not governed by the imperatives of a competitive marketplace, they presented a similar conundrum. Despite a repeated emphasis on the need for realistic models and especially lifelike hands, in the Soviet context, as in the capitalist West, an artificial limb was successful only in so far as it hid an impairment and called attention to it at the same time. War heroes Mares'ev and Petrov (to be discussed further below) were famous not solely for their military exploits, stunning though these may have been, but because the men were disabled when they achieved them. Thus in Mares'ev's case, it was not his courageous eighteen-day crawl back to friendly territory but the return to battle on prosthetic legs—his victory over his own disability combined with his victory over the enemy—that would define him. Similarly, the prostheticized shock workers who overfulfilled production quotas were recognized as "heroes of labor" in part because they overcame the obstacles of their own bodies to achieve these remarkable industrial feats.

VIKTOR KONONOV AND THE ARTICULATION OF PROSTHETIC MANHOOD

The imperatives outlined above explain the enthusiasm with which prostheses were greeted as a solution to the "problem" of the disabled veteran in their capacity to counter the vast material, psychological, and symbolic damages wrought by the war through a simple technological fix. Hailed for enabling ex-soldiers' return to work and their ability to function independently, prostheses thereby served as a potent agent of

---

[34] McCallum, "Fate of the New Man" (cit. n. 21), 141. See also Natal'ia Konradova and Anna Ryleva, "Geroi i zhertvy: Memorialy Velikoi Otechestvennoi," *Neprikosnovennyi zapas* 2–3 (2005): 134–48, http://magazines.russ.ru/nz/2005/2/ (accessed 10 June 2013).

[35] Marquard Smith, "The Vulnerable Articulate: James Gillingham, Aimee Mullins, and Matthew Barney," in *The Prosthetic Impulse: From a Posthuman Present to a Biocultural Future*, ed. Marquard Smith and Joanne Morra (Cambridge, 2005), 43–72.

remasculinization.[36] The cultural and political importance attached to them can be seen in the conferral of Stalin Prizes (the country's highest award) on several inventors of artificial limbs, all disabled themselves, in the years following the war, an honor they shared with such notables as Andrei Tupolev (the designer of the medium-range bomber Tu-2, 1948), Mstislav Rostopovich (1950), and Andrei Sakharov (1954).[37] According to Aleksei Iugov, a medical doctor turned journalist for the prominent *Literaturnaia gazeta*, the Soviet Union equated developments in prosthetics with such essential inventions as jet-propelled aircraft or the newest metallurgical technology.[38] Given the paucity of visual images of the war disabled, this trend is especially noteworthy.[39]

Among the Stalin Prize winners, one recipient in particular came to signify the transformational potential of artificial limbs and of those who employed them. Viktor Kononov had been working as a mechanic when he designed for his own use an artificial hand that would become the most heralded and one of the most widely supplied prosthetics of the war and postwar years.[40] Made initially for below-the-elbow amputees,[41] "Kononov's arm" was the first completely active hand produced in the Soviet Union, meaning all five fingers could bend to grasp and, through a lock mechanism, hold an object. Lightweight and easy to put on and take off without assistance, the device enabled a high degree of self-sufficiency.[42] And as the most lifelike of any prosthesis available, it was deemed to be of great psychological benefit as well (fig. 3).[43]

The attention Kononov's arm received also derived from its inventor's biography. A poster child for the transformational power of Soviet technology, he was an unschooled autodidact devoted to self-improvement, and thus quite literally a self-made man. Moreover, he achieved his great success only after he became disabled. The subject of countless publications, including a biography, Kononov's story follows the typical socialist realist narrative, in which the hero progresses from adversity through consciousness to success and happiness, thanks to the opportunities avail-

[36] Of course, an emotional and financial investment in science and technology as a means of bolstering a masculinity at risk was not exclusive to either the Soviet Union or the postwar era. See, e.g., the contributions by Michael S. Reidy, "Mountaineering, Masculinity, and the Male Body in Mid-Victorian England," and Michael Robinson, "Manliness and Exploration: The Discovery of the North Pole," both in this volume.

[37] B. P. Popov and D. I. Gritskevich, "Protezirovanie v RSFSR," *Ortopediia, travmatologiia, i protezirovanie* 5 (1956): 3–8, on 7. The Stalin Prize was awarded annually to honor the country's highest achievements in the fields of science, engineering and technology, literature, and the arts. Archive of the Academy of Medical Sciences (AMN), f. 9120, op. 3, d. 2, l. 156; RGASPI, f. 17, op. 125 d. 585, ll. 120, 177; GARF, f. 413, op. 1, d. 1939, l. 59.

[38] Aleksei Iugov, "Chelovek prevyshe vsego!" *Literaturnaia gazeta*, 7 August 1948, 4.

[39] For other examples, see N. Shenk, "Izobretateli-invalidy," *Sotsial'noe obespechenie* 2 (1941): 16–8; "Konstruktor-izobretatel'," *Sotsial'noe obespechenie* 5 (1941): 12–3; Reshchikov, *S pomoshch'iu rabochego proteza*; Treskov, "O protezakh" (both cit. n. 15).

[40] Before the Second World War, the prosthetics industry in the Soviet Union was embryonic, with institutes (where limbs were designed) and factories (where they were fabricated) limited to a few major urban centers. For a brief history of the development of the industry and an assessment of the range of prostheses available at the time, see L. P. Nikolaev and I. A. Shumilin, "Otechestvennye aktivnye protezy," in *Aktivnye protezy verkhnikh konechnostei*, ed. A. K. Prikhod'ko and A. M. Veger (Kharkiv, 1949), 69–94.

[41] Eventually the arm was modified for above-elbow and bilateral amputees.

[42] A. R. Kreslin and K. I. Ivanov, *Pamiatka po osvoeniiu, pol'zovaniiu i ukhodu za protezom predplech'ia s aktivnoi kist'iu konstruktsii laureata stalinskoi premii V. E. Kononova* (Moscow, 1952).

[43] O. S. Dobrova, "Opyt protezirovaniia bezrukikh," *Ortopediia travmatologiia i protezirovanie* 1 (1956): 35–8, on 35.

Figure 3. *Kononov in his workshop. Source: Russian State Archive of Scientific-Technological Documentation (Samara), f. 180, d. 552, l. 87. I am grateful to Doctor of Medical Science Mikhail Anatol'evich Dymochka, chief federal expert for medical-social expertise, Federal State Budgetary Division of the Federal Bureau of Medical-Social Expertise of the Russian Ministry of Labor, and to the Central Research Institute of Prosthesis and Prosthesis Construction for permission to use figures 3 and 4.*

able only in the USSR.[44] Born into a poor peasant family with many mouths to feed, his father's death when Viktor was ten forced the boy into the workforce, toiling long hours for very little money. Eventually he took up a variety of professions, including joiner, locksmith, blacksmith, and boat mechanic's assistant. In 1915 he was drafted, wounded, and, when he had recovered, sent to the automobile division, transporting the injured from the front lines. He fought during the Civil War and became a me-

[44] B. Azbukin, *Chelovek idet k tseli* (Moscow, 1950).

chanic for the secret police, eventually and fatefully joining the transport division of the Red Army stationed in Mongolia.[45]

On 28 October 1928, while driving his superior to headquarters by motorcycle, Kononov was in a serious accident in which he lost his right arm. Having recovered enough to return to work a few months later, he no longer drove "masterfully, with strength and confidence," as one of the many articles written about him related.[46] The artificial limb he ordered from Germany, praised for its high quality, turned out to be a standard cosmetic prosthesis, and on his very first business trip three of the fingers broke off.

In 1929 Kononov designed a crude prosthetic that he used for the next four years, though he remained dissatisfied with its limitations. To achieve his goal of an artificial hand that bent at the joints and was capable of grasping and holding, he inserted steel cables in the fingers (originally encased in leather, eventually replaced by rubber hosing), which were attached inside the wooden palm to a single cable wound around a drum. At the end of the hand was a metal band connected to a sleeve, itself fastened to a shoulder harness. A movement of the shoulder caused the fingers to close and, by activating the lock mechanism located on the forearm, allowed the grasp to be maintained. The hand could be removed easily and replaced with any number of work- and living-related attachments (clamp, hook, hammer, pencil holder, spoon, etc.). Given the range of movements it permitted, when sheathed by a leather glove it was almost indistinguishable from a natural hand.[47] In 1932 the Moscow Prosthetics Factory produced a mock-up of his device, and Kononov was hired as a designer by the Scientific Research Clinic of Prosthetics and Orthopedics of the Russian Republic—the premier prosthetics institute in the country—to continue improvements on his invention.[48] In 1941 his device was put into limited production, and by 1946 Kononov's arm was manufactured in factories and workshops across the country.

Both the invention and the inventor served as useful propaganda in the country's developing Cold War narrative, further evidence of the purportedly huge disparity between the treatment of veterans in the United States and the Soviet Union.[49] Soviet prosthetists and health workers hailed Kononov's arm as far superior to American models, and in particular to the much-celebrated "Wonder Arm," in appearance, function, simplicity of design, and cost.[50] As they explained, this was not surprising, given

[45] M. Kirsanov, "Ruka kononova," *Sotsial'noe obespechenie* 10 (1940): 16.

[46] Ibid.

[47] L. G. Kapachnikova, "Protezirovanie invalidov s odnostornnei kul'tei predplech'ia v srednei treti protezom kononova" (Kandidat. diss., TsNIIPiP MSO RSFSR, 1952).

[48] Ibid., 17. See also M. Vapshevich, "Ruka kononova," *Sotsial'noe obespechenie* 2 (1941): 14–5; M. Mikhailov, "Pervoe soveshchanie rabotnikov proteznoi promyshlennosti," *Sotsial'noe obespechenie* 7–8 (1940): 24–7. Eventually the institution was renamed the Central Scientific Research Institute of Prosthetics and Prostheticization of the Social Welfare Ministry of the Russian Soviet Federation of Socialist Republics.

[49] Iugov, "Chelovek prevyshe vsego!" (cit. n. 38); I. A. Shumilin, "Raznovidnosti aktivnykh protezov," in Prikhod'ko and Veger, *Aktivnye protezy verkhnikh konechnostei* (cit. n. 40), 6. Despite such jingoistic pronouncements, Soviet orthopedists and prosthetists followed American developments closely; on two occasions specialists traveled to the United States on extended fact-finding missions, visiting leading specialists, treatment centers, and manufacturers. Soviet specialists purchased American equipment, specialist literature, and devices, which they sought to replicate domestically (typically unsuccessfully, given the expense and complexity of the American technology). Russian State Archive of Technical-Scientific Documentation (hereafter cited as RGANTD), f. 146, op. 1, d. 21, ll. 22-23, 33; GARF, f. 413, op. 1, d. 252, l. 20 ob.; d. 560, l. 95.

[50] While a number of foreign specialists and organizations, most notably the Paris-based International Federation of Veterans, were aware of the Kononov arm and sought information on its manufacture,

the superior nature of Soviet technology and the special attention the state showed its disabled veterans: in place of the vast opportunities for fulfillment available in the USSR, only neglect awaited American war invalids, whose economic, physical, and psychological needs remained unmet.[51]

Articulated in Kononov's story was the fantasy that not only could a profound injury be erased, but so too could the psychological and emotional baggage that went with it. Thus, in the words of D. Reshchikov, a double amputee, prosthetics take invalids "from the most depressed condition, from severe psychological suffering, to the point at which no trace remains."[52] Moreover, artificial limbs made it possible to move up the professional ladder, from blue- to white-collar labor: in Kononov's case, from driver to designer, and in Reshchikov's, from factory technician to teacher. In his address to the Prosthetics Institute after winning the Stalin Prize, Kononov spoke to this ability directly: "In our country, a country building communism, the boundary between intellectual and physical work is fading. A vivid example of this is the conferral of the highest title on me, a former worker and now a master designer."[53]

The achievements of the invalid-inventors provided much-needed reassurance, despite or perhaps because of the overwhelming evidence to the contrary, that the disabled could become more self-reliant and competent than before they were injured. Yet regardless of the grave labor shortage due to the country's astronomical wartime losses (over 20 million perished, including both civilians and combatants), veterans with disabilities encountered numerous impediments to work.[54] Ignoring a law requiring them to do so, employers often refused to hire the war wounded outright to make room for demobilized nondisabled servicemen or offered them low-status and low-paying jobs, such as watchman. Moreover, because disability status was calculated according to loss of labor capacity, by facilitating a return to work, prostheses served to disqualify veterans from the meager benefits to which they would have been entitled otherwise.[55]

Already the subject of numerous professional publications, Kononov's invention

---

I have not come across any mention of the prosthesis by their American counterparts. RGANTD, f. 146, d. 127, ll. 3-4.

[51] On the development of the prosthetics industry in the United States, see Serlin, *Replaceable You* (cit. n. 20), chap. 1. Interestingly, the enthusiastic response of the amateur inventors to the state's call for proposals marks another point of discrepancy between the American and Soviet experiences: in the United States, the government sought to enlist disabled veterans in the development of new prostheses and was surprised by their lack of interest (personal communication with Audra Jennings).

[52] D. Reshchikov, "Moi chudesnye ruki: kak ia vladeiu rabochimi protezami," *Sotsial'noe obespechenie* 8 (1957): 31.

[53] RGANTD, f. 146, op. 1, d. 50, l. 46.

[54] GARF, f. 413, op. 1, d. 651, ll. 7–18. GARF, Social Welfare Ministry: 413/1/651/7–18; Beate Fieseler, "Soviet-Style Welfare: The Disabled Soldiers of the Great Patriotic War," in *Disability in Eastern Europe and the Former Soviet Union*, ed. Michael Rasell and Elena Iarskaia-Smirnova (London, 2014), 18–41, esp. 28–30.

[55] Before the early 1930s, the Soviet Union, like tsarist Russia before it and like most of its Western counterparts, calculated disability benefits according to the nature of one's injury. In 1932, in the midst of the country's mass-industrialization campaign, disability categories were tied instead to one's occupation. Thus a right-handed accountant who lost his left hand was entitled to no compensation. Fieseler, "Soviet-Style Welfare" (cit. n. 54), 24–5. See also Edele, *Soviet Veterans* (cit. n. 5), 82; Bernice Madison, *Social Welfare in the Soviet Union* (Stanford, Calif., 1968); Sally Ewing, "The Science and Politics of Soviet Insurance Medicine," in *Health and Science in Revolutionary Russia*, ed. Susan Gross Solomon and John Hutchinson (Bloomington, Ind., 1990), 69–96; and Ethel Dunn, "Disabled Russian War Veterans: Surviving the Collapse of the Soviet Union," in Gerber, *Disabled Veterans* (cit. n. 8), 251–71.

was first introduced to the general public in a January 1943 article in *Trud*, the chief newspaper of the Soviet trade unions.[56] That was where Captain Chizhov read about it. While serving on the Western front, he was injured by a shell fragment and lost several fingers. His hand had to be amputated, and, remembering the article, he requested and was fitted with a Kononov arm. With his new limb, Chizhov wrote to express his gratitude to the inventor: now he could smoke a cigarette, light a match, chop wood, and, most importantly, hold and shoot a heavy machine gun, thereby allowing him to rejoin his unit in "the destruction of German fascism."[57]

Another invalid-inventor who received a fair amount of attention was Boris Efremov, a Stalin Prize winner for his above-knee prosthesis. Efremov staged a particularly elaborate and public demonstration of his leg's capabilities, accompanied by witnesses testifying (in writing) to his accomplishment.[58] As reported in the 15 November 1944 edition of *Krasnaia zvezda*, the Red Army newspaper:

> On an overcast autumn morning a car stopped in Maiakovskii Square. The door opened and a man wearing a blue mackintosh emerged. This was Efremov. He quickly walked across the sidewalk. In the crowd it was difficult to distinguish him from other pedestrians. He didn't even have a stick in his hands, typically used by those with prostheses. He crossed Gor'kii Street, crossed the square at Belorusskii Station and continued his way along the Leningrad Highway.
>   In the first 29 minutes he walked almost two and a half kilometers. Later he slowed down a little bit. A strong wind interfered with his movement, and he encountered steep inclines and slopes, but the designer Efremov continued on in his effort. Without stopping to rest, he arrived at the end of his planned journey (Rechnoi Vokzal). In two hours and 23 minutes Efremov walked 11 km 600 m on prosthetics.[59]

In spite of this display of manly vigor and endurance, Efremov's leg was ultimately a failure, roundly rejected by veterans for its "ugliness." Their response underlines the importance of appearance, beyond mere functionality, to conceptions of prosthetic masculinity—features Kononov's arm ostensibly was able to fulfill. There were also institutional factors at work: while Efremov received substantial financial support and encouragement from the army, prosthetists and social workers affiliated with the Ministry of Social Welfare (under whose jurisdiction prostheses were developed and fabricated) were adamantly opposed to his invention's manufacture.[60]

EXEMPLARY INVALIDS: MODELING PROSTHETIC MANHOOD

Kononov and the other invalid-inventors were not the only soldiers heralded in the press whose achievements relied on prostheses. Without question the most celebrated of this group was Alexander Mares'ev, the combat pilot made famous in *Pravda* by the war correspondent Boris Polevoi and later immortalized in his novel *The Story of*

[56] K. Golitsinskii, "O proteze Kononova," *Trud* 11 (6655) (14 January 1943): 4.

[57] RGANTD filial, f. 180, op. 1, d. 552, l. 103.

[58] Central Archive of the Ministry of Defense (hereafter cited as TsAMO), f. 75, op. 12328, d. 236, l. 9.

[59] Ibid., l. 8: "Vypiski iz gazety *Krasnaia zvedzda* ot 15 november 1944 No. 270/5950/ o proteze konstruktsii B. F. Efremova." See also F. Tokarev, "Eshche o protezakh," *Pravda*, 16 August 1945.

[60] Whether this was owing to its appearance or to an interagency scramble for resources or authority, the industry's leading specialists were hostile to the invention and its inventor. GARF, f. 413, op. 1, d. 754, ll. 19, 137, 141.

a Real Man (1946).[61] On 4 April 1942, Mares'ev's I-16 fighter plane was hit during an air fight over the northwestern front, and he was forced to parachute into the forest behind enemy lines, severely damaging his legs. For eighteen days, unable to stand, he pulled and crawled his way over frozen ground back to friendly territory before being rescued. Both limbs had to be amputated below the knee, and Mares'ev spent the next several months in a military hospital undergoing treatment and rehabilitative therapy. In the hospital he was given a pair of prosthetic legs (not Efremov's, as they were designed for transfemoral amputations) and, resolving to return to the front as a pilot, spent countless grueling and painful hours learning to walk again. Eventually he mastered the limbs so well that he was not only walking but dancing.[62] The battle to conquer his own body was only one of the obstacles Mares'ev faced; he next had to overcome opposition from the medical review board and the military command, as permission from both groups was required before he would be allowed to fly. A little over a year later, Mares'ev resumed active duty, downing several enemy planes in the battle of Kursk.[63] For his actions he was awarded the title of Hero of the Soviet Union.

Similarly celebrated was Vasilii Petrov, a war hero who seemingly rose from the dead to fight again. In 1943, as deputy commander of the 1850th Anti-tank Artillery Regiment on the Ukrainian front, Petrov was seriously injured, losing both arms while holding the Bukrinsk bridgehead on the Dnieper River, an action for which he won his first Hero of the Soviet Union award. Believed dead, Petrov was taken to the morgue along with the other battle casualties. When the brigade commander learned about the death, he sent two men to retrieve the body in order to bury Petrov with military honors. They searched for close to a day before discovering him just barely alive in a shed amongst countless corpses awaiting burial. They brought him to the battalion medical station and demanded that the base surgeon operate immediately. The doctor refused, arguing that his time would be better spent on those more likely to survive, until the men held a pistol to his head and insisted. Petrov survived and after his recovery petitioned to be sent back to the field, a request personally approved by Stalin. Equipped with prosthetic arms,[64] Petrov served as deputy commander and then commander of the 248th Anti-tank Regiment. He was awarded his second Hero of the Soviet Union decoration for holding a beachhead on the Oder River.[65]

While there were other isolated cases of soldiers with artificial limbs returning to active duty (like Captain Chizhov), few were capable of doing so, much less perform-

---

[61] Boris Polevoi, *Povest' o nastoiashchem cheloveke* (Moscow, 1966). For Mares'ev's "real" biography, see Aleksandr Abramov, "Krasnye sokoly: Russkie aviatory letchiki-asy, 1914–1953: Mares'ev Aleksei Petrovich," http://airaces.narod.ru/all6/maresyev.htm (accessed 9 September 2013); "Aleksei Mares'ev, 'Ia ne iz legendy,'" interview by Anatolii Dokuchaev, October 2003, http://www.bratishka .ru/archiv/2003/9/2003_9_6.php (accessed 5 May 2015); Dinara Al'bertovna Borisova, "'V nebo iz Ibres'. Aleksei Mares'ev," www.edu.cap.ru/home/9239/dokument/maresev.doc (accessed 5 May 2015). Note the different spelling of the pilot's real (Mares'ev) and fictional (Meres'ev) surnames. For the purposes of this discussion, I will refer to Polevoi's narrative, as that is the version with which the public was familiar, either through the novel itself or through the film, released two years later (Aleksandr Stolper, 1948). The book was also the basis of Sergei Prokofiev's final opera, banned from being performed after a single, closed-door performance in 1948. Francis Maes, *A History of Russian Music*, trans. Arnold Pomerans and Erica Pomerans (Berkeley and Los Angeles, 2005), 339.

[62] Polevoi, *Povest'* (cit. n. 61), 228, 238.

[63] Abramov, "Krasnye sokoly" (cit. n. 61).

[64] GARF, f. 146, op. 1, d. 13, l. 17.

[65] Nikolai Nikitich Skorokhodov, *Chelovek-legenda* (Saransk, 1970). See also Vladimir Gorishniak and Iakov Nagnoiny, "My rodom: Vasilii Petrov," http://yarodom.livejournal.com/790481.html (accessed 6 July 2013).

ing the sort of extraordinary feats achieved by decorated war heroes or Stalin Prize winners. Though this level of distinction was out of reach for most everyone, the press also was filled with examples of little heroes to emulate. With the right character and attitude anyone could become a shock worker, exceeding production goals and displaying selfless commitment to the country through whatever kind of work one could do. The highest pinnacle of such achievement was to be named a Stakhanovite, in honor of miner Aleksei Stakhanov, who surpassed his work quota fourteen times over by extracting 102 tons of coal in less than six hours.[66]

In a series of pamphlets intended for disabled soldiers, readers were introduced to many such little heroes. A Stakhanovite and machine repairman before the war, senior sergeant Mikhail Kuliabin lost his right arm in battle. From an article in *Trud* he learned about the possibility of adapting machines for amputees like himself and was inspired to do the same. Kuliabin returned to his factory and, with a similar modification to the equipment, overfulfilled his quota by 150 percent. The author noted that he looked no different from anyone else on the shop floor, concluding: "In short: it is impossible to call this the work of an invalid."[67] While these articles never claimed so outright, readers might easily conclude that artificial limbs produced improved specimens, capable of the sort of achievements not possible in their predisabled, preengineered bodies.[68] Hence the large number of war disabled who became Stakhanovites or moved up the professional ladder. If this was not enough of an incentive, little hero narratives emphasized a moral equivalency between front line and factory line: "Yesterday a fighter battling the enemy, today [Kuliabin] is a fighter on the labor front. The only difference is that he has exchanged his rifle for a repair tool."[69]

MEN AND THEIR MACHINES

With his experience more readily accessible as a model of prosthetic manhood than those of a Mares'ev or a Petrov, Kononov remained the archetype for this new improved breed of Soviet man. Sources depicted him as leading an exceptionally well-rounded and fulfilling life. Kononov practiced gymnastics, played the xylophone, filled and lit his own pipe, tinkered with car engines, rode a bike and a motorcycle. Just as his detachable hand facilitated his migration from one form of work to another, Kononov's liminal status enabled him to shift between categories of identity: he could engage factory workers and rub elbows with the most highly esteemed academics. He traveled the country to personally demonstrate his invention (and his own life) to others with disabilities. Yet there was one significant exception to this picture of the perfect well-rounded life: his self-sufficiency was quite literal since the one realm in which he seemed to be lacking was romantic companionship. Whatever the relational status and history of the real Kononov, the celebrated invalid-inventor Kononov was, as far as the coverage of him was concerned, single. His primary "attachment" was to his arm.

---

[66] This heavily promoted movement celebrated and richly rewarded those who surpassed production targets. On the phenomenon, see Aleksei Grigorevich Stakhanov, *The Stakhanov Movement Explained* (Moscow, 1939); Lewis H. Siegelbaum, *Stakhanovism and the Politics of Productivity in the USSR, 1935–1941* (New York, 1988).

[67] E. Kharitonovich, *Chelovek nashel sebia* (Moscow, 1944), 28–31.

[68] M. F. Iashonkov, "A. I. Titovu i drugim," *Sotsial'noe obespechenie* 10 (1956): 50–3, esp. 51.

[69] Kharitonovich, *Chelovek nashel* (cit. n. 67), 21.

Attention to this issue may have been deemed unnecessary, given the vastly unequal ratio of women to men after the war and a general discomfort with overt attention to sexual matters characteristic of the culture.[70] Nonetheless, Kononov's biography underlines the very distinct nature of late Stalinist masculinity, and prosthetic masculinity, in particular. In this respect, as in others, the Soviet model differed from its wartime counterparts. In the United States during this era, addressing impaired veterans' anxieties regarding sexuality and their relationships with wives or girlfriends was seen as fundamental to the rehabilitation agenda, which sought to help men reestablish their prewar status as the head of the family.[71] In Japan, state policy encouraged single women to marry disabled soldiers.[72]

In contrast, rehabilitation in the USSR was depicted as a solitary endeavor; as Kaganovsky writes, "the Stalinist text leaves little room for the conventional marriage plot."[73] These popular narratives showed that war invalids capable of achieving true independence by means of their artificial limbs did not need to depend on women. Popular illustrations as well as those appearing in professional publications reinforced this theme of male self-reliance. With their Kononov arm, men could chop wood and hammer nails, write and shave, pour tea and eat soup, as well as perform the sort of domestic tasks that would have been relegated to the woman of the household, such as making the bed and brushing one's shoes and clothing (fig. 4). Notably, this muddling of gender roles did not work in reverse; in the few illustrations showing women using the device, they are engaged in such strictly feminine pursuits as sewing and knitting. Nor does the problem of male sexual performance seem to have garnered much attention in the psychiatric literature, as it did in the West.[74] In those instances in which the disabled veteran does get the girl, it occurs only after his recovery is complete—a reward for his efforts, as it were. The process of both psychological and physical rehabilitation was something to be tackled and endured alone.

The role that would have been played by the loving wife or girlfriend in another context was filled instead by the lathe, loom, or drill, the nuclear family by the greater

[70] According to Catherine Merridale, there were 20 million fewer men than women at the end of the war; Merridale, "The Collective Mind: Trauma and Shell-Shock in Twentieth Century Russia," *J. Contemp. Hist.* 35 (2000): 39–55, on 52. On sexual asceticism in the postrevolutionary context, see my *The Dictatorship of Sex: Lifestyle Advice for the Soviet Masses* (DeKalb, Ill., 2007). For the postwar era, see Deborah Field, *Private Life and Communist Morality in Khrushchev's Russia* (New York, 2007), chap. 4; and Bucher, *Women* (cit. n. 10).

[71] David Gerber, "Heroes and Misfits: The Troubled Social Reintegration of Disabled Veterans in *The Best Years of Our Lives,*" in Gerber, *Disabled Veterans* (cit. n. 8), 70–95. On a similar understanding of successful rehabilitation after the First World War, see Beth Linker, "Shooting Disabled Soldiers: Medicine and Photography in World War I America," *J. Hist. Med. Allied Sci.* 66 (2011): 313–46, http://jhmas.oxfordjournals.org (accessed 9 July 2013).

[72] Lee Pennington, "Occupational Therapy: Japanese Disabled Veterans and the Postwar Occupation of Japan" (paper presented at the American Historical Association Annual Conference, New York, January 2009).

[73] Kaganovsky, *How the Soviet Man Was Unmade* (cit. n. 32), 89; Clark, *The Soviet Novel* (cit. n. 30), 69.

[74] The mental and sexual health of disabled veterans was a low priority given the pervasive trauma affecting every corner of society. As Paul Wanke, a historian of Soviet military psychiatry, notes, "the Soviet Union was simply unable and unwilling to provide such services"; see Wanke, *Russian/Soviet Military Psychiatry, 1904–1945* (New York, 2005), 94, 112. On American attention to such issues, see Beth Linker and Whitney Laemmli, "Half a Man: The Symbolism and Science of Paraplegic Impotence in World War II America," in this volume. On the understanding and treatment of trauma in Russia throughout the twentieth century, see Merridale, "The Collective Mind" (cit. n. 70); and Julia Furst, "Introduction," *Late Stalinist Russia* (cit. n. 5), 1–20, on 5.

Рис.49. Удержание стака-
на и пользование им с по-
мощью протеза Кононова
/инв.Р-Н, сл.З/.

Рис.50. Пользование расческой
с помощью протеза Кононова.
Тот же случай.

Рис.51.Чистка щеткой
одежды с помощью протеза
Кононова.
/Инв.Б-Н,сл.25/.

Рис.52. Чистка щеткой обуви
с помощью протеза Кононова.
Случай тот же.

Figure 4. *Using the Kononov arm to hold a glass of water, comb hair, brush clothes, and clean shoes. Source: Kapachnikova, "Protezirovanie" (cit. n. 47), 151.*

collective Soviet family. It was their mastery of feminized machines that signaled the disabled men's true recovery. Before the war, Aleksandr Chibirikin had been a lathe operator, a Stakhanovite who overachieved norms by 400 percent. As the commander of his unit, he lost his right arm rescuing his men from three different tanks under enemy fire. Returning to his factory after his recovery, Chibirikin became an instructor, teaching other men how to "love their machines."

> After this conversation the machines became for the youths living, complex, thinking beings. They needed to be fed, watered, kept free of dirt and the metal dust that covered everything. The students had to be smarter in order to control without error such compli-
> cated organisms. Chibirikin told the workers about when he started, his helpless bewilder-

ment when she, this mysterious machine, for some inexplicable reason broke down. . . .
"To get them to work demands that we are also precise. The machine loves precision,
attention." Within three months every student was surpassing target norms.[75]

Chibirikin's address to the young men sounds more like courtship or wedding night
advice than technical manual.

Despite the country's looming demographic crisis, reintegration into the collective family was a higher priority than reintegration into the individual family.[76] While
Mares'ev had a love interest whom he later married, his recovery was inspired by a
newspaper article about a one-legged World War I pilot and by the example of the
fellow patient who gave it to him. As the party-appointed political commissar of his
regiment, Semen Vorob'ev stood in for the state; indeed, he was the "real man" of the
title. Similarly, one of the little heroes of the prosthetic narrative, Izrail Kheifets, lost
not only his legs and all his fingers to the Germans but also his father, wife, and three
children. Contemplating suicide, Kheifets was saved thanks to the solicitude of the
state and the intervention of social welfare employees. The ministry provided food,
firewood, a new suit, and prostheses. As the author concludes, "Kheifets imagined
for himself loneliness. The motherland opened wide for him the doors to the greater
Soviet family."[77]

CONCLUSION

Whatever expectations the ex-soldiers had for attaining the model of prosthetic manhood detailed above were disappointed by the system-wide failures of the prosthetics
industry. The small number of research institutes and manufacturers operating at the
start of the war were unprepared for and overwhelmed by the demand, made even
greater by the loss of facilities located in the occupied western areas of the country.
The ministries of social welfare and health were embroiled in a pronounced and pro-
longed turf battle for control, with ultimate jurisdiction shifting from one to the other
and back again; as noted above, there were also clashes between the welfare ministry
and the army medical establishment. There was conflict between the orthopedists who
prepared the stumps, the engineers who designed the devices to go over them, and the
workers who fabricated them. Within the factories there was little coordination and
frequently downright hostility between the medical and manufacturing divisions. The
archival record brims with evidence of manifold breakdown at all stages of the process. Materials, when available, were of poor quality and subject to frequent delays in
delivery. Labor shortages and extremely low salaries meant that factories were often
understaffed and the people who worked there had received limited training, resulting in shoddy construction and requiring a return of the devices for repairs several
times a year (according to one report it was unusual if a device did not break during
the first three months). Inadequate instruction and supervision extended to those who
measured the intended recipients, resulting in poor fit and ensuing complications from
stumps reopening, strains on compensating body parts, and other problems.[78]

[75] V. I. Cherevkov, *Snova v stroiu* (Moscow, 1945), 13–4.
[76] On this point as it relates to literature, see Krylova, "'Healers of Wounded Souls'" (cit. n. 10), 317.
[77] Cherevkov, *Snova v stroiu* (cit. n. 75), 32–6.
[78] For a detailed examination of the industry's failures, see my "Prosthetic Promise and Potemkin
Limbs in Late-Stalinist Russia," in Rasell and Iarskaia-Smirnova, *Disability* (cit. n. 54), 42–66.

Disabled veterans complained bitterly of the entire process. They were required to make multiple trips, in many cases traveling great distances, for fittings and then to receive the devices, and they endured long delays for their appointments. Because there were no seats in the waiting rooms, amputees had to sit on the floor or stand for several hours while they waited to be seen. For those obliged to stay overnight or longer, no accommodations were provided, forcing amputees to sleep in train stations. War heroes or no, they encountered the same rude service and bureaucratic intransigence for which the Soviet Union was notorious. Once they received their limbs, they were given little instruction in how to care for or properly use them. When, inevitably, the devices broke, delays in repairs sometimes stretched for many months and even years.[79] Not surprisingly, many chose to forgo wearing their prosthetics entirely or use them only cosmetically. These inadequacies continued well into the late 1950s.[80] In her address to the Russian congress of physicians employed in the prosthetics industry, Deputy Social Welfare Minister M. F. Aleksashina commented on the absurdity of the state's continued failure to accommodate upper limb amputees:

> We live in an age when the enormous growth of technical culture, when the technical apparatus employed in such powerful hydrostations as Kuybyshev and Stalingrad, the likes of which history has never seen before, inspires the entire scientific and technical world. And to this day we still can't manage upper limb prostheses.[81]

The accomplishments of a Mares'ev or Petrov remained out of reach for most Soviet citizens, whether disabled or not. Similarly, prosthetic manhood, while theoretically possible, was unattainable by all but the very few amputees fortunate enough to have had their limbs made and fitted individually: the mechanical arms of Kononov and Petrov by the Prosthetics Institute and Mares'ev's legs by an elderly craftsman trained before the revolution who worked by hand.[82] None of these were manufactured via the system of industrial production so central to the country's proletarian identity. The disconnect between Kononov's arm as totemic promise and those mass-produced in factories was especially striking. Work attachments were often unavailable or unusable with "improved" models.[83] When it was determined by the early 1950s that the grip was too weak and the arm too heavy for industrial labor, it was recommended that the device be given only to white-collar employees.[84] Despite repeated promises to expand the range of available options, hands continued to be produced in only one (masculine) size, as opposed to the six originally planned, and in one skin shade, rather than four.[85]

The failure to deliver on prosthetic masculinity amounts to a double amputation. The state, which promised to step in and take the place of the missing limb by ex-

---

[79] See, e.g., RGANTD, f. 146, op. 1, d. 39, ll. 6–7.

[80] Miftakhov, "Bol'she vnimaniia" (cit. n. 16); S. Pozniakov, "Pochemu invalidy vozrashchaiut protezy?" *Sotsial'noe obespechenie* 10 (1956): 43–6.

[81] GARF, f. 413, op. 1, d. 2056, l. 28; and similarly Central State Archive of the City of Moscow (TsGA Moskvy), f. 1046, op. 1, d. 4, l. 237.

[82] Polevoi, *Povest'* (cit. n. 61), 152, 155.

[83] RGANTD, f. 146, op. 1, d. 98, l. 213.

[84] Kapachnikova, "Protezirovanie" (cit. n. 47), 198.

[85] In this case, the Rubber and Chemical Industry Ministries, which bore responsibility for the outer rubber coverings, were at fault. GARF, f. 413, op. 1, d. 1631, l. 12; f. 438, op. 1, d. 615, l. 2; RGANTD, f. 146, op. 1, d. 98, l. 130. On race as a factor in the determination of limb shades in the United States, see Beth Linker, *War's Waste: Rehabilitation in World War I America* (Chicago, 2011), 115.

changing it for a mechanical one, only made matters worse. Ultimately, the only prosthetic masculinity delivered was discursive. It allowed the Soviet system to claim that it was making up for the mutilation of its soldiers without actually doing the real work to help their bodies, thus creating phantom limbs to compensate for phantom pains.

After Stalin's death and the initiation of the cult of World War II, visual representations of the Great Patriotic War invalids became more prevalent.[86] By this time many of the most seriously impaired had passed away. In her analysis of Cold War masculinities, Erica Fraser argues that the wartime disruptions of gender norms allowed for the postwar emergence of multiple sites of maleness.[87] The moment of prosthetic men passed, to be replaced by astronauts, scientists, and athletes: all far less problematic embodiments of Soviet manhood.

[86] McCallum, "Fate of the New Man" (cit. n. 21), chap. 2. Interestingly, Vasilii Petrov served as the prototype for sculptor Valentin Znoba's Soldiers of Victory Monument, created for the Great Patriotic War Museum in Moscow.

[87] Fraser, "Masculinities" (cit. n. 8), 4, 56. On cosmonauts as a symbolic model of masculinity, see also Slava Gerovitch, "'New Soviet Man' Inside Machine: Human Engineering, Spacecraft Design, and the Construction of Communism," *Osiris* 22 (2007): 135–57.

Recipes for Men:
Manufacturing Makeup and the Politics
of Production in 1910s China

by Eugenia Lean*

ABSTRACT

In the first decade of Republican China (1911–49), masculinity was explored in writings on how to manufacture makeup that appeared in women's magazines. Male authors and editors of these writings—some of whom were connoisseurs of technology, some of whom were would-be manufacturers—appropriated the tropes of the domestic and feminine to elevate hands-on work and explore industry and manufacturing as legitimate masculine pursuits. Tapping into time-honored discourses of virtuous productivity in the inner chambers and employing practices of appropriating the woman's voice to promote unorthodox sentiment, these recipes "feminized" production to valorize a new masculine agenda, which included chemistry and manufacturing, for building a new China.

In the 1910s, a curious print culture phenomenon appeared in China's urban areas. Journals such as *Funü zazhi* (Ladies' journal) and *Nüzi shijie* (Women's world) began to run columns and articles that provided highly detailed, technical information on soap, hair tonic, perfume, and rouge. These columns did not describe how to consume but instead how to produce these items, offering the home as the perfect space for such cosmetic production. In 1915, *Funü zazhi* ran several such articles in its "Technologies" [*xueyi*] section. Titles include "A Brief Explanation on Methods for Making Cosmetics" in the January issue, "Method for Making Rouge" in the March issue, and "Method for Producing Cosmetics" in the May issue.[1] Also in 1915, *Nüzi shijie*

* Columbia University, 925 IAB MC 3333, 420 West 118th Street, New York, NY 10027; eyl2006@columbia.edu.

This research was supported by the Charles A. Ryskamp Fellowship from the American Council of Learned Societies, the Taiwan Fellowship from the Ministry of Foreign Affairs and the Center for Chinese Studies, National Central Library, and funding from the Weatherhead East Asia Institute at Columbia University. Thanks to Erika Lorraine Milam, Robert A. Nye, Debbie Coen, Daniel Asen, Liza Lawrence, Alexander Des Forges, Jacob Eyferth, and two anonymous readers for reading drafts of this essay. I would also like to thank participants at the "Masculinities in Science/Sciences of Masculinity" conference, sponsored by the Philadelphia Area Center for History of Science, upon which this volume is based, and the audiences at related talks that I have given at Princeton University; Harvard University; Columbia University; Ohio State University; Triangle East Asia Colloquium, North Carolina; Fudan University, Shanghai; and the Academia Sinica in Taiwan.

[1] Ling Ruizhu, "A Brief Explanation of the Methods to Make Cosmetics," *Funü zazhi* 1 (January 1915): 15–8; Hui Xia, "Method for Making Rouge," *Funü zazhi* 1 (March 1915): 15–6; Shen Ruiqing, "Method for Manufacturing Cosmetics," *Funü zazhi* 1 (May 1915): 18–25.

featured a new column from January to May, "The Warehouse for Cosmetic Production," which appeared regularly under the "Industrial Arts" [*gongyi*] section. What makes these writings notable is that they deem their detailed technical manufacturing information as highly appropriate for women to apply in the domestic realm. And yet, despite the claim that these were tasteful recipes for women, the production of this discourse was not just for women, but also—if not primarily—by and for men. By employing gendered rhetorical strategies, male authors and editors inflected the discourse of domestic production with moral legitimacy. They sought to "cleanse" the manual work of technology of any problematic class connotation so that it could be reclaimed by new-style urban men, including lettered connoisseurs of technology, amateur scientists, and would-be manufacturers.

A typical example of the gendered portrayal of domestic manufacturing in these writings can be found in the first run of the *Nüzi shijie* column, "The Warehouse for Cosmetic Production" (hereafter, "The Warehouse"). The piece appeared in the January 1915 issue of this women's magazine and was titled "An Exquisite Method for Manufacturing Hair Oil." The editor of the column, as well as the overall editor of *Nüzi shijie* in 1915, was Tianxuwosheng (Heaven Bore Me in Vain), the pseudonym for Chen Diexian (1878–1940), a highly influential editor who was also a romance novelist and, later, an industrial captain. In the column that appeared in the February issue, Tianxuwosheng wrote that the first appearance of the column had elicited much interest and a reader by the name of Mme Xi Meng had already sent in a request asking him to divulge more tips.[2] Highly amenable to this request, Tianxuwosheng provided information on how to produce some of the basic ingredients of hair tonic. Listed in both Chinese and Latin,[3] these ingredients included:

純粹硫酸	Acidum Sulphruicum [*sic*]
檸檬油	Oleum Limonis
. . .	
玫瑰精	Spiritus Rosae
硼砂	Borax
橙花水	Aqua Aurantii Florum
油精	Spiritus
. . .	
丁香油	Nelkeuöl [*sic*]
肉桂油	Oleum Cinnamomi
橙皮油	Oleum Aurantii Corticis
屈里設林	Glycerin
. . .	
白米澱粉	[Rice Starch]
白檀油	Oleum Santali[4]

---

[2] Tianxuwosheng, "Huazhuangpin zhizao ku," *Nüzi shijie* 2 (February 1915): 3.

[3] Note that a few ingredients are not listed in Latin. Glycerin is English, and Nelkeuöl [*recte* Nelkenöl] is the German word for oil of cloves.

[4] Tianxuwosheng, "Huazhuangpin zhizao ku" (cit. n. 2), 4. These ingredients are better known as sulfuric acid, oil of lemon, the essence of roses (i.e., the scent of roses), borax or hydrated sodium borate, orange flower water, alcohol, oil of cloves, oil of cinnamon, oil of orange peel, sugar alcohol, rice starch, and oil of sandalwood.

Many of the items, he explained, could be purchased in Shanghai's pharmacies, but since *spiritus rosae* was particularly expensive, he wanted to make its recipe readily available. His instructions read:

> Extract the fragrance of fresh flowers, and attach it onto something solid, so that it lasts and does not disappear. There are many ways of doing this. One can use a method for suction; the method for squeezing, the method for steaming, the method for soaking. None of these are as ideal as the method for absorption. To make *spiritus rosae*, use the method for absorption.[5]

What follows is a detailed and technical description of how to achieve this method for absorption at home. The tools, instruments, and materials needed include bottles, tubes, alcohol burners, hydrochloric acid, and no less than five pounds of marble.

A simple review of the first two entries of "The Warehouse" column shows how these writings struck a curious balance between presenting highly technical detail and a practical "how-to" sensibility. Moreover, this column presented the information as appropriate for household use by women. Household manufacturing and production were thus elevated over the consumption of mass-produced items from new-style pharmacies, and experimentation and lab work were promoted as appropriate for the home. Finally, the editor seemed driven by a mission to unveil heretofore secret recipes, formulas, and how-to knowledge in the new urban press. By doing so, he sought to make such information available to an engaged reading public, ideally made up of female readers like Mme Xi Meng.[6] The presentation of manufacturing and production knowledge as appropriate for women and located in the domestic arena was explicit.

And yet participants in this discourse included men, ranging from male connoisseurs of technology to dabblers in chemistry and from cosmopolitan lettered elites to budding industrialists. Often closely linked to the rising industrial sector in which advances in chemistry and manufacturing were promoted as necessary for the betterment of Chinese society, male authors and editors were key producers of this discourse on virtuous domestic production. They rendered the how-to pieces in reformist women's periodicals as sites where experiential engagement with chemistry and manufacturing was promoted as crucial in strengthening China, as well as a sign of good taste and bearing. In a period when men were obsessively writing about the plight of women in Chinese society and promoting models of new womanhood, these instructive pieces, which were also written and compiled by men, featured idealized visions of manufacturing women in the inner chamber in order to promote an agenda for men that wedded modern chemistry, technology, and hygienic domesticity.[7] As we will see, they drew on a long history of male literati appropriating the more ethically powerful woman's voice to levy criticism against the political orthodoxy and promote alternative male personae. By 1915, a transitional period when masculinity was in flux, the pure "inner chambers" were deemed a site where virtuous agendas

---

[5] Ibid., 6.

[6] Some writers claimed to be women, and I will discuss the significance of the gender of both writers and consumers of such writings below.

[7] A great deal of work has been done on the male-dominated discourse of New Womanhood in early twentieth-century China, and how such a discourse on women served male interests to define themselves and their masculine agendas of reforming the family, society, and the Chinese nation. See, e.g., Susan Glosser, *Chinese Visions of Family and State, 1915–1953* (Berkeley and Los Angeles, 2003).

such as building industry, strengthening the nation, and laying a strong manufacturing base were articulated as alternatives to the traditional route of government service or the contemporary option of republican politics, which were mired in paralysis and dysfunction. Such a view complicates the more conventional narrative drawn from modern Western experiences in which the masculinization of science and industry marginalized and excluded women in both practice and rhetoric. In this volume, for example, several essays examining the rise of modern science in the United States and Europe provide detailed analyses of the contexts in which forms of technology, science, and related endeavors have been declared manly.[8] From the perspective of 1915 China, on the other hand, we see that at least in rhetoric, women were identified as the primary agents in the construction of new masculinist agendas.

POLITICAL DISARRAY, KNOWLEDGE PRODUCTION, AND THE REMAKING OF MASCULINITY

To grasp the full significance of the gendered presentation in these recipes, we need to take into consideration how by the end of the nineteenth century, educated men had become increasingly disenchanted with serving as officials in the imperial bureaucracy or engaging in other traditional activities of China's lettered elite. Weakened by half a century of internal rebellion and Western imperialist aggression, the Qing dynasty (1644–1911) had in the early twentieth century begun to engage in eleventh-hour reforms geared toward saving the moribund dynasty. These reforms had a profound impact on elite strategies of cultural, social, and political reproduction. Most significant was the 1905 dismantling of the examination system, the bureaucratic mechanism long used by the imperial state to recruit bureaucrats and ensure their ideological and institutional loyalty. The institutionalized ties between educated men and the political center were fundamentally severed, and state privileging of the Confucian canon and its moral text-based knowledge came to an abrupt end. The political disenfranchisement of educated Chinese men deepened with the fall of the empire in 1911. The first decade of the new Republic (1911–49) started with great promise but quickly disintegrated into a period of political disarray. The goals of the 1911 revolution—including the establishment of a legitimate constitution and parliamentary government—were proving elusive. By 1915, China's experiment with republicanism was effectively in shambles when militarist Yuan Shikai threatened to turn China from a republic back into an empire with himself at the helm. In a context in which "traditional" modes of Confucian knowledge and long-vaunted careers in the imperial bureaucracy were no longer feasible, many of China's lettered men started to explore new regimes of knowledge and experiment with new social and occupational roles, as well as reconsider fundamentally what it meant to be an elite man. With the chaos of republican politics threatening national strength, especially in the north, these men often traveled to China's new treaty ports, where modern print industries,

[8] See esp. the essays by Michael S. Reidy on late Victorian male alpinists ("Mountaineering, Masculinity, and the Male Body in Mid-Victorian Britain"), Alexandra Rutherford on mid-twentieth-century American psychology ("Maintaining Masculinity in Mid-Twentieth-Century American Psychology: Edwin Boring, Scientific Eminence, and the 'Woman Problem'"), Erika Lorraine Milam on the gendering of human nature in the 1960s and 1970s ("Men in Groups: Anthropology and Aggression, 1965–84"), and Nathan Ensmenger on gender competition in the professionalization of computer programming in the United States ("'Beards, Sandals, and Other Signs of Rugged Individualism': Masculine Culture within the Computing Professions"), all in this volume.

growing consumer cultures, and mercantile and light manufacturing were flourishing.[9] In this new context, industry and manufacturing, along with the once taboo realm of commerce, increasingly became legitimate ways of strengthening Chinese society and nation outside of the political arena.

Writings such as those found in "The Warehouse" helped promote chemistry, industrial technology, and manufacturing knowledge as foundational pillars of national strength. It was far from "natural" for well-educated men to turn toward production and manufacturing. The literati had long felt a severe distaste for hands-on engagement with things for subsistence or commercial purposes, which, despite the many reforms taking place, persisted into the twentieth century. Suspicion of certain skills and forms of knowledge long associated with toiling artisans and profit-pursuing merchants remained strong among the well educated. To change such entrenched views, publications highlighted new fields of knowledge, most prominently, chemistry and physics, which had experienced advances in nineteenth-century Europe's second industrial revolution and fueled industrial development in Europe, the United States, and, after the 1870s, Japan. They also incorporated insights from related fields, like industrial technology and manufacturing. The writings also strongly emphasized the making of things (*zhizuo* or *zhizao*), demanded physical engagement with material objects, and exhorted experimentation and experiential knowledge. These skills and virtues—long the purview of the artisan class—were now identified as appropriate for the urban educated elite.

To mobilize support for this agenda of redefining elite masculine endeavors, writers of these recipes adopted the woman's voice as an effective vehicle with which to present the information on manufacturing and chemistry. To "domesticate" the technical know-how, editors and writers of these pieces—who were primarily men—portrayed technical knowledge and manufacturing practice as feminine and located production squarely in the domestic realm. By promoting the manufacture of goods such as cosmetics and beauty items, these articles not only tapped into a growing global market that revolved around the care of the self and the cultivation of health and beauty but also drew from globally circulating discourses that endorsed the idea that the health of nations depended on the health of their national citizens.[10] In a context where China was subject to unfair treaties in the hands of imperialist powers and had earned the reputation of being the "Sick Man of Asia," advancing the health and beauty of its female citizens gained symbolic significance.[11]

To understand fully why editors like Tianxuwosheng decided to appropriate the female voice, we also need to keep in mind the longer historical trend of gendering virtuous production as feminine in late imperial China. In the late imperial period,

[9] Disenfranchised literati had already started to find new opportunities in treaty ports in the latter part of the nineteenth century, when moribund court politics and widespread factionalism and corruption drove many away from Beijing. For more on how literati translated their cultural skills into profit in the new circles of print, entertainment, and leisure in cities such as Shanghai, see, e.g., Christopher Reed, *Gutenberg in Shanghai: Chinese Print Capitalism, 1876–1937* (Honolulu, 2004).

[10] See, e.g., Mary Lynn Stewart, *For Health and Beauty: Physical Culture for Frenchwomen, 1880s–1930s* (Baltimore, 2000), on the physical culture of French women in the late nineteenth and early twentieth centuries, which focused on their health and beauty within the context of the French state's concern for healthy families and the reproductive fitness of the nation. As we see here, in these recipes, the concern for female health and beauty lurked behind the agenda of promoting manufacturing knowledge among progressive readers.

[11] For more on the term "Sick Man of Asia," see the introduction to Ruth Rogaski, *Hygienic Modernity: Meanings of Health and Disease in Treaty-Port China* (Berkeley and Los Angeles, 2004).

there had been considerably more room for upper-class women to engage in handwork and manufacturing. Francesca Bray identifies a gendered distribution of elite skills in the late imperial period in terms of the concept of *qiao* (craft, cunning, skill).[12] Manual *qiao* was an attainment to which educated men or even farmers never really aspired. Instead, it was a characteristic primarily of male artisans. Yet, as a female attribute, *qiao* transcended class. Associated with "womanly work," especially in textile production, it functioned to denote a relation to the material world through which women crafted a path to virtue.[13] The Confucian slogan, "men till and women weave" (*nangeng nüzhi*) alluded to this gendered economy of *qiao*, suggesting that virtuous male skill resided in the agricultural arena, while female skill was linked to the production of textiles. In these how-to pieces from the 1910s, the gendered presentation of manufacturing invoked these earlier discourses of Confucian household management and virtuous female production.

There was also a long-standing tradition in Chinese literary history of male literati assuming a woman's sentimental voice, especially when alienated from the political center. As research on the last decades of the Ming dynasty (1368–1644) has shown, in an earlier era of political crisis, disenfranchised literati appropriated the symbolically powerful voices of concubines in their writings to express discontent with orthodox knowledge and politics.[14] A woman's voice was seen as more sentimental, more genuine for expressing inner emotions, such as frustration, pathos, and anger, and hence more effective in articulating criticism of the failure of orthodoxy and statecraft. In the final decades of the Qing, lettered men echoed their late Ming counterparts. The man-of-feeling persona gained currency as a vehicle to legitimize new endeavors and pursuits such as literary projects that included assuming the highly sentimental voices of the effete heroes of popular romance novels, most notably the famous *Dream of the Red Chamber*.[15] With the masculinity of China's lettered men in crisis during the fragile years of the early Republic, male editors could also appropriate the female persona to claim a new form of moral authority. In the writings here, editors identified a female subject as the ideal producer, promoted household production of womanly things, and tapped into the morally charged trope of the domestic. They evoked classical discourses on production and the investigation of things to render chemistry, manufacturing, and bodily engagement with production as legitimate alternatives to politics and statecraft for elite men.

In a period when institutions of knowledge production, social occupations, and masculinity were all in flux, the "ownership" of science, technology, and industry was also not yet fixed. As lettered men turned to new forms of knowledge and engaged in new endeavors, they did so before systematic industrialization, the professionalization of occupational organizations related to industry and manufacturing, and the founding

---

[12] Francesca Bray, *Technology and Gender: Fabrics of Power in Late Imperial China* (Berkeley and Los Angeles, 1997).

[13] To be sure, talented gentry women [*cainü*] were highly invested in expressing their virtue and class identity through poetry and textual engagement like their male counterparts. Yet female activities such as embroidery, painting, and textile production were considered equally respectable and a crucial sign of female virtue and classical education.

[14] Ellen Widmer, "Xiaoqing's Literary Legacy and the Place of the Woman Writer in Late Imperial China," *Late Imperial China* 13 (1992): 111–55.

[15] For more on the ritualized role-play based on novels, such as *Dream of the Red Chamber,* see Catherine Yeh, *Shanghai Love: Courtesans, Intellectuals, and Entertainment Culture, 1850–1910* (Seattle, 2006).

of related modern scientific academic disciplines such as chemistry and physics. As a result, the questions of who owned "science," who was responsible for adapting and producing new knowledge on manufacturing, where such knowledge was to be applied, and for what purposes, were under debate. To be sure, in practice artisanal production and indigenous merchant activities from salt manufacturing to papermaking continued to thrive.[16] But, intellectuals faced with the Qing state's weakness in the latter part of the nineteenth century started to put indigenous regimes of knowing the natural world and "traditional" forms of technology under harsh scrutiny. In response to this perceived political and technological crisis, an array of Chinese actors had started to adapt new forms of production knowledge and apply them to strategic ends. During the Self-Strengthening Movement (ca. 1861–95), the Qing state had sponsored the establishment of arsenals where Chinese provincial statesmen worked closely with Western missionaries and experts, along with Chinese artisans, to translate and adapt Western technological knowledge. With the express goal of building China's military armaments and national strength, the Self-Strengthening Movement came to an end following China's defeat in the Sino-Japanese War in 1895. The ensuing early Republican state proved far less effective in sponsoring industrial development in a systematic manner. In the absence of strong state action, scholarly societies of the late Qing, and new scientific organizations of the 1910s, such as the Science Society of China (SSC), emerged from this vacuum, taking initial steps toward professionalizing and providing an institutional identity to modern science.[17] Missionary schools beginning in the first decades of the twentieth century started offering courses on chemistry, engineering, and physics, but it was not until the 1920s that the institutionalization and professionalization of science developed systematically with the creation of academic disciplines in Chinese universities and related professional "fields" more broadly.[18]

If the period was in flux, it was not merely a "transitional" stage in China's inevitable march toward modern industry, a characterization that smacks strongly of teleology. Nor was it an inferior, less substantial engagement with science and manufacturing than similar situations in other periods or places. Rather, we should consider that China in the 1910s represented a time of multiple opportunities and possibilities as industry and manufacturing developed in various informal spaces and sites. Industrial endeavors were not yet firmly established in factories, laboratories, or the research halls of academia but were pursued in locations such as the studios of literati, reading rooms, domestic and private spaces, and the how-to columns in women's magazines emerging in China's burgeoning consumer culture. Statesmen no longer

---

[16] For work on Chinese salt merchants, see Madeleine Zelin, *The Merchants of Zigong: Industrial Enterprise in Early Modern China* (New York, 2005). For work on paper makers, see Jacob Eyferth, *Eating Rice from Bamboo Roots: The Social History of a Community of Handicraft Papermakers in Rural Sichuan, 1920s–2000* (Cambridge, Mass., 2009).

[17] James Reardon-Anderson, *The Study of Change: Chemistry in China, 1840–1949* (Cambridge, 1991), 93–101.

[18] For more on the development of modern chemistry education, see ibid., esp. chap. 5. For work on the emergence of the modern field of geology, see Grace Shen, *Unearthing the Nation: Modern Geology and Nationalism in Republican China, 1911–1949* (Chicago, 2013). For forensics, see Daniel Asen, "Dead Bodies and Forensic Science: Cultures of Expertise in China, 1800–1949" (PhD diss., Columbia Univ., 2012). The professionalization of science coincided with the rise in the professionalization of modern occupations such as medicine, law, and journalism during the 1920s. See Xiaoqun Xu, *Chinese Professionals and the Republican State: The Rise of Professional Associations in Shanghai, 1912–1937* (Cambridge, 2000).

spearheaded these pursuits as they did during the Self-Strengthening Movement, and academics and professional industrialists did not yet monopolize them as they were to do in later periods. Key players were regional elites, male and female urban connoisseurs, maverick entrepreneurs, and amateur scientists, many of whom appropriated and applied this knowledge and these endeavors in a leisurely manner rather than as academic or professional specialists. Knowledge about chemistry and industry was increasingly commodified, published in commercial journals in burgeoning print markets like Shanghai, where experts of "taste" promoted industrial know-how for self-fashioning urban consumers. In such a context, the recipes discussed in this article functioned in multiple ways. Rhetorically they urged readers to develop an interest in scientific know-how, while supplying tips and information for practical application. The pieces could thus appeal to connoisseurs of technology and to urban readers eager to establish their cosmopolitanism. They were also attractive to amateur scientists and would-be manufacturers who sought concrete knowledge that would advance their hands-on endeavors in industry and manufacturing.

THE WOMEN'S PRESS

Before we turn to the articles themselves, it is worth discussing the nature of China's women's press. The gendering of this knowledge began with the appearance of these pieces in women's press journals. Despite being known as women's journals, early twentieth-century women's press titles offered unique editorial space where both male and female readers could explore new ideas and epistemological configurations.[19] It is important to appreciate the extent of the mixed-gender nature of these journals in considering the "woman's voice" as a powerful tool in which the modern subjectivities of men and women alike could be expressed. In early twentieth-century China, womanhood had become increasingly politicized as a site for the articulation of modernity. Identities such as the new female student and the modern bourgeois housewife were powerful rhetorical tropes, which any writer, regardless of gender, could employ to dispense information or explore new regimes of knowledge. Male contributors frequently assumed the moniker of "Lady" [nüshi] so-and-so to enjoin male and female readers to be a part of the "women's world."[20]

As a particularly lively sector of the commercial press, women's journals quickly came to provide an exciting forum for exploring new ideas and ways of thinking, including information we would now classify as related to the fields of modern science and technology. The journals were hardly uniform, and their exploration of new forms of knowledge assumed different political connotations. Nüzi shijie and Funü zazhi,

[19] For an excellent treatment of the mixed-gender nature of authors and readers of Funü zazhi, see Hsu-Chi Chou, "Yuedu yu shenghuo—Yun Daiying de jiating shenghuo yu 'Funü zazhi' de guanxi" [Reading and lifestyle—Yun Daiying's family life and its relationship with "The Ladies' Journal"], Si yu yan: Renwen yu shehui kexue zazhi 43 (September 2005): 107–90.

[20] By identifying themselves as female students at a particular school, and/or by referring explicitly to their targeted (female) readers as "comrades in the women's world," several of the writers seemed to signal that they were women by using nüshi 女士 ("Lady" or "Mme"). However, we should be careful not to take such authorial claims of speaking in a woman's voice too literally in every single instance. Lesser-known or unknown male writers often assumed names ending with nüshi in order to have their work published in the women's press. See Jacqueline Nivard, "Women and the Women's Press: The Case of the Ladies' Journal (Funü zazhi) 1915–1931," Republican China 10 (1984): 37–55, on 48. That said, women writers—while in the minority—adopted these same conventions and wielded them effectively to legitimate their growing participation in the public literary realm.

the two journals that featured these how-to writings, were important women's press titles in early twentieth-century China. *Nüzi shijie* had a brief run, published from December 1914 to July 1915. In contrast, *Funü zazhi* was one of the longest-running women's journals. Unlike missionary-produced science magazines of the end of the nineteenth century, neither title exclusively featured writings on science and technology. Founded in December 1914 by Tianxuwosheng, the editor of the "Warehouse" column, *Nüzi shijie* was actually primarily devoted to poetry and fiction, writings to be consumed by the refined and genteel female [*guixiu*] reader (fig. 1). Information falling under the categories of industrial arts [*gongyi*], household [*jiating*], fine arts [*meishu*], and hygiene [*weisheng*] only appeared toward the back of the magazine.

Whereas *Nüzi shijie*'s targeted reader was the genteel female or *guixiu*, *Funü zazhi* explicitly identified modern female students as its intended audience and featured progressive opinion pieces and social issue essays or articles [*lunshuo*]. After an initial set of illustrations and photographs (often of respectable modern women), the leading content section of the journal featured opinion essays, including pieces that dealt with topics such as female education. In the back part of the journal, sections such as technologies [*Xueyi*], home economics [*Jiazheng*], fiction, literary selections (literally, "a garden of literature" or *Wenyuan*), the arts, and other miscellaneous items were found. Articles on how to make cosmetics and toiletries were included in the "technologies" [*Xueyi*] section, along with other "popular science" articles, including "A brief discussion of daily physics and chemistry," "Method to eliminate stains from cloth," "The way to measure Chinese weights," "Animals' self-defense," and "The consciousness of plants."[21]

The way women's journals classified these writings reveals how these pieces sought to prescribe new knowledge and information and, more specifically, redistribute the ownership of technical knowledge and skill from a regime of production once associated with artisans to a more lettered audience. Notably, the how-to writings were not classified under categories of traditional elite knowledge such as poetry and belles lettres, which constituted separate sections of at least the *Nüzi shijie*. Rather, the magazines classified the writings under different yet related terms, with *Nüzi shijie* categorizing them as *gongyi*, or "industrial arts," whereas in *Funü zazhi* they fell under the category of *xueyi*, or "technologies." The overlap and fuzziness of the two terms speak to the instability of the categories in the 1910s.[22] The term *gongyi* appeared more specifically focused on industrial knowledge of manufacturing through chemical processes, and *xueyi* suggested a broader subject matter. *Nüzi shijie*'s usage of *gongyi* more narrowly focuses on forms of knowledge and practices involving the use of original ingredients to make something new, often with chemical processes, such as food recipes or dye making. In contrast, the *Funü zazhi*'s use of *xueyi* was more expansive and included a range of topics including domestic production of soap, 100 simple cures for common maladies, and an introduction to a Western-style chessboard.

The readership of both *Nüzi shijie* and *Funü zazhi* was somewhat exclusive.

[21] Some of these were multi-issue articles, and they are chosen from a review of the first four years of the journal, from 1915 to 1919.

[22] These terms started to accrue these meanings in an earlier context of science and technology translation. In the 1895 technological treatise *Xixue fuqiang congshu* [Collectanea of Western learning and political economy] based on missionary translations of Western technology, technologies such as soap making were classified under the heading of *xueyi*, which, in turn, was classified under *gongyi*, or "industrial arts."

Figure 1. *The cover of the January 1915 issue of* Nüzi shijie. *The cover visually evokes the inner chambers of the* guixiu *by featuring a well-groomed, respectable woman, posing demurely in a graceful sitting position. On other covers, the woman of the inner chambers is framed by part of a doorway or wall, accompanied in some cases by a cat or a teakettle, items of the household domain.*

Whereas the major dailies would reach far more readers in the next few decades, these journals nonetheless had a decent distribution. Literary historian Perry Link estimates that 3,000 copies of the *Nüzi shijie* were being published and that copies were then shared or read collectively, with an estimated 10,000 readers per issue.[23] The geographic reach of these journals was also impressive. Examining the letters to

[23] Perry Link, personal communication with Eugenia Lean, 27 March 2006, Princeton University, Princeton, N.J.

the editor that were sent into a regular column in the *Funü zazhi* on medical advice that ran from 1929 to 1931, historian Chang Che-chia argues that while the majority of the letters were from Shanghai, Guangzhou, and the Jiangnan area, letters came from as far as Japan and Chongqing.[24] While the medical column examined by Chang was published later than the *Funü zazhi* issues I study here, its reach suggests that even in the 1910s, the geographic distribution of the *Funü zazhi* was not necessarily confined to the Shanghai area.

Articles in these journals were frequently direct, if unattributed, translations of pieces in the Western or Japanese press, or translated modules of texts, recipes, and images taken from a host of different sources and pasted together. It was common practice in the publishing world of early twentieth-century China for authors and editors of urban presses to draw from their publishers' libraries for sources and engage in the wholesale importation of images, text passages, and recipes when compiling articles.[25] Producing an article involved cutting and pasting blocks of text with imported images. Editors would then add introductory remarks, or commentary on the images, to give the article a sense of cohesiveness. Some of the pieces under consideration here may have been produced in this manner. For example, in the recipe to make *spiritus rosae* that appears earlier in this essay, there is no mention of the exact amount of the flower to use, a missing detail that may point to the recipe's piecemeal production. Furthermore, the name of the purported female reader who wrote in demanding more recipes, Mme Xi Meng, sounds very much like a transliteration of a Western name (possibly Simone?), which suggests that the article may have been translated.[26] The particular manner in which these articles were produced thus poses interesting challenges and opportunities for the historian. Even if we cannot identify a single authorial intent, these pieces nonetheless remain fascinating precisely because of how the modular texts or images were linked together to make a compelling "read." Introductions, prefaces, and other editorial touches reveal the logic of compilation. Thus, translation, appropriation, and local adaptation rather than original authorship of these recipes must be central to our analysis.

THE ARTICLES: HOUSEHOLD TIPS OR MORE?

Written in an accessible form of classical Chinese, the January 1915 *Funü zazhi* article, "A Brief Explanation on Methods for Making Cosmetics," was a typical howto piece from a women's magazine. The author, Ling Ruizhu, identified herself as a third-year student at the #2 Girls School of Jiangsu. The article offered instructions on how to make several different kinds of cosmetics. One entry was for "Soap," identified by the technical Chinese term *shijian*. The entry reads:

---

[24] Chang Che-chia, "*Funü zazhi* zhong de 'yishi weisheng guwen'" ["Consultation on medicine and hygiene" in the *Ladies' Journal*], *Jindai Zhongguo funü shi yanjiu* 12 (December 2004): 145–68, on 153. Since Chang's analysis focuses on the final years of the publication, the journal's reach may have been somewhat differently configured geographically in the earlier years.

[25] Wang Fei-hsien, personal communication with Eugenia Lean, 1 December 2008, University of Chicago, Chicago.

[26] Xi Meng was used as a transliteration for the name of a female English poet, who was introduced in first issue of *Nüzi shijie* along with other foreign poets. See mention of her in the table of contents of volume 1, issue 1 of *Nüzi shijie*. As an editor of the journal, Tianxuwosheng no doubt knew of this transliteration.

[*Shijian*] is commonly referred to as *feizao*. When it provides a cosmetic function, it is called *xiangzao*. There is more than one kind. There is cassia-scented soap, *zhulan*-scented soap,[27] sandalwood soap, etc. For ingredients, one must collect (caustic) soda and cow's fat, pig's fat or coconut oil, etc. Then, add perfume. Its quality depends upon the expensiveness of the perfume, and doesn't have anything to do with the difficulty of making it. . . . When you use it, first dissolve the soap in water, and the soda will slightly float apart. Then, you lather it in water and use it to clean your skin or body hair of filth. Its efficacy is remarkable.[28]

Information was imparted in a straightforward, prescriptive tone. Other entries detailed information on making freckle juice, face powder, and perfume.

Upon closer inspection, however, the writings curiously included what was arguably technical information and chemical knowledge superfluous in the actual production process, which belies any characterization that they are merely "practical advice" or "domestic hints." New forms of technology, including modern chemical principles of saponification, the reaction of alkali with fat or oil, both of which are needed to make modern soap, were described explicitly and in detail. Some pieces purposefully highlighted the language and conceptual categories of modern chemistry. Take, for example, the March 1915 article, "Method for Making Rouge":

Rouge is made from the flower petals of the red flower 紅花 (Carthamus tinctorius L. [*sic*]) and is one of the important items among cosmetics. Red flower is part of the chrysanthemum family. . . . Its height is around 2–3 *chi*. Its flower petals contain red and yellow pigment. The red pigment named Karthamin ($C_{14}H_{16}O_7$) is the main component of rouge. . . . When [the petal pigment] comes into contact with an acid type then it precipitates into sedimentation. If you take the dissolved sediment part, then high quality manufacturing is possible.[29]

In explaining the ingredients of rouge, this passage explicitly used scientific terms. *Carthamus tinctorius L.* was the Latin name for safflower used in the Linnaean taxonomic system, referring to the flower needed to make the red dye for rouge. In the original article, the term was written in the Romanized alphabet and thus stood out sharply in the otherwise primarily Chinese text. The Latin term for red dye, *Karthamin*, was similarly Romanized. So too was the chemical compound, $C_{14}H_{16}O_7$. The text also employed the discourse of modern chemistry in describing the rouge-making process. Sodium carbonate (sodium salt of carbonic acid) was specified as the preferred acid-type for precipitating the petal's pigment into sedimentation.

These writings emphasized experimentation, sensory know-how, and bodily engagement. "A Brief Explanation on Methods for Making Cosmetics" was typical in that it featured a high degree of prescriptive technical knowledge in the form of practical advice. The author, Ling Ruizhu, not only displayed expertise in chemistry and manufacturing but also served as a practical advisor, whose "tips" resulted from her own hands-on experience in making and experimenting with things. In the technical explication of manufacturing soap, Ling advised readers that the best way to tell whether the soap was high quality was through one's senses. "With the tip of the tongue, taste it. If there is no peppery, acrid taste, then it is a quality commodity

---

[27] In modern botany, *zhulan* is *Chloranthus spicatus*, or makino, a tree with fragrant yellow beadlike seeds.

[28] Ling, "A Brief Explanation" (cit. n. 1), 17–8.

[29] Hui, "Method for Making Rouge" (cit. n. 1), 15.

[*shangpin*]."[30] This advice, while terse, nonetheless spoke volumes in alluding to a new and practical way of knowing and by promoting ingenuity and inventiveness as virtues. Experiential knowledge and experimentation were valued. Chen Diexian exhorted his readers to experiment with basic ingredients in the January issue of "The Warehouse," the entry with which we started this essay. Technical and scientific information could be obtained from mundane ways of knowing, including hands-on experience, trial and error, and physical engagement with ingredients.

While these writings were more than just "domestic hints," their emphasis on the domestic sought to convince readers of the moral relevance of industry and technology through the trope of the domestic. It was no accident that they promoted the ideal domestic site for chemistry and manufacturing as the well-to-do space of the genteel inner chambers. In his May 1915 contribution to "The Warehouse" column, Chen Diexian, writing under the name Xu Yuan, identifies women's inner chambers as ideal sites for chemical experimentation:

> Use [the method for absorption] to make gifts for your friends in the inner chambers (*gui-you*). It is rather enjoyable. Moreover, you can use seasonal flowers and make different kinds of solid fragrances; you are not restricted to using rose essence [the only fragrance available on the market]. The chemical method . . . of absorption . . . has become the strategy of experimentation (*shiyan ji*) in the inner chambers. All you need to do is obtain some simple tools and follow the [instructions below].[31]

The inner chambers had become a place where one's senses (in this case, one's olfactory senses) could be deployed as a site for scientific experimentation. In the rest of the article, Xu described the exact kind of jar to use at home, provided instructions on how to make a copper sheet into a thin tube, and how to place it in the jar, with the precise measurement of how far away from the four sides the tube needed to be. He also provided directions for how to sterilize the jar without cracking the glass and how to use Vaseline and scents to make solid perfumes. The end of the article narrated the exact process by which to calibrate the strength of fragrances in the making of perfume with different kinds of flower petals. Xu concluded by advising, "There is only one matter one must really pay attention to. If the fragrance is too pungent, then the result is not beautiful. Fragrance is like color. If a color is too strong, it appears murky and dark. If a scent is too pungent, it causes unpleasant olfactory senses. If you use less, the scent will unfold and be alluringly fragrant. Thus, those making scents must understand the physics (behind the process)."[32]

By identifying the well-to-do space of the inner chambers as the ideal domestic site for chemistry and experimentation, Xu Yuan underscored the exclusivity of such productive practices. These were not the quarters of more modest homes or parlors of modern treaty-port homes. They were the refined inner chambers of elite households. References to gift exchange among genteel women, the seemingly disinterested act of "enjoyment," and an emphasis on "friendship" among like-minded people further evoked the image of literati cultivation long associated with such fine spaces. At the same time, Xu made sure to present this space as fully modern. The inner chambers did not resemble the traditional literati study, where reflective moral contemplation

---

[30] Ling, "A Brief Explanation" (cit. n. 1), 18.
[31] Xu Yuan, "Huazhuangpin zhizao ku," *Nüzi shijie* 5 (May 1915): 1.
[32] Ibid., 2–3.

took place through the study of texts. Nor were they a space for talented women to embroider, exchange poetry, or participate in a separate "woman's culture." They differed considerably from imperial period households that had been engaged in textile production.[33] In these early Republic writings, the inner chambers had become a space where things were manufactured through chemical means, a place where experiments abounded and modern production practices could be enjoyed.

Xu proceeded in one entry to tout domestic production as superior to items available in stores. He writes triumphantly,

> Now, in the inner chambers, it is a common practice to take hair tonic as suitable for use. That which is sold on the market is also called hair gel and uses fat. It is sold in small glass bottles or stored in small porcelain boxes. Its weight is no more than an English tael and its selling price is regularly above a silver *yuan* and two *jiao*. If you buy the original ingredients and make it yourself, [however], it is far more inexpensive. Try what is explained [below]. . . . You can use it whenever you like, and it is hardly different from what you can buy on the market. [Furthermore] . . . that which you can buy on the market is all made into a rose scent. . . . [At home], if you like other kinds of scents, you can use . . . other kinds of perfumes and add it into the mixture.[34]

Making an appeal to frugality, Xu promoted the homemade version as the same in quality as the store-bought type, and far less expensive. He stressed how home manufacturing allows for more flexibility and thus the production of more scents. He then warned that store-bought fragrances ran the risk of being "infected with dirt" as they were mixed with different kinds of inferior fat. Household production was heralded as superior to market consumption.

To underscore the relevance and urgency of this technical know-how, these writings not only gendered the know-how but also invoked several powerful moral discourses. The classical belief in the importance of the household management of womanly virtue and production for the larger polity is, for example, an important concept for grasping the significance of the *Funü zazhi* article, "The Method for Producing Cosmetics."[35] Its author, a Mme Shen Ruiqing, declared that women's knowledge of how to make cosmetics, and thus the ability to truly know their nature, was at the crux of improving the state of the household (*jiating*). She began with a warning: "Cosmetics are things [*pin*] that women need. Their price is rather expensive. When women are being thrifty, cosmetics count as a kind of expendable item. Is it better then not to use them?" To answer the question, Mme Shen invoked the traditional belief that women's proper appearance and deportment was one of four womanly virtues. "Having balding hair and not treating it, and having teeth plaque and not cleaning it, all obstruct the hygienic welfare of the household [*jiating weisheng*] . . . if we use [cosmetics] without knowing their nature [*pinxing*], this is lacking in household knowledge [*jiating zhishi*]." By placing such a premium on household order, Mme Shen strongly evoked the classical Neo-Confucian [*lixue*] discourse that linked the harmony of the domestic feminized realm [*nei*] with the moral harmony of the masculinized outer [*wai*] realm

---

[33] With the imperial-era discourse of "men till and women weave" [*nangeng nüzhi*] identifying female production of cloth as the crux of a productive empire, women of all classes actively engaged in their homes in the "womanly work" of weaving, textile making, and managing the household production of things. See Bray, *Technology and Gender* (cit. n. 12), esp. pt. 2.

[34] Xu, "Huazhuangpin zhizao ku" (cit. n. 31), 1.

[35] Shen, "Method for Manufacturing Cosmetics" (cit. n. 1), 18–25.

of political cosmology. Family relations had metaphorical implications for political relations; wifely chastity and filial piety were metaphors for political loyalty. If men's pursuit of textual knowledge had tended to exclude them from the realm of practical manual labor, women's work in the household was deemed crucial for a productive state of affairs beyond the domestic domain.

In identifying the production of women's things to be the crux of household harmony and emphasizing knowing the nature of objects through their production, the *Funü zazhi* article also resonated strongly with the imperial discourse on *gewu zhizhi*, the "investigation of things" (abbreviated hereafter as *gewu*). The philosophical discourse on *gewu* was primarily seen as an act of textual classification rather than an actual engagement with material objects. As the subject of broad learning [*bowu*], the concept of things [*wu*] in *gewu* was defined expansively, referring to "objects, events, mental and physical phenomena, the unknown, and the anomalous."[36] As such, things were to be known, or decoded, through the words and language of philosophers and connoisseurs. Accordingly, *gewu* practitioners expended a great deal of textual energy etymologizing the world and life of things to probe universal principles and seek the moral harmony of the larger sociopolitical cosmos.[37] As the classical Confucian text "The Great Learning" [*Daxue*] explained, it was by grasping the nature of things that individuals could complete their knowledge, and by extension, that an imperial state could achieve its harmony.[38]

By taking as its basic premise that obtaining knowledge (especially textual knowledge) of material things such as women's cosmetics and toiletry items will result in moral harmony, this *Funü zazhi* article appeared to be drawing on the tradition of *gewu*. For Mme Shen, full understanding of women's objects (in part, achieved through textual study of her article) was the key to cultivating household harmony. Vanishing face cream and hair grease were presented as at once necessary and potentially destabilizing. They helped women uphold the virtue of appearance and deportment and yet were fraught with potential danger. If their nature was not understood properly, they could potentially lead women and, by extension, the household to dissolute, wasteful consumption. The piece thus provided a prescription against a scenario of household dissolution by detailing production know-how and the proper understanding of the nature of such things. The rest of the article recounted the chemical makeup of different cosmetic items, and how to manufacture and properly use them. It included a list of original chemical ingredients in the section on the "dissection of [the item's] various parts," the manufacturing method [*zhifa*], the method for using the item [*yongfa*], and supplementary notes that included a warning about how to handle the glycerin properly when making lotion that removes freckles.[39] Readers

---

[36] Benjamin Elman, *On Their Own Terms: Science in China, 1550–1900* (Cambridge, Mass., 2005), xxix.

[37] For more on the rich and complex history of the discourse of *gezhi* in imperial China, see ibid.

[38] *Daxue* was one of the four canonical Confucian classics, institutionalized in the Song dynasty as part of the civil service examination curriculum. Originally part of the *Book of the Rites, Daxue* was authored by Confucius (551–479 BCE) in the sixth century BCE.

[39] Glycerin is a neutral, colorless, thick liquid that freezes to a gummy paste and has a high boiling point. It can be dissolved into water or alcohol, but not oils. Since many things will dissolve into glycerin more easily than they do into water or alcohol, it functions well as a solvent. Soap fats already contain glycerin, and it is when the fats and lye interact in the formation of soap that glycerin is left out as the "by-product." The above warning about handling glycerin is probably because glycerin is often

also learned how to make hair dye, face powder, l'eau de toilette, liquid rose rouge, face lotion, perfume glycerin, toilet powder, lavender water, hair tonic, pimple caustic, rose-scented hair grease, lavender-scented hair tonic, camphor tooth powder, and "Hazeline Snow" vanishing face cream.[40]

While the relationship between the moral and the material might not have been new, what was unprecedented was the crucial subtext behind the 1915 interest in the material nature of "women's things." In a postimperial order, modern China had moved away from a Confucian cosmology to become—at least in principle—a Republican nation-state competing in the international arena of capitalism. The focus in these articles on broadening the knowledge of the constituency of certain commodities was meant to ensure China's ability to manufacture goods and its success in this global realm of commerce, rather than harmonize a Confucian social order, despite their palpable concern with China's increasingly commercialized society. The rise of global capitalism in the nineteenth century had resulted in Chinese markets being flooded with foreign commodities [pin 品], ranging from opium to industrial goods. The humiliating military defeats by Western powers and modernized Meiji Japan seemed to confirm China's technological and material inferiority. This anxiety ensured that materiality loomed large in these articles and notably often in a language that arose from the context of international trade.

The choice of toiletry items, such as soap and hair tonic, was moreover hardly accidental or frivolous. As Mme Shen's reference to the need for a hygienic welfare of the household suggests, the international discourse of China's deficiency in hygiene and health conditions that had gained currency by the latter part of the nineteenth century cast a large shadow in these pieces and was invariably linked to the perception that China's weakened state led to increasingly aggressive and violent imperialist subjugation. Chinese intellectuals, reformists, and administrators quickly internalized this discourse. The need to modernize the "sick man of Asia" implied that the Chinese body had to be made anew into a clean and healthy one. Reformists and intellectuals were thus actively promoting new regimes of "hygienic modernity" in the reformist press, through anti-foot-binding associations, and in the creation of physical education curricula and institutions.[41] The *Funü zazhi* and *Nüzi shijie* writers allied themselves to this cause by calling for the hygienic welfare of the household and were charged with a political and moral imperative. The focus on the production of these objects served to contain the object of cosmetics from a potentially dangerous consumerist attraction into something that would make the nation strong.

Concern with China's manufacturing competitiveness was also apparent in the article, "Method for Making Rouge":

> Our country's rouge has been long famous. But as a beauty product of the (traditional) inner chambers, [our rouge] has had no significance for the fate of the nation and the lives of citizens. However, if we acquire the (proper) methods, we can manufacture it

---

used to make nitroglycerin. But this warning may in fact be moot since glycerin is not an explosive substance by itself and has to be turned into nitroglycerin before it becomes explosive.

[40] Hazeline Snow was a British brand and one of the most popular vanishing face creams marketed in China's treaty ports at the time.

[41] For more on "hygienic modernity" in China's treaty ports, such as Tianjin, see Rogaski, *Hygienic Modernity* (cit. n. 11).

[like] recent imports, including the so-called foreign rouge. . . . [Indeed] without quality manufacturing techniques [*zhizao jingqiao*], how could these [foreign] countries have robbed us of our power? Thus, with the production technique I describe below, I speak to my comrades in the women's world and contribute for all to examine and study; with this production technique, we can obtain true mastery![42]

In this passage, the potential wealth of the nation was directly linked to technical expertise and proper knowledge. The segment also illustrates the metaphorical quality of the "woman's world" and the domestic sphere. The author identified products of the old inner chambers as being dangerously outdated, and because of their lack of relevance for the polity, a potential hazard to the fate of the nation. An appeal is made for the author's "comrades in the woman's world" to take seriously the competitive new production techniques outlined in the article. The urgent plea to improve methods of manufacturing was meant not only for actual housewives. "Comrades in the women's world" referred to new-style lettered and entrepreneurial elites, including men. By engaging in chemistry, experimentation, and production (whether through the production or consumption of these writings, or in actual practice), participants in this discourse could articulate an unprecedented urban identity that turned on a newfound zeal to reform the nation through commerce and manufacturing.

In short, these writings adapted a variety of strategies to legitimate the chemical and manufacturing knowledge they were promoting to China's urban readers. By identifying the feminized inner chambers as the spatial site of scientific activity, these pieces wove together a moral discourse defining virtuous production as relevant and necessary for the new age. Columns like "The Warehouse" rendered ideal reader-producers like Mme Xi Meng as exemplars, who embodied new forms of technical knowledge and scientific production to which both male and female readers would ideally aspire. The pieces also mobilized an array of moral discourses. The classical Neo-Confucian discourse on the management of the household served to establish that the activities described on these pages had implications beyond the domestic arena. The *gewu* legacy shaped the emphasis placed on the need to know thoroughly the nature of things, including soap and cosmetics. And, the late nineteenth-century imperialist discourses on hygiene made the writings relevant to the contemporary world where China's national fate was at stake. Taken together, the message delivered by these pieces was clear. Productive self-sufficiency in the feminine domain of the inner chambers was to be both a metaphor for and a means to material self-sufficiency for China. Such writings were part of a larger contemporary discourse on virtuous women's work [*nügong*] that filled the pages of women's journals such as *Funü zazhi,* discussed by Constance Orliski. Yet, whereas Orliski argues that this broader discourse engaged specifically in the construction of middle-class ideals of productive women, I argue that the feminized discourse of production examined here was just as much about the construction of new norms of masculinity.[43] As we see next, these writings—despite being gendered feminine—were sources of inspiration and concrete information for male dabblers in chemistry, budding manufacturers, editors and authors of domestic science columns, and leisurely connoisseurs of technology.

[42] Hui, "Method for Making Rouge" (cit. n. 1), 15.
[43] Constance Orliski, "The Bourgeois Housewife as Laborer in Late Qing and Early Republican Shanghai," *Nan Nü* 5 (2003): 43–68.

AMATEUR SCIENTISTS AND WOULD-BE INDUSTRIALISTS

Editors, authors, and readers of these pieces included connoisseurs of chemistry—men and women—as well as male would-be industrialists and amateur science experimenters. All were invested in advancing domestic chemical production and the industriousness of the virtuous *guixiu* as a means to disseminate and endorse the practical information and chemical know-how provided in these recipes as a legitimate source of male knowledge and endeavors. The early twentieth century saw the flourishing of light industry and manufacturing in China. While a stable and mature industry may not have existed until the Nanjing decade (1927–37),[44] Chinese companies had started to open soap and cosmetic factories as early as the first decade of the twentieth century. One of the earliest domestic soap factories, the Nanyang Soap and Candle Factory of Jiangsu Province, was established in 1910.[45] Another early cosmetic company was the China Chemical Industries Company, which was established in 1911 and whose founder, Fang Yexian, I discuss below.[46] By 1915, the Ministry of Agriculture and Commerce established in Beijing the Bureau of Industrial Research, which was dedicated to the analysis of native products, including cosmetics and dyes, and played a crucial role in promoting and building native industry.[47] A number of cosmetic and pharmaceutical companies were founded soon thereafter. In 1917, two years after his tenure as the editor of the *Nüzi shijie* and its "Warehouse" column, Chen Diexian himself established the pharmaceutical company Household Industries, with $10,000 of capital he had earned from his fiction and editing work.[48] The most popular product produced by Household Industries was Chen's own invention, the "Butterfly" brand (*Wudipai*) tooth powder that could double as face powder.[49]

In this context, it is not difficult to imagine how would-be industrialists might have turned to these 1915 recipes for inspiration, if not actual manufacturing tips. A strong political undercurrent that would appeal to such readers regarding the need to domesticate production not just literally, but figuratively, can be found in the content of

[44] James Reardon-Anderson attributes this to the Nationalists for providing enough direct investment in research and industry, a degree of political and social stability, as well as a dose of benign neglect. See Reardon-Anderson, *The Study of Change* (cit. n. 17).

[45] For more information on this company, see *Zhongguo guohuo gongchang shilue* [A historical sketch of China's national product factories] (Shanghai, 1935), 23–4.

[46] For more on this company, see ibid., 63–4. See also *China Industrial Handbooks Kiangsu* (Shanghai, 1933), 507–11.

[47] In the late Qing under the New Policies reforms, Yuan Shikai had established a similar bureau in Zhili, known as the Bureau of Industry, which lasted from 1903 to 1907 and was meant to develop modern industry in the Zhili region. It established a higher education facility in industry, organized displays, ran workshops, lectures, and night classes to teach trade and industry, printed educational materials, and visited factories to encourage and develop industry. For this information and more, see the Bureau's 1907 gazetteer, *Zhili gongyi zhi chubian* [Zhili industrial gazetteer], 1st ed. For histories on the role of government—national and regional—in developing industry, see relevant sections in Yutang Sun, *Zhongguo jindai gongshi ziliao* [Source materials on the history of industry in modern China] (Beijing, 1957), and Jun Gong, *Zhongguo xin gongye fazhang shi dagang* [The historical synopsis of the development of new industry in China] (Shanghai, 1935).

[48] For more information on this company, see *Zhongguo guohuo gongchang shilue* (cit. n. 45), 23–4, 117–8, and 127–8. See also *China Industrial Handbooks Kiangsu* (cit. n. 46), 508.

[49] The official English name of the tooth powder was "Butterfly," which was a translation of the Shanghaiese pronunciation of the Chinese name of the product, *Wudi*. However, the name brand when read or pronounced in Mandarin connoted another meaning. The compound *wudi* literally means "peerless," or "without enemies," and *Wudipai* means "Peerless Brand." In the context of the National Products Movement, of which Chen was a leader, the name thus gained great significance.

several pieces. In "Method for Producing Cosmetic Soap with Care," Kuang Yu, the author of an article in the March 1915 "Warehouse" column of *Nüzi shijie*, wrote,

> High-quality soap can . . . make the skin smooth and glossy (like jade). Thus, among cosmetic products, soap occupies an important place. Today, our countrymen establish factories to manufacture soap and attend to its production. Generally, most of the products are inferior. One reason is because the original ingredients are poor quality. Another reason is that the manufacturing method is not ideal. We cannot be surprised that the soap used in our inner chambers cannot compete with imported products. Trace the cause of this . . . not to a *thorough* enjoyment of using foreign products [per se], but to the fact that in our country there are no good products.[50]

Included in a column dedicated primarily to domestic production, this piece is notable for its abiding concern with China's fledgling manufacturing industry and its inferior products. As if retorting to critics who might say that it is the fault of consumers who enjoy using imports, the passage underscored that it was not lavish attachment to foreign items, but rather substandard production of domestic products that was at fault. Moreover, the reference to inner chambers in the piece speaks to the metaphorical meaning of the domestic trope in which to bolster the products of the inner chambers was to improve the industrial strength and competitiveness of China. In an era when patriotic consumption and production were gaining traction in campaigns such as the National Product Movement, such writings resonated with patriotic readers, some of whom certainly included actual or would-be manufacturers.[51]

The rest of the article offered concrete tips and manufacturing methods to improve China's ability to compete internationally.

> To make this, you first need to take 1,000 *liang* of high-quality cow fat (you can also use the highest quality olive oil), throw it into a cauldron, and melt it until the fat is completely dissolved. Then, take 150 *liang* of high-quality caustic soda, and dissolve it into clear water until it becomes a thin liquid. . . . When adding the caustic soda, you must use a wooden paddle and stir without stopping. After cooking it to the point of boiling for approximately 4 hours, the caustic soda and the cow fat will have thoroughly combined. That which floats on top is pure white, fine and delicate soap paste. . . . Choose the best fragrance . . . add it in; thoroughly mix . . . pour it into a mold, and . . . an exquisitely fragrant colored soap results.[52]

Where the piece merits particular attention is its mention of 1,000 *liang* of high-quality cow fat, which is around 83.33 pounds of cow fat, a considerable amount that would produce a large amount of soap.[53] This recipe was clearly not intended for mere household production, but for large-scale manufacturing.

Biographies reveal that would-be manufacturers and dabblers in chemistry were indeed active participants in this discourse. First and foremost is Chen Diexian, the

[50] Kuang Yu, "Huazhuangpin zhizao ku," *Nüzi shijie* 3 (March 1915): 3.

[51] For more on China's early twentieth-century National Product Movement, see Karl Gerth, *China Made: Consumer Culture and the Creation of the Nation* (Cambridge, Mass., 2003).

[52] Kuang, "Huazhuangpin zhizao ku" (cit. n. 50), 3.

[53] According to a "Chinese Weights and Measures Table" published in the "Chinese Maritime Customs, 1922–1931 Decennial Reports," a *liang* is a tael, and sixteen *liang* is equivalent to one *jin* (catty), and 1,000 *liang* would be around 62.5 *jin*. With 100 *jin* (catties) the equivalent of 133.33 pounds, 62.5 *jin* would be around 83.33 pounds.

editor of the how-to column of the *Nüzi shijie* that we are examining here. His bio-graphical accounts portray him as an amateur chemist who turned his literati studio into a chemistry lab in his early years in Hangzhou.[54] He also opened a shop to sell books and imported scientific instruments and appliances, as well as a public library-cum-reading room, which featured translated scientific texts. In early 1913, while serving in a staff post in Zhenhai, a city on the Zhejiang coast near Ningbo, Chen spent his days experimenting with cuttlefish and coastal brine to locally source calcium and magnesium carbonate, key ingredients in tooth powder. The result of his experiments was the above-mentioned versatile tooth powder that could double as face powder. It was with this dual-functioning tooth powder, along with proceeds he made from being a professional writer and editor in Shanghai, that Chen went on to found House-hold Industries, which became a leading domestic pharmaceutical company that com-peted with major international companies for market share in urban China's com-petitive hygiene and cosmetic market. As an industrial captain, powerful Shanghai editor, and advocate for manufacturing know-how, Chen was a key participant in using the new genres of women's journals to promote new forms of knowledge that could serve as the foundation of the new masculine ideal of serving the nation through industry.

While we cannot be sure he read these exact pieces, Fang Yexian (1893–1940), the above-mentioned cofounder of one of Republican China's most successful light chemical industry companies, China Chemical Industries Company (hereafter, China Chemical), read similar material to procure the knowledge necessary for his manu-facturing endeavors. Born in Shanghai into a business family (family enterprises in-cluded banking houses, pawn shops, jewelry shops, and sundry goods stores),[55] Fang was sent to the Anglo-Chinese College, a missionary educational facility founded in 1882 in Shanghai, where he developed a particular interest in chemistry. There he studied with a German chemistry instructor who had been hired by the Shanghai Municipal Council of the International Settlement.[56] He pursued his interests outside of the classroom, too, by taking advantage of the late nineteenth-century missionary-related network of public science and reading rooms. He avidly consumed journalistic writings of precisely the sort considered here, along with specialized books, to create a household laboratory, engage in chemical experiments, and make everyday items such as cosmetics at home. As a result of such activities, Fang founded China Chemical with his mother's support in 1912 (his father was against the enterprise). It started with a few individuals and disciples but in twenty years became the leading Chinese manu-facturer of tooth powder, with one of its most popular items being the Sanxing brand tooth powder. The company was also successful in manufacturing food-flavoring powder, including the Boddhisattva powder (*Guanyin fen*), and Jiandao soap. Later,

[54] For biographical information on Chen Diexian, see Chen Dingshan, "Wode fuqin Tianxuwosheng—Guohuo zhi yinzhe" [My father, Tianxuwosheng—recluse of the National Products (Movement)], in *Chun Shen jiuwen* (Taipei, 1967), 180–204; and Chen Xiaocui, Fan Yinqiao, and Zhou Shoujuan, "Tianxuwosheng yu Wudipai yafen" [Tianxuwosheng and Wudipai tooth powder], in *Wenshi Ziliao Selections*, ed. Zhongguo renmin zhengzhi xiehuiyi quanguo weiyuanhui wenshi ziliao yanjiu wei-yuanhui [The National Association of the Chinese People's Political Society's Association for the Research of Historical and Literary Materials], vol. 80 (Beijing, 1982).

[55] For biographical information on Fang Yiexian, see, e.g., Pan Junxiang, ed., *Zhongguo Jindai Guohuo Yundong* [China's modern National Products Movement] (Beijing, 1996).

[56] The Chinese name of the instructor was Dou Bolie, most likely a rendition of the German name.

Fang became known as the King of National Products because of his anti-Japanese activism in the National Products Movement.[57]

As Chen's and Fang's biographies indicate, young men often dabbled in science in informal spaces and venues at the turn of the century. Some of these amateur scientists went on to build successful careers in the industry. Others, however, did not have the financial support to be able to start their own pharmaceutical empires or build industries but found other uses for such formulas and recipes. A colorful example is the famous Daoist practitioner Chen Yingning (1880–1969). Like Fang, Chen was an amateur scientist who practiced in informal settings and among literati friends and collaborators. He built a home laboratory to engage in chemical experimentation in pursuit of Daoist goals of longevity through external alchemy. As Xun Liu details in his study of modern Daoism in urban Shanghai, Chen's pursuit of domestic experiments was a team effort.[58] Chen's wife, a modern gynecologist, and four other fellow practitioners of alchemic self-cultivation, financed the endeavor. Located near the entertainment center of the Yu Garden, Chen's urban home set aside two rooms for his experiments, most of which involved smelting metals in attempts to test the veracity of secret formulas from ancient alchemical recipes. His laboratory was stocked with key alchemical minerals such as cinnabar, mercury, silver, and lead and was equipped with heating furnaces and refining crucibles. He would invite friends to his laboratory, perform experiments in front of them, and discuss for hours into the night various formulas and their efficacy for self-cultivation. Xun Liu suggests as well that this group, many of whom were doctors, scientists, and scholars educated abroad or in modern universities in China, were exposed to popular science journals featuring how-to pieces similar to the ones under consideration here and that their alchemical experiments were clearly informed by modern chemistry.

While the Fang Yexians and Chen Yingnings of the world provide profiles of potential male participants in this discourse, it is entirely possible that women participated for similar reasons. Actual examples are hard to come by, but Wu Yizhu (1882–1945), Chen Yingning's wife, is the kind of woman who might have put these writings to use. Wu was one of several new-style women who took advantage of the array of new ideas and opportunities available in treaty ports and in the urban press, along with their male counterparts. She was a well-educated female who traveled in the circles of these new urban elites. In addition to having trained at the Sino-Occidental Medical College, she became a successful modern gynecologist and was an active participant in building modern Daoist circles and supporting, if not actively engaging in, the domestic chemical experimentation integral to Daoist pursuits of self-cultivation.[59]

From these biographies, we can see how these articles were part of a larger exploration of manufacturing and chemistry that was being pursued in a variety of settings. This was a moment before the professionalization of industry and science in China,

---

[57] It was this same activism and his unwillingness to work in collaboration with the pro-Japanese Wang Jingwei government that finally led to Fang's assassination by secret agents in 1940. For an English-language account of his involvement in the National Products Movement, see Gerth, *China Made* (cit. n. 51), esp. 180–1.

[58] Xun Liu, *Daoist Modern: Innovation, Lay Practice and the Community of Inner Alchemy in Republican Shanghai* (Cambridge, Mass., 2009), 71, 299n68. Liu suggests that they were most likely aware of contemporaneous attempts by Sinologists and historians of science in the West and Japan to link Daoist alchemy to the origins and history of modern chemistry, by viewing Daoist alchemic experiments as iatrochemistry, the forerunner to modern chemistry.

[59] For more on Wu Yizhu, see ibid., 56–8.

which would function to discipline and draw boundaries around associated activities and knowledge, locating them, for example, in formal sites such as the university, the science lab, or the modern factory. The 1910s saw practices associated with modern science, including experimentation, observation, building laboratories, and using modern scientific equipment, often taking place in the home, in literati studios, and in storefronts. While textbooks and technological compendia were sources for technical information and modern science, so too were missionary reading rooms, science displays, and how-to writings in the women's press. Chemistry was pursued out of a variety of motives, whether to increase manufacturing, for leisurely connoisseurship, or as part of other cultural pursuits, such as updating Daoism for the modern age.

CONCLUSION

By 1915, the promises of the 1911 revolution seemed increasingly undeliverable, and Western imperialists and Japan continued to compromise Chinese sovereignty, humiliating China internationally and persisting with invasive forms of extraterritoriality. In the greater Shanghai and Guangdong areas, where women's journals were widely circulated, more commercially oriented elites readily disengaged from the political scene. The "domestic hints" literature examined here provided these disenfranchised editors and authors the opportunity to appropriate the domestic realm to define an alternative site of elite moral activity. The household was redefined as a feminine site where chemistry and manufacturing were to be enjoyed and exalted above mass manufacturing and the crass market. As avid participants in this discourse, male writers and authors appropriated the feminine to explore new forms of knowledge and endeavors. The domestic and, indeed, the feminine itself became metaphors for the productivity of a commercial and scientific culture that thrived, in contrast to the (masculine) political arena, where the failure to establish the republic and build a new nation was all too painfully obvious.

The domestication of manufacturing knowledge in these recipes resonated with the burgeoning National Products Movement and eventually acquired the meaning of domesticating foreign technology for purposes of increasing native manufacturing and strengthening the nation. This movement had already started with boycotts of American goods starting in 1905. By the 1910s, Chinese patriotic sentiment had only deepened and turned against the Japanese in particular. Having declared war on Germany in 1914, Japan followed up not by participating substantively in the European theater but by attacking Germany's concession in Shandong, China. This served as a pretext for Japan to issue a humiliating set of demands, better known as the Twenty-One Demands, to the Chinese government. It was at this point that figures like Chen Diexian became politicized. As biographical accounts later noted, his animus toward the Japanese had been key in motivating him to found his company, Household Industries, in 1918 and develop his Wudipai (Peerless) brand tooth powder to defeat Japan in the market place.[60] Chen went on to become a leading National Products Movement leader, promoting virtuous copying of foreign technology for the purpose of establishing China's manufacturing autonomy. The 1915 recipes featuring female producers in the inner chambers—some of which Chen compiled—set the stage for

[60] See, e.g., Chen, "Wode fuqin Tianxuwosheng," 184, and Chen, Fan, and Zhou, "Tianxuwosheng yu Wudipai yafen," 217–8 (both cit. n. 54).

anti-imperialist activities and patriotic manufacturing that served as alternative ways for men to serve the country.

In only a few years, the promotion of "Mr. Science" by Beijing-based intellectuals would substantially challenge the gendered portrayal of technology found in the 1915 how-to pieces, as well as the politics of such knowledge. In 1919 in what is now known as the May Fourth Movement, students and citizens of Beijing marched to protest China's humiliating treatment in the signing of the Versailles Treaty and called for the need to strengthen the Chinese nation domestically against warlords and internationally against imperialists. Central to the goals of the movement's participants was the rejection of old ways and the introduction of "new culture," including Western science and political philosophies, which intellectuals had begun to explore and promote during the New Culture Movement (ca. 1915–9). In a *New Youth* essay published in January 1919, Chen Duxiu, the dean of Peking University and a key leader of the New Culture Movement, promoted "Mr. Science," along with "Mr. Democracy," as the antidote to the superstition of the past and the oppressive shackles of Confucianism and traditionalism.[61] Adopted by May Fourth activists as a slogan, "Mr. Democracy and Mr. Science" were soon consecrated as essential parts of China's modernity. Through these icons, cosmopolitan intellectuals associated with the iconoclastic New Culture movement sought to claim ownership over "science" and define its purpose in society, as well as define proper norms of elite masculinity. They ardently argued that men truly concerned with China's national fate had to engage in "modern science" [*kexue*] for nation-state building purposes and the pursuit of modernity. The earlier depictions of manufacturing and chemistry as activities to be pursued leisurely and for enjoyment in the inner chambers by genteel women quickly came to be seen as illegitimate and frivolous.

Historians of China have internalized this canonical May Fourth narrative that "masculine" science deserves attention rather than "feminine" domestic know-how. Accordingly, much attention has been paid to how state-sponsored and "formal" science in modern China developed in direct response to modern China's political plight. Scholarship has tended to focus on the earlier self-strengthening arsenals built by the Qing state, and the founding of formal academic scientific fields like geography and archaeology that occurred in the early twentieth century.[62] What has not yet been considered, however, is how China's national plight spurred exploration of industry, chemistry, and manufacturing in seemingly apolitical and inconsequential spaces such as China's burgeoning consumer culture and its women's magazines. Not surprisingly, the writings explored here have tended to be either completely overlooked or treated too modestly in the past, misleadingly referred to as "domestic hints" or mere "practical knowledge" for housewives. To remedy such omissions, this essay takes these how-to writings seriously. We have seen how politics formed a crucial undercurrent in these women's press pieces and how these recipes helped legitimate new knowledge,

---

[61] Chen Duxiu, "Xin qingnian zui'an zhi dabianshu" [*New Youth's* reply to charges against the magazine], *Xin Qingnian* [New Youth] 6 (15 January 1919): 10–1.

[62] For recent work on the arsenals, see Yue Meng, "Hybrid Science versus Modernity: The Practice of the Jiangnan Arsenal, 1864–1897," *EASTM* 16 (1999): 13–52, and Elman, *On Their Own Terms* (cit. n. 36). For scholarship on the formation of modern geology, see Shen, *Unearthing the Nation* (cit. n. 18), and for archeology, Fa-ti Fan, "Circulating Material Objects: The International Controversy over Antiquities and Fossils in Twentieth-Century China," in *The Circulation of Knowledge between Britain, India and China: the Early Modern World to the Twentieth Century,* ed. Bernard Lightman, Gordon McQuat, and Larry Stewart (Leiden, 2013), 209–36.

shape urban male identity and taste, as well as sow the seeds for elite male activity in manufacturing and production in the early part of the twentieth century.

Rather than the more familiar story that emerges from studies focused on the West, where masculine norms are mobilized to legitimate the production of knowledge about technology or science, we have seen here how feminine ideals served to authenticate industrial and manufacturing know-how in China's burgeoning consumer culture in 1915. Indeed, paying attention to these Chinese recipes raises questions about the way scholarship in the history of science has thought about the relationship between modern science and technology and masculinity. Philosophers of science have powerfully challenged the ideal that science is "pure" and "objective" knowledge, while feminist theorists identified the processes by which women have been systematically excluded from the domain of modern science.[63] Historical case studies based on examples from the West have tended to follow suit by illuminating the process by which women have been excluded from the domains of science and technology. To challenge essentialist assumptions of science and technology as inherently male endeavors, many have sought to historicize the process by which science and technology have been gendered masculine.[64] Pioneering and rich as these studies are, a global perspective helps complicate the picture. Not simply a curious exception to the rule, 1915 China serves to remind us that as actors in different societies mobilized gender to legitimate the authority of scientific and technological fields and knowledge, they drew from a wide variety of histories of gendering knowledge and practice. Indicative of a broader epistemological reordering of elite knowledge taking place in the 1910s for both men and women, these writings deployed feminine tropes to ensure that chemistry and manufacturing were appealing and desirable to a group of new urban readers. The "genteel woman's" voice rather than any masculine trope proved most conducive to help articulate the contours of masculinity and make these recipes for the manufacturing of new men in early twentieth-century China so potent.

[63] See, e.g., Evelyn Fox Keller, *Reflections on Gender and Science* (New Haven, Conn., 1985); and Sandra Harding, *The Science Question in Feminism* (Ithaca, N.Y., 1986).

[64] Classic examples include Robert A. Nye, "Medicine and Science as Masculine 'Fields of Honor,'" *Osiris* 12 (1997): 60–79; Ruth Oldenziel, *Making Technology Masculine: Men, Women, and Modern Machines in America, 1870–1945* (Amsterdam, 1999).

Mountaineering, Masculinity, and the Male Body in Mid-Victorian Britain

by Michael S. Reidy*

ABSTRACT

Golden-age mountaineers attempted to codify gender, like flora and fauna, by altitude. They zoned the high Alps masculine. As women also reached into the highest regions, male alpinists increasingly turned to their bodies, and the bodies of their guides, to give scientific validity to their all-male preserve. Edward Whymper traveled to the Andes in 1879, where he transformed Chimborazo into a laboratory and his own body and those of his guides into scientific objects. His work helped spearhead a field-based, vertical approach to human physiology that proliferated after the turn of the century. By viewing gender through a spatial lens and using the sides of mountains to map it, this essay highlights the gendered notions that directed early research in high-altitude physiology.

> The fact is—and it cannot be too strongly insisted on—that there really exist three distinct Switzerlands, suspended one over the other at different altitudes. The first—the Switzerland of ladies, children, elderly gentlemen, and ordinary folk in general—includes all the valleys and lakes traversed by railways, highway roads, and steamers. . . . The second region . . . takes in the localities which cannot be reached in carriages, but to which prudent lads and lasses may roam on foot or horseback. . . . Our third and uppermost Switzerland supplies the Alpine Club with spots where a human foot has never trod, or where the number of its footprints may be counted. It furnishes peaks ascended only by scientific men and human donkeys.
> —Charles Dickens, "Foreign Climbs," 1865[1]

Edward Whymper's successful ascent of the Matterhorn on his eighth attempt in July 1865 mired the burgeoning sport of mountaineering in an ethical debate. During the descent, on a rather tricky section of mixed ice and rock, a quick slip, one false move, and four men fell thousands of feet, their clothes later found fused to their flesh,

* Department of History and Philosophy, 2-170 Wilson Hall, Montana State University, Bozeman, MT 59717; mreidy@montana.edu.

I would like to thank David Agruss for valuable discussions, and James H. Meyer, Joseph Taylor, and Peter Hansen for close readings and helpful comments on earlier versions of this essay. I also owe an enormous debt to Erika Lorraine Milam, Robert A. Nye, and the anonymous reviewers for insightful suggestions.
[1] Dickens, "Foreign Climbs," *All the Year Round* 14 (2 September 1865): 135–7.

with arms and legs missing from their mangled bodies. Three were British, including Charles Hudson, vicar of Killington, the most promising young alpinist of the age, and Lord Francis Douglas, a member of the British landed elite. "Why is the best blood of England to waste itself in scaling hitherto inaccessible peaks," the *Times* mocked, "in staining the eternal snow, and reaching the unfathomable abyss, never to return?"[2] The nascent climbing fraternity pleaded with Whymper to respond. The "bitter attacks on the Alpine Club and all mountain climbers," they feared, had produced the general sentiment that climbing was "morally wrong."[3]

Several such attacks came from Charles Dickens. He scoffed at the "salubrious excitement of mountaineering for over-worked men; the proud preeminence of England in manly courage."[4] It was a dismal sign of the suffocating state of British society. No wonder, he frothed, that the weary businessman, benumbed lawyer, or bored professor, "if he have a fibre of manhood in him, rejoices in the change, rejoices in the adventure, rejoices (this largely enters into the Englishman's account) in his power of proving to himself that he is neither effete nor effeminate."[5] During a time when working-class bodies were being mangled in London factories in the name of industrial progress, that the inner "fibre" of England's best and brightest was being mutilated in Swiss crevasses seemed beyond the pale.

Leslie Stephen—literary critic and mountaineering apologist—insisted that climbing was a sport, nothing more, but the raging debate in the public press suggested otherwise. So, why all the fuss? What was at stake? In this essay, I argue that definitions of masculinity were at the bottom of the Matterhorn tragedy. I show how golden-age mountaineers attempted to codify gender, like flora and fauna, by altitude. How this worked and why it worked is part of a fascinating story of alpine heights, of death and self-definition, of guides and gentlemen, and of pioneering alpinists' ability to use their bodies to define the masculine in what Dickens pejoratively referred to as the third Switzerland.

Dickens's use of a spatial metaphor of "distinct" zones should come as no surprise. He had a knack for feeling the pulse of his age. His move through the vertical—from civilization to wilderness, from common sense to hubris, and from the feminine to the masculine—mimicked the spatial perspective that had taken hold in the sciences by midcentury. Vertical zonation had become a steady guide to research in the geological, botanical, and zoological sciences. It was a guiding organizational force behind biogeography and early oceanography, atmospheric studies of light and heat, and the Humboldtian sciences of terrestrial magnetism, meteorology, and radiation physics. As scientists began exploring the ocean depths and the upper atmosphere, attempting to fathom the oceanic and aerial realms through technologies such as deep-ocean soundings and hot air balloons, a vertical orientation enabled researchers to answer questions that had previously been beyond their reach.

In the middle decades of the nineteenth century, mountaineers used this same

---

[2] As quoted in David Robertson, "Mid-Victorians amongst the Alps," in *Nature and the Victorian Imagination*, ed. U. C. Knoepflmacher and G. G. Tennyson (Berkeley and Los Angeles, 1978), 113–36, on 114.

[3] Alfred Wills to Edward Whymper, 6 August 1865, and Howel Buxton to Edward Whymper, 9 August 1865, MS 822/35, "Letters Concerning the Matterhorn," Scott Polar Research Institute, Cambridge, England.

[4] Dickens, "Hardihood and Foolhardihood," *All the Year Round* 14 (19 August 1865): 85–7.

[5] Ibid.

guiding principle to advance the study of human physiology. A wide range of cultural forces influenced this new interest in the human body. One was the Darwinian emphasis on the natural mechanisms responsible for how the body worked, including its physical and mental limits. Another was the cultural fascination with the consequences of the laws of thermodynamics, especially the seemingly irreversible deterioration associated with entropy. By reducing the body to a machine, which took in food and expended energy through physical labor, nineteenth-century physiologists could focus on fatigue, that onerous quality responsible for limiting productivity, be it on the factory floor or in the trenches of war.[6] The quest to determine the limits of the human body in extreme environments was part of a larger pursuit to reduce human physiology to an exact science. Mountaineers were in a propitious position to advance these questions. By placing their own bodies in extreme environments, mid-Victorian alpinists helped spearhead a scientific, field-based approach to high-altitude physiology that flourished after the turn of the century.[7]

A focus on the human body linked Darwinian evolutionary theory and the sport of mountaineering from the very beginnings of alpinism. The first edited volume on climbing, *Peaks, Passes, and Glaciers*, appeared in 1859, the same year as Darwin's *Origin*. The three foundational mountaineering texts in Britain—Tyndall's *Hours of Exercise*, Stephen's *Playground of Europe*, and Whymper's *Scrambles amongst the Alps*—were all published in 1871, the same year as Darwin's *Descent of Man*. Victorian mountaineers were well versed in Darwin's powerful new theory, acutely aware of its implications for morality and the focus it placed on the human body: as an animal, a specimen, and an experimental object.[8]

These foundational mountaineering narratives move consistently upward, physically and metaphorically: from civilization to isolation, from domestication to wilderness, and from the feminine to the masculine. Mountaineering narratives have never been simply about climbing, especially during the mid-Victorian golden age, when the sport was invented and debated.[9] The mountains provided the perfect physical geography to discuss issues of race, class, nationalism, civilization, modernity, morality, and physical ability.[10] But the most highly charged topic of debate

[6] Anson Rabinbach, *The Human Motor: Energy, Fatigue, and the Origins of Modernity* (New York, 1990), 4, 63. For the manner in which Victorians used the metaphor of the machine to determine the differences between the sexes, see Cynthia Eagle Russett, *Sexual Science: The Victorian Construction of Womanhood* (Cambridge, Mass., 1989).

[7] The historiographies of both sport and human physiology place this type of scientific analysis in the late nineteenth and early twentieth centuries, attributing it primarily to the work of French and German physiologists. See, e.g., Richard Holt, *Sport and the British: A Modern History* (Oxford, 1989); Vanessa Heggie, *A History of British Sports Medicine* (Manchester, 2011); John B. West, *High Life: A History of High-Altitude Physiology and Medicine* (Oxford, 1998); John M. Hoberman, *Mortal Engines: The Science of Performance and the Dehumanization of Sport* (New York, 1992); and Rabinbach, *The Human Motor* (cit. n. 6).

[8] For Tyndall and Stephen, two of the leading evolutionary naturalists of their day, these questions were paramount. See Michael S. Reidy, "Scientific Naturalism on High: The X-Club Sequesters the Alps," in *Victorian Scientific Naturalism: Community, Identity, Continuity*, ed. Gowan Dawson and Bernard Lightman (Chicago, 2014), 55–78; and Reidy, "Cosmic Emotion," *Alpinist Magazine*, Summer 2012, 93–6.

[9] Susan Schrepfer, *Nature's Altars: Mountains, Gender, and American Environmentalism* (Lawrence, Kans., 2005), 3.

[10] See the recent spate of secondary material covering the history of mountaineering, including Peter H. Hansen, *The Summits of Modern Man: Mountaineering after the Enlightenment* (Cambridge, Mass., 2011); Ann C. Colley, *Victorians in the Mountains: Sinking the Sublime* (Burlington, Vt., 2010); Joseph E. Taylor III, *Pilgrims of the Vertical: Yosemite Rock Climbers and Nature at Risk* (Cambridge,

was gender. Mountains became a preferred site for the cultivation of all that was considered masculine and the expulsion of all that was deemed "effete and effeminate." The body became the tool, the instrument that contemporaries employed to fight their battles over these contested concepts. Thus, this essay analyzes broad geographical trends in the sciences that Victorians used to divide landscapes into zones in order to underscore the gendered notions that directed pioneering research in high-altitude physiology.

WHY CLIMB?

> The people beneath our feet inhabit a different world. They don't belong to us, nor we to them. We have risen above the latitudes where battles are fought and crowns are lost; we breathe a purer ether and serener air; and, like the gods from Olympus, look down complacently upon the races of men who make haste to destruction.
>
> —Member of the Alpine Club, 1873[11]

Three climbers, in particular, structured the narrative of mountaineering in the mid-Victorian era. Edward Whymper towered over the sport, much like the mountain with which he will forever be associated. His rival on the Matterhorn, John Tyndall, infused the sport with science, scaling the mighty Weisshorn in 1861. And Leslie Stephen turned the Alps into a playground, especially for British agnostics. They were the elite of the climbing community. Whymper's *Scrambles*, Tyndall's *Hours,* and Stephen's *Playground* helped define the culture of climbing in Europe.[12] One might assume these adventuresome texts to be filled with hairbreadth (and harebrained) escapes and dangerous fights to the death with the horror of the heights. Literary scholar Francis O'Gorman, for instance, pinpoints the masculinity in both Stephen's and Tyndall's work primarily in their "bravery in facing the physical dangers of the mountains."[13] Historian of science Bruce Hevly likewise argues that scientists established authority primarily through the "physical discomfort, if not immediate danger" they experienced as they climbed.[14]

Yet the Matterhorn disaster produced an outpouring of anger, encouraging all three alpinists to downplay the dangers of climbing in their publications. "Our great climbers are really getting modest," one reviewer quipped. "For some reason or other it is becoming the fashion to represent one's own performances in the gross with a tone of gentle disparagement."[15] The lack of brag and bravado smelled of a conspiracy,

---

Mass., 2010); and Maurice Isserman and Stewart Weaver, *Fallen Giants: A History of Himalayan Mountaineering from the Age of Empire to the Age of Extremes* (New Haven, Conn., 2008).

[11] "The Doctor Abroad," *Blackwood's Edinburgh Magazine* 113 (1873): 657–77, on 664.

[12] See, e.g., Claire-Elaine Engel, *A History of Mountaineering in the Alps* (London, 1977); James Ramsey Ullman, *The Age of Mountaineering* (Philadelphia, 1964); and, for a popular account, Fergus Fleming, *Killing Dragons: The Conquest of the Alps* (New York, 2000).

[13] Francis O'Gorman, "'The Mightiest Evangel of the Alpine Club': Masculinity and Agnosticism in the Alpine Writing of John Tyndall," in *Masculinity and Spirituality in Victorian Culture*, ed. Andrew Bradstock, Sean Gill, Anne Hogan, and Sue Morgan (New York, 2000), 134–48, on 135.

[14] Bruce Hevly, "The Heroic Science of Glacier Motion," *Osiris* 11 (1996): 66–86, on 66, 84. For a nuanced discussion of Victorian mountaineering and risk, see R. D. Eaton, "In the 'World of Death and Beauty': Risk, Control and John Tyndall as Alpinist," *Vict. Lit. Cult.* 41 (2013): 55–73.

[15] "Hours of Scrambling Exercise—Tyndall and Whymper," *Saturday Review of Politics, Literature, Science and Art* 32 (1871): 59–60, on 59.

another reviewer noted, as alpinists "gloss over the dangers of mountaineering and give the fiercest Alps a quiet, comfortable character, — in fact, to warrant them one and all to carry a lady if required."[16]

Whymper, Tyndall, and Stephen all attained hero-like status among mountaineers for their passionate defense of climbing in ways other than dallying with danger. Danger was certainly involved, but they spoke of this in terms of a benign struggle with nature that ultimately benefited their bodies. Climbing was safe if one took precautions, hired a trustworthy guide, and knew one's natural limits. Then, it provided an "education in bravery, in self-collectedness, in self control, and in the power of acting in sudden emergencies."[17] The sport exemplified the Victorian fascination with the assertion of physical and mental control of one's own body.

Victorian mountaineers also spoke of their yearning to escape the monotony and drudgery of a mundane, urbanized existence, fearing that increased cultivation had eroded a healthy mental and physical equilibrium.[18] Tyndall said he climbed to "restore that balance between mind and body which the purely intellectual discipline of London is calculated to destroy."[19] For Stephen, the mountains offered "strong stimulants" to reinvigorate his "sluggish imagination."[20] The mountains represented a liminal space blending mind and body, a vertical arena where imaginations could be rejuvenated by the body being pressed to its physical limits. This could not be accomplished in the flat, sullied streets of London. It required a flight to the heights. As one critic rather curiously put it in the *Westminster Review* in 1864, "We must rise now and then, like the whales, to a purer medium."[21] This purer medium enabled mountaineers to break down distinctions between physical and intellectual labor, to stress their mental motivations while experiencing hearty, physical activity.

But most of all—and like Dickens, I must stress that this theme appeared throughout British accounts—mountaineers spoke of the manly nature of climbing, of ridding the bored Englishmen of everything that was "effeminate and effete." These mountaineers drew upon well-established models of masculinity circulating at the time in novels and weeklies, from the books of Charles Kingsley and Thomas Hughes to the heated discussions of women's education in the *Times*.[22] Alpinists believed that they were in a good position to advance this discourse, to approach questions of gender from a privileged position on high. One reviewer noted that each author had his merits—Stephen his buoyancy, Tyndall his imagination, and Whymper his artistic excellence. "But they agree in one virtue, that of manliness."[23]

Physical and mental qualities attained through climbing were always conflated with questions of manliness and the male body. "To attain the perfect balance of body required for difficult glacier and rock work," one commentator noted, "brings one

---

[16] "The Alps; or, Sketches of Life and Nature in the Mountains," *Brit. Quart. Rev.*, July 1862, 71–102, on 98–9.

[17] "Switzerland in Summer and Autumn. Part I," *Blackwood's Edinburgh Magazine* 98 (1865): 323–45, on 344.

[18] Robert Nye, *Masculinity and Male Codes of Honor in Modern France* (Oxford, 1993); Taylor, *Pilgrims of the Vertical* (cit. n. 10).

[19] John Tyndall, *Mountaineering in 1861* (London, 1862), vi.

[20] Leslie Stephen, *Playground of Europe* (London, 1871), 180.

[21] "Mountaineering," *Westminster Rev.*, October 1864, 276–90, on 279.

[22] Charles Kingsley, *Two Years Ago: A Novel* (1857); Thomas Hughes, *Tom Brown's Schooldays* (1857); and Hughes, *Tom Brown at Oxford* (1861).

[23] "Hours of Scrambling Exercise" (cit. n. 15), 59–60.

very near, at least, to the perfection of physical manhood."[24] Another argued that the "qualities of manhood" were all refined through the physical exertions required on steep cliffs: "a reasonable disregard of pain and of life, that insensibility to physical privation, that lightning readiness of hand and eye, that dogged temper of endurance which men have called manliness ever since the days of the Trojan war."[25] Mountains provided Tyndall with a laboratory for his scientific research, the perfect backdrop for his lack of faith, and an outlet for his agnosticism, but most importantly, he stressed, "they have made me feel in all my fibres the blessedness of perfect manhood."[26] In the mountains, physical privation and mental acuity were rarely disassociated from questions of gender.

ESTABLISHING GENDERED ALTITUDINAL ZONES

> Between the tropic or table d'hôte zone, inhabited by elderly pa-
> pas, and the glacial regions devoted to the chamois and to climb-
> ers who can equal or surpass the chamois, comes the temperate
> or middle-aged and ladylike zone of mountaineering.
> —"Mountaineering," *Saturday Review*, 1862[27]

There is nothing inherently masculine about either mountains or the sport of mountaineering. Historically, mountains were the province of the gods, a place to experience ghastliness or to contemplate the sublime, none of which had ever been strictly codified as masculine in character. Women could see godliness in the mountains, could fear them or see the sublime in them, or even want to conquer them.[28] In fact, the mountaineering narratives of 1871 were as popular among women as men, as were the debates about the merits of climbing.[29] Women, it seems, were banished from the heights primarily in Britain. As historian Susan Schrepfer demonstrates, a group of New England women founded the first mountaineering club in the United States in 1863, and few subsequent clubs in the United States, New Zealand, or Canada were exclusively male.[30] Women routinely undertook serious alpine excursions, often attaining the summit first.[31]

Even in Britain, as climbing grew in popularity, it became increasingly difficult to sustain the process of gendering the sport masculine.[32] Women climbers, including Lucy Walker and Marguerite Claudia "Meta" Brevoort, the aunt of William Coolidge, were celebrated climbers in their own right.[33] In 1871, Walker became the first

---

[24] "Switzerland" (cit. n. 17), 343.

[25] "Mountaineering" (cit. n. 21), 281.

[26] As quoted in "A Scientific Climber," *Critic* 24 (April 1862): 413–5, on 414.

[27] "Mountaineering," *Saturday Review of Politics, Literature, Science and Art* 13 (1862): 627–8.

[28] Ruth Oldenziel, *Making Technology Masculine: Men, Women and Modern Machines in America 1870–1945* (Amsterdam, 1999), 10.

[29] Schrepfer, *Nature's Altars* (cit. n. 9), 13. This is similar to the Arctic exploration narratives discussed in Michael Robinson's essay, "Manliness and Exploration: The Discovery of the North Pole," in this volume.

[30] Schrepfer, *Nature's Altars* (cit. n. 9), 69. Women were not allowed as members of the British Alpine Club until 1976.

[31] Ibid., 72.

[32] "Mountaineering," *London Review of Politics, Society, Literature, Art, and Science* 109 (1862): 98–9.

[33] Rebecca A. Brown, *Women on High: Pioneers of Mountaineering* (Boston, 2002); Colley, *Victorians* (cit. n. 10), 133.

woman to summit the Matterhorn. That same year, Brevoort climbed the Weisshorn, a mountain that had overpowered both Tyndall and Stephen on their first attempts, and which Tyndall had finally summited only a decade earlier. Women climbers, more-over, exhibited all the qualities of their male counterparts, good and bad: they were competitive, self-aggrandizing, self-promoting, and they cared about when they got to the summit and where they would fit in mountaineering history. And they were obviously great athletes. Climbing the Weisshorn or the Matterhorn is no easy feat. As literary scholar Ann Colley has recently pointed out, the sheer number of women mountaineers by the last decades of the nineteenth century suggests that all did not share the idea of an exclusively male domain. Yet, though women were not "summar-ily discouraged" from climbing,[34] gender still operated as a primary means by which Tyndall, Stephen, Whymper, and members of the all-male Alpine Club codified the mountain landscape.

Just as men erected barriers against women entering the labor market or gaining access to education and the professions, so too did a parallel process of spatial bound-ary making occur in the mountains. Because women were regularly climbing to high altitude, members of the male climbing fraternity focused more diligently on finding zones where women and lesser men could be excluded. Similar to the boundary-making processes seen in the essays by Nathan Ensmenger, Erika Lorraine Milam, and Alexandra Rutherford in this volume, male alpinists attempted to invent an all-male upper zone that could be distinguished from the supposedly demasculinized glaciers and valleys.[35] The Alps, especially its glaciers, were becoming uncomfortably accessible, prompting the famous British mountaineer, Alfred Mummery, to describe the history of every mountain in Switzerland in three stages: "An Inaccessible Peak.—The most difficult ascent in the Alps—An easy day for a lady."[36] It is useful here to note that the masculinization of the higher realms took place at exactly the same time as the masculinization of technology, and for similar reasons.[37] Just as machines be-came the measure of men, so too could the ability to reach high altitudes (and lati-tudes) be used to measure masculinity.

Thus, not all mountain regions were gendered masculine. Titles of texts from the period also included references to a "temperate zone" that was "ladylike." Compare, for instance, Charles Hudson and Edward Kennedy's *Where There's a Will There's a Way: An Ascent of Mont Blanc by a New Route*, with Mrs. Freshfield's *Alpine Byways or Light Leaves Gathered in 1859 and 1860*, or contrast Alfred Wills's *The High Alps* with Mrs. Cole's *A Lady's Tour Round Monte Rosa*. Or, if we simply study the iconic paintings and photographs of the era, we find scenes of summits highlighting males in the act of climbing (fig. 1) while pictures of glaciers feature women in skirts in the act of walking.

Duplicating similar moves in the counting-houses and colleges of London, women began to rebel against what they rightfully perceived as a repression of their physical

[34] Colley, *Victorians* (cit. n. 10), 5.

[35] Nathan Ensmenger, "'Beards, Sandals, and Other Signs of Rugged Individualism': Masculine Culture within the Computing Professions," Erika Lorraine Milam, "Men in Groups: Anthropology and Aggression, 1965–84," and Alexandra Rutherford, "Maintaining Masculinity in Mid-Twentieth-Century American Psychology: Edwin Boring, Scientific Eminence, and the 'Woman Problem,'" all in this volume.

[36] As quoted in "A Happy Mountaineer," *Bookman* 8 (1895): 85–6, on 85.

[37] Oldenziel, *Making Technology Masculine* (cit. n. 28), 11.

Figure 1. The First Ascent of the Matterhorn *by Gustave Doré, 1865.*

abilities.[38] They took to the high peaks. Marie Paradis had climbed Mont Blanc very early, in 1808, but she was a "peasant" from Chamonix and never published her own story. Published accounts stressed how guides literally carried her to the top as she swooned, semiconscious and fearful to the point of hallucination. There was less to mock, however, in the second female ascent. Henriette d'Angeville, a woman of noble rather than peasant birth, summited the peak in 1838 after a decade of planning. She attributed her success primarily to the triumph of her will. "My physical powers were abandoning me," she recalled near the summit. "I heard without hearing and saw without seeing."[39] The first British woman to scale Mont Blanc represents a similar triumph over both the mountain and established gender roles. Accompanied by her husband, nine guides, and a sixteen-year-old porter, Mrs. Hamilton passed two nights on the ascent, attained the summit, and returned to Chamonix where she joined the celebration, dancing and toasting her guides with sherry.[40]

The most accomplished British women climbers such as Lucy Walker rarely published their accounts of summiting high peaks. The narratives of those who did publish were often confined to the lower regions. One of the earliest and most popular was *A Lady's Tour Round Monte Rosa*, unsigned, but known to be the work of Mrs. Henry Warwick Cole. Her aim in writing the book, she related, was the hope of inducing others, "especially members of my own sex," to follow in her footsteps.[41] The narrative is gendered distinctly feminine.[42] Cole focused on the inns and hotels, the proper feminine dress, the ability to hire mules or donkeys, and whether "recent numbers of the 'Times' and 'Quarterly Review'" were readily accessible.[43] She represented the local guides as a lazy lot, selfish, ignorant, often timid, and always drunk. In the climax of the text, Cole walks around the Chamonix valley rather than conquering one of the surrounding peaks. She begins and ends in civilization.

The text posed little threat, as Cole did not challenge the upper male zone. It was a ride on a mule "round" Monte Rosa, not a climb up, so most reviewers treated her book with a wink and a nod. The one negative review mistakenly believed "the bold lady" actually summited the peak. The reviewer castigated Cole for forfeiting the "gentler and more loving characteristics of her kind." In fact, the reviewer questioned what "kind" she was. "In daring, in physical strength, and in closeness and accuracy of thought she seems as much a man as Semiramis or Lady Macbeth."[44] Similarly, in "A Lady's Ascent of the Breithorn," published in *Chambers's Journal* in 1877, an alpinist also described herself as "the hapless object of half a woman."[45] The Breithorn is an extremely high but relatively accessible peak overlooking Zermatt. Its accessibility

---

[38] David Rosen, "The Volcano and the Cathedral: Muscular Christianity and the Origins of Primal Manliness," in *Muscular Christianity: Embodying the Victorian Age*, ed. Donald E. Hall (Cambridge, 1994), 17–44, on 20.

[39] As quoted in Hansen, *Summits* (cit. n. 10), 172.

[40] "A Lady's Ascent of Mont Blanc," *Leader* 5 (1854): 845; "A Lady's Ascent of Mont Blanc," *Reynold's Miscellany of Romance, General Literature, Science and Art* 13 (1854): 212.

[41] [Mrs. Henry Warwick Cole], *A Lady's Tour Round Monte Rosa* (London, 1860), 2.

[42] For "proper" and "improper" females, see J. B. Bullen, *The Pre-Raphaelite Body: Fear and Desire in Painting, Poetry, and Criticism* (New York, 1998), 166.

[43] [Cole], *A Lady's Tour* (cit. n. 41), 58.

[44] "Alpine Tours," *Dublin University Magazine* 54 (1859): 475–86, on 475.

[45] "A Lady's Ascent of the Breithorn," *Chambers's Journal*, 1877, 535. See also R. A. E., "A Lady's Ascent of a Snow Mountain," *Golden Hours: A Monthly Magazine for Family and General Reading* 15 (1882): 441–5, for a description of the ascent of the Cima di Jazi, a lofty snow-covered peak of the Monte Rosa range, reaching 12,500 feet.

lessened its value for mountaineers, and tellingly, it became known as "the Ladies' Mountain."[46]

Other climbing narratives written by women seemed to follow a similar course. *Alpine Byways* by "A Lady" appeared in 1861, the same year that Tyndall climbed the Weisshorn. Its frontispiece featured a magnificent woodcut of the mountain, but only as seen from below (fig. 2). In fact, none of the seven illustrations depict icy crags or rocky pinnacles. The "lady" seemed to know her place. However, "without aspiring to exploits which may be deemed unfeminine," the author explained that the pursuit of "manly amusement" had spread far and wide, "making wives and sisters seek participation in the pleasures which they hear so vividly described."[47]

The text then subtly undermined the all-male vertical zonation of mountain regions. While always with her husband and young son, she consistently left the "well-known routes" and "(accompanied by a lady friend) sought to extend our acquaintance with the by-ways and higher passes of the Alps."[48] While climbing the Schilthorn, she stressed how "there were no defined tracks, and we had to climb up steep, rough rocks . . . affording precarious footholds, where a slip would have had unpleasant consequences."[49] At one point, they had to cross a knife-edged ridge, "skirting the snow on one side, while on the other the rocks descended precipitously below us."[50] This sounds very much like the accounts given by Whymper, Stephen, and Tyndall. As she ascended ever upward, a decidedly masculine narrative follows, with the crossing of extensive snowfields embedded with crevasses, snow bridges, the cutting of steps, and the incessant dallying with danger through deep snow to the summit. Her husband was enfeebled by the climb, both on the way up, when his "breathing became oppressed and painful," and on the way down, when he "sprained his knee and nearly exhausted his powers."[51]

The author took more chances with her second book, *A Summer Tour in the Grisons and Italian Valleys of the Bernina*, published in 1862. Both the Grisons and Bernina Alps had rarely been visited, so the title alone suggested experiences beyond the ordinary. This time, she signed her name, Mrs. Henry Freshfield, to the title page and acknowledged her authorship of *Alpine Byways*. From the beginning, she consciously focused on heights and included a table of "Mountain Heights in English Feet."[52] Following a common trope of alpine narratives, she began by comparing most travelers to sheep, never wandering from the beaten path. She then asked the reader to follow her as she broke new trails, scrambled across knife-edge ridges, glissaded down snowfields, and climbed "high above the glacier."[53] In her first book, her husband often had trouble following her on the mountain; in this bolder text, he is purposefully and unceremoniously left behind at the hotel.

---

[46] See Annie S. Pec, "A Woman's Ascent of the Matterhorn," *English Illustrated Magazine* 157 (1896): 53–62, on 56. This language still prevails. Today, most climbing areas have "girlfriend friendly routes," usually the warm-up climb. The irony, of course, is that some of the greatest climbers today are female, including the greatest rock climber of my generation, Lynn Hill, the first to free climb "The Nose" on El Capitan in Yosemite National Park, as well as the first to free climb it in a day.

[47] A Lady, *Alpine Byways, or Light Leaves Gathered in 1859 and 1860* (London 1861), 2.

[48] Ibid. Her son, Douglas Freshfield, became a world-renowned climber, the first to open up the Caucasus to climbing, one of the first to visit Kanchenjunga, and a future president of the Alpine Club.

[49] Ibid., 10.

[50] Ibid., 12.

[51] Ibid., 28, 32.

[52] Mrs. Henry Freshfield, *A Summer Tour in the Grisons and Italian Valleys of the Bernina* (London, 1862).

[53] Ibid., 102.

Figure 2. *The Weisshorn from above Ronda (from a drawing by Edward Whymper). From [Mrs. Henry Freshfield]*, Alpine Byways *(cit. n. 47), frontispiece.*

Mrs. Freshfield's mountaineering texts predated the narratives of Stephen, Tyndall, and Whymper, who focused exclusively on the heights, insisted on the courage and trustworthiness of guides, and defined their own regions as the place where the *Times* could never be found. They realized the boundary they were attempting to create was much more porous than they liked to admit. The higher Mrs. Freshfield climbed, the higher their exclusive zone needed to be. Most of all, they required a boundary-making instrument. For this, they turned to their bodies, and when that proved insufficient, they co-opted the bodies of their guides, the real natives of the third Switzerland.

THE BODY AND THE GUIDE

> The mountaineers as a rule seem to think that their consumption and assimilation of food is a subject of very general interest, not only to themselves, but to the public. If there could be removed from the various Alpine chronicles all that bears upon breakfasts and suppers, the sustaining character of concentrated meat, the vicious consequences of excess in bread and milk, the use and abuse of cold tea, and the invigorating properties of chocolate, there would vanish all that lends glow, so to speak, and substance to a considerable number of narratives.
> —"Guides or No Guides," *Saturday Review*, 1870[54]

Whymper, Stephen, and Tyndall focused incessantly on their own bodies: their physical health, the amount they could walk, the fitness of their stomachs, and especially on everything that they ate. Tyndall's solo ascent of Monte Rosa, the first time the mountain had been climbed alone and without a guide, is representative of other narratives of high-altitude climbing in its focus not on where he went, but on what he ingested. He commented on the "cowardly and apathetic" nature for the need of "incessant 'refreshing'" found in more touristy narratives, and then related what he ate at each and every point. He had "neither brandy nor wine," merely "four ounces of bread and ham."[55] Near the summit, "I had eaten nothing. I had two mouthfuls of sandwich and nearly the whole of the tea that remained."[56] To lighten his load, he left the rest of the sandwich behind. We know little about the route he took, but we know, practically to the ounce, what he ate the entire day.

Tyndall was religious about what he consumed. As one commentator noted, he viewed "physical existence as the mystic substratus of men's moral nature" and was particularly fixated on eating wholesome food. "The self-same atmosphere forced through one instrument produces music; through another noise," Tyndall thought, "and thus the spirit of life, acting through the human organism, is rendered demoniac or angelic by the health or the disease which originate in what we eat."[57] More to the point, Tyndall's caloric intake was repeated in almost every review of his climbing accomplishments. "It is not every man" noted one commentator, "who can, like Professor Tyndall, climb the Weisshorn on a box of meat lozenges, and Monte Rosa on a couple of ham sandwiches."[58] On the Monte Rosa, noted the *Athenaeum*, Tyndall

---

[54] "Guides or No Guides in the Swiss Mountains?" *Saturday Review of Politics, Literature, Science and Art* 29 (1870): 681–2, on 681.

[55] John Tyndall, *The Glaciers of the Alps* (London, 1861), 151.

[56] Ibid., 155.

[57] As quoted in "A Scientific Climber" (cit. n. 26), 413.

[58] "Mountaineering in 1861: A Vacation Tour," *Athenaeum*, no. 1801 (1862): 589–91.

persevered "without a guide, without a coat, and without a neckcloth, provided only with a small bottle of tea and a ham sandwich."[59] If Tyndall and others didn't speak of food, reviewers superadded it to the account of the alpinists' own narratives, underscoring bodily exertion, the need for sustenance, and the health of the body.[60] As historian of sport Richard Holt put it, "The Victorians were much preoccupied with matters of health."[61]

The focus on food enabled Tyndall and others to demonstrate their physical endurance under extreme privation, directing the focus specifically on the body itself.[62] As Cynthia Eagle Russett has shown, "the idea that great thinkers are hearty eaters" also enabled the Victorians to equate a healthy body with a healthy mind, the one leading to the other.[63] The self-discipline required to deny bodily wants was also the key to both intellectual and physical control. They did not distinguish between the rigorous self-discipline needed to conquer peaks and the same manly discipline displayed by "truth-seekers" in their quest for intellectual or spiritual attainment.[64] Tyndall, Stephen, and Whymper all dwelled on the physicality involved in climbing in the higher regions of the Alps, the simian contortions of their movements, and the resulting inscriptions produced on their flesh. "Blood shot eyes, burnt cheeks, and blistered lips," Tyndall recounted, "are the results of the journey."[65]

A focus on the male body was a defining aspect of the cultural milieu of "muscular Christianity," the unification of physical strength and evangelicalism that became so widespread in the second half of the nineteenth century. It was Charles Kingsley, argues historian James Eli Adams, who "placed the male body into widespread circulation as an object of celebration and desire."[66] Though muscular Christianity exerted a profound influence on Victorian mountaineers, it is actually a rather strange moniker. First, following the work of Rousseau, Humboldt, de Saussure, and others, many Victorian alpinists viewed the high Alps in decidedly secular terms. Stephen and Tyndall were among the most outspoken agnostics of their day, their views formulated in the early 1860s, at the height of their climbing careers.[67] High Alpine environments provided the ideal backdrop to establish their new creed and to prove through physical exertion what they popularized in print.

Second, Tyndall, Stephen, and Whymper were not muscular. They all had a similar physique: tall and slender, certainly tough, but more wiry than strong. Tyndall spoke of the "fibres" of his frame, a tenacity that he portrayed as more moral than physical.[68]

---

[59] "Our Weekly Gossip," *Athenaeum*, no. 1719 (1860): 453–4.

[60] See, e.g., "The Alps" (cit. n. 16), 91.

[61] Holt, *Sport* (cit. n. 7), 87.

[62] Like a host of other students at Cambridge, Stephen mingled athleticism and intellectual study. For the close connection between athletics and the Mathematical Tripos Exams, see Andrew Warwick, "Exercising the Student Body: Mathematics and Athleticism in Victorian Cambridge," in *Science Incarnate: Historical Embodiments of Natural Knowledge*, ed. Christopher Lawrence and Steven Shapin (Chicago, 1998), 288–323.

[63] Russett, *Sexual Science* (cit. n. 6), 104–5.

[64] Stephen Shapin, "The Philosopher and the Chicken: On the Dietetics of Disembodied Knowledge," in Lawrence and Shapin, *Science Incarnate* (cit. n. 62), 21–50.

[65] John Tyndall, *Journals*, 25 August 1861, JT.2.13c, Royal Institution of Great Britain, London, England (hereafter cited as RI).

[66] James Eli Adams, *Dandies and Desert Saints: Styles of Victorian Masculinity* (Ithaca, N.Y., 1995), 150.

[67] Reidy, "Scientific Naturalism" (cit. n. 8).

[68] As quoted in "A Scientific Climber" (cit. n. 26), 413.

Stephen was often described as striding "from peak to peak like a pair of compasses."[69] With their tall, lanky frames, physically diminished from lack of food and physical exertion, they looked more like Christ on the Cross than an athlete on the rugby field. And they had the ascetic control to prove it. Through temperament and training, they limited their bodily needs and desires. Their accounts echo a precious narrative of Western culture: the suffering, death, and rebirth of Christ. Christ's body—its actual physical form—his suffering and eventual resurrection, dominated thinking at the time,[70] and mountaineering narratives recapitulated the same narrative. Alpinists also suffered. They too died on their way to the summit, often one toe or finger at a time. And, in the end, they too were resurrected—physically, emotionally, and spiritually. The mountain life was an ascetic life, and like Christ bearing a burden for humanity, suffering physically on the cross for our sins, so too did climbing inscribe suffering directly on the bodies of climbers.[71]

So, if the elite mountaineers were thin and wiry, where were the muscles to be found? The answer leads us to Dickens's "human donkeys," the bodies in the mountains that actually did all the work, the brawn connected to the brain. Stephen was explicit about this point: "It is the guide who precedes the rest of his party, cuts out the steps, supports the exhausted, and carries the provisions; while the traveller only carries himself and his brandy flask."[72] Stephen idolized Ulrich Lauener and Melchior Anderegg, his guides of many years (fig. 3). He described Lauener as "square-shouldered," "gigantic," "the most picturesque of guides," and "the very model of a true mountaineer." Tyndall, likewise, described his guide, Joseph Johann Bennen, as giving the "impression of great strength and great decision." Other guides are "not so strong as Bennen; and Bennen knows no fear: what man can accomplish he will do."[73]

The guides were not discussed ambivalently, as we might expect. They were not invisible, making up the "silent majority who never made it into the world of documents."[74] They were always named—Christian Kaufmann, Johann Auer, Eduard Balmat, Anderegg, Lauener, Bennen, Jean Antoine Carrel. They were present because their bodies made them so.

In some respects, the guides were obviously subordinate to their employers. Their bodies were objectified commodities belonging to the climber, an ownership that was functionally hierarchical.[75] They were paid for their services, and they climbed at the behest of their employer. The amateur planned the entire climb, directed it, decided

---

[69] Frederic William Maitland, *The Life and Letters of Leslie Stephen* (London, 1906), 143.

[70] Norman Vance, *Sinews of the Spirit: The Ideal of Christian Manliness in Victorian Literature and Religious Thought* (New York, 1985), 6.

[71] Stephen D. Moore, *God's Gym: Divine Male Bodies of the Bible* (New York, 1996), 81; Adams, *Dandies* (cit. n. 66), 4–16.

[72] As quoted in "The Playground of Europe," *Chambers's Journal*, issue 384 (1871): 276–80.

[73] Tyndall went on to write of Bennen, "He is proud of his calling and will show no danger. He is conscientious, and depend upon it, if you lose your life in his company he will lose his in yours, for he will die to save the man he leads." John Tyndall to Heinrich Debus, 4 August 1858, Tyndall Correspondence, RI. Tyndall was unfortunately correct. Bennen died leading three British amateurs up the Haut de Cry in 1864.

[74] Nicholas Jardine and Emma Spary, "The Natures of Cultural History," in *Cultures of Natural History*, ed. N. Jardine, J. A. Secord, and E. C. Spary (Cambridge, 1996), 3–13, on 9.

[75] Donald E. Hall, "On the Making and Unmaking of Monsters: Christian Socialism, Muscular Christianity, and the Metaphorization of Class Conflict," in Hall, *Muscular Christianity* (cit. n. 38), 45–65, on 50.

Figure 3. *Leslie Stephen with his longtime guide, Melchior Anderegg, circa 1870.*

where they should go and when they ought to turn back. But this was true only to a point. Verticality reigned supreme here as well, dictating the power relationship between a guide and "his" amateur. The guide led the party. It was his decision that constrained the choices of the amateur, defining where they could go and when they had to turn back. Up to a certain height, the guide was at the behest of the amateur.

Past that point, the amateur was at the behest of the guide. The decision-making and power structure was thus mapped onto a vertical scale. Although the "amateur" was referred to as "Herr," which retained a class distinction, the guides were the Bergführers, or mountain masters.[76]

The relationship that middle-class Victorian alpinists had with their working-class guides is fascinating in its complexity. Everything strikes a historian's ears as slightly amiss, partly because the class distinctions that were so apparent in other venues were lessened in the high Alps. The mountain ascent became one of the few places where men could transcend class divisions, at least rhetorically, erasing the social stratification between worker and gentleman.[77] One of the peculiar merits of mountaineering, one commentator noted, was the "human fellowship" that it provided between different classes of companions. Gentlemen usually interact with their "laboring fellow men" only in artificial circumstances, making true brotherhood impossible. "Men of education and of wealth meet their toiling brothers only as employers, as rulers, as teachers,—never, by the nature of things, as friends." Yet, the mountain environment changed all of that, for reasons that were clear enough: "Conventional reserve, however thickly coated, shrivels off from men who owe each other their lives several times a day."[78] With the creation of climbing came the advent of climbing partners, and a bond formed between men irrespective of social class or education. Then and now, this is reverently referred to as a "fellowship of the rope."[79]

As Dennis W. Allen has argued with respect to Charles Kingsley and others, "the emphasis on physical vigor in muscular Christianity is also an act of co-optation, the ideological annexation by the middle classes of the body itself, which had formerly been associated with the working classes."[80] In the high Alps, the guide's body literally became an extension of the climber, and its qualities—paid for by the British amateur—an extension of middle-class British masculinity. Physically tethered by a rope, their fellowship united their bodies into one form, with the same fate and the same destiny, be it triumph or defeat, courage or weakness.

Though the relationship between guides and amateurs often blurred common distinctions of class and authority, it actually helped solidify contemporary notions of masculinity. The metaphorical figure of Christ that some saw in the bodies of the British amateur, when combined with the muscular bodies and physical strength of

[76] For insightful discussions of the relationship between Victorian mountaineers and their guides, see Hansen, *Summits* (cit. n. 10); Eaton, "In the 'World of Death and Beauty'" (cit. n. 14), 64–73; and Helen Blackman, "A Spiritual Leader? Cambridge Zoology, Mountaineering, and the Death of F. M. Balfour," *Stud. Hist. Phil. Biol. Biomed. Sci.* 35 (2004): 93–117. In the use of "Bergführers," note the patriarchy implied. There were women climbers but no women guides. The same phrasing was at work with the introduction of the "master mechanic" in the 1880s. See Oldenziel, *Making Technology Masculine* (cit. n. 28), 108.

[77] As "H" wrote in the *Alpine Journal*, "Dangers and difficulties shared, and the exchange of thoughts and opinions . . . wonderfully diminish, for the time, at least, the gulf that exists, socially, between them while the courage, presence of mind, endurance and unselfishness which is so often displayed on behalf of the traveller, makes him feel that his advantages of birth and education do not weigh so very heavily against native worth." "Letter to Editor," *Alpine Journal* 44 (1863–4): 132.

[78] "Mountaineering" (cit. n. 21), 285.

[79] There is also a complicated connection between the creation of a muscular Christian fellowship in the mountains, and the rugged, cross-class brotherhood formed in socialist circles in the nineteenth century. See esp. Hansen, *Summits* (cit. n. 10); and Sheila Rowbotham, *Edward Carpenter: A Life of Liberty and Love* (London, 2009).

[80] Dennis W. Allen, "Young England: Muscular Christianity and the Politics of the Body in 'Tom Brown's Schooldays,'" in Hall, *Muscular Christianity* (cit. n. 38), 114–32, on 128.

the guides, offered a means of uniting seemingly disparate forms of manhood, the ascetic and the muscular, that in himself the middle-class amateur often lacked. By co-opting the guide's body in this way, the Victorian mountaineer could combine the elements of both working-class and middle-class masculinity, mixing autonomy, skill, and physical strength with judgment, restraint, and mental prowess, a swaggering masculinity with a respectable one.[81] The mountains are thus a fruitful place to look for the transition from early Victorian conceptualizations of masculinity, based on ascetic self-discipline and moral strength of conduct, to the late Victorian rhetoric of courage, bodily exertion, and physical strength.[82]

WHYMPER AND THE PHYSIOLOGY OF MASCULINITY

> As water finds its own level, so does the Alpine tourist, after a short experience, know his own range on the Alpine scale. A little training may enable him to reach the extreme limits of that range. . . . But the range itself is fixed by tastes and habits as well as physical powers.
> —*Saturday Review*, 1862[83]

Mountaineers brought seemingly contradictory visions of masculinity together in the mountains, the muscular and the ascetic, the working and the genteel. Their focus on their own bodies and those of their guides also enabled them to tease out the differing physical and mental capacities of men and women, imparting scientific validity to the all-male vertical zone. The problem for women—and the advantage for men—could be found in the body.

The transition from identifying the masculine in the male's mind to situating it in the body, from the moral to the physical, is a dominant theme in the mountaineering literature of the mid-nineteenth century. Thus, in addition to the domestic manliness outlined by historian John Tosh, where the "Victorian codes of manliness made scant acknowledgement of the body,"[84] we also see the creation of a rugged, outdoor manliness that focused on the body itself. The high alpine environment provided an ideal geography to discuss, debate, and ultimately define the physiology of the male body and the characteristics that might flow from it.[85] Where a climber's body could go was not simply a matter of physical training or mental preparedness. The range was fixed by nature, like water finding its level.

Edward Whymper was obsessed with finding just that fixed limit. All mountaineers know his *Scrambles amongst the Alps*, perhaps the seminal text in the history of mountaineering. It contains chapter after chapter of exciting explorations in the heart of the Swiss heights, climaxing with a firsthand recounting of the Matterhorn tragedy. Few today, however, read his *Travels amongst the Great Andes of the Equator*, even though it contains an account of the first European to climb in the Andes to over

[81] Stephen Meyer, "Work, Play, and Power: Masculine Culture on the Automotive Shop Floor, 1930–1960," in *Boys and Their Toys? Masculinity, Technology and Class in America*, ed. Roger Horowitz (New York, 2001), 13–32.

[82] Adams, *Dandies* (cit. n. 66), 230, though Adams does not discuss mountaineering.

[83] "Mountaineering" (cit. n. 27), 627–8.

[84] John Tosh, "What Should Historians Do with Masculinity? Reflections on Nineteenth-Century Britain," *Hist. Workshop J.* 38 (1994): 179–202, on 182.

[85] Ibid., 180. Tosh acknowledges this point as well, noting how the Imperial Service enabled Victorian young men to move beyond the domestic sphere and thus domestic masculinity.

20,000 feet, the first to spend the night on Cotopaxi, and the first to climb Chimborazo (not once but twice, by two different routes).[86] Alpinists don't read it today because it's not really a mountaineering narrative; it is primarily a contribution to high-altitude physiology. That is certainly how Whymper's contemporaries viewed his work.[87]

Whymper focused incessantly on his own body. His stamina was the stuff of legends. He often walked thirty-five miles in a day, including twenty-seven miles nonstop during a walk from Edinburgh to London. He had his pedestrianism down to a science, calculating his pace at about eleven minutes per mile.[88] In 1865, the year of the Matterhorn tragedy, he calculated that he had climbed over 100,000 vertical feet in eighteen days.[89] He transformed his walking prowess into both climbing proficiency and scientific achievement. His focus on his own body propelled him to a unique, if largely forgotten, place in the history of physiology.

Whymper had long dreamed of making a name for himself through science and exploration, eventually making two exploratory voyages to Greenland in 1867 and 1872, to the glaciers of South America in the mid-1880s, and to the Rocky Mountains in the early 1900s. His love for mountains began in the Swiss Alps in the 1860s, a result of his training as an engraver. William Longman sent him to the mountains to make woodcuts for the first edited volume on climbing, *Peaks, Passes, and Glaciers*, and Whymper became entranced by high-altitude environments. The Matterhorn seemed to control him like a spell. He made numerous attempts on the mountain, all from the Italian side, before successfully climbing the Hornli Ridge from Switzerland. "There seemed to be a cordon drawn around it," he said of the deadly mountain, "up to which one might go, but no further."[90]

After the solemn events of 1865, Whymper turned his attention to defining the actual physical cordons that limited human movement in the high altitudes. His work fit into a larger contemporary cultural fascination with how the human body worked, a nascent scientific approach to understanding the body's natural limits.[91] By searching for the limits of the human body, he helped spearhead a scientific, field-based approach to high-altitude physiology that would flourish only after the turn of the twentieth century.

The limits of the human body in extreme environments, especially the sickness brought on by high altitude, had become a popular topic among researchers, particularly on the Continent. The symptoms of mountain sickness seemed to correspond almost perfectly to the presumed physiology of the "effete and effeminate," suggesting a topic of study that could highlight the inner workings of the human body and thus a useful means for distinguishing scientifically between males and females. Scientists

---

[86] Indeed, it contains the first ascent of some of the world's greatest mountains, including Chimborazo (20,564 ft.), Cayambe (18,996 ft.), Antisana (18,714 ft.), Carihuairozo (16,750 ft.), Sincholagua (16,360 ft.), Cotacachi (16,290 ft.)—all higher than the highest mountain in Europe.

[87] See the numerous reviews of the text in all the major journals of the day, including, e.g., "Travels amongst the Great Andes," *Athenaeum*, no. 3366 (30 April 1892): 557–8; "The Andes of the Equator," *All the Year Round* 7 (21 May 1892): 486–92; and "On the Great Andes; Or Life at Low Pressure," *Chambers's Journal of Popular Literature, Science and Arts* 9 (28 May 1892): 343–5. And see the reviews in the scientific literature of the day, such as "Mountain Sickness," *Brit. Med. J.*, 16 April 1895, 829.

[88] Edward Whymper, *Travels amongst the Great Andes of the Equator*, ed. Loren McIntyre (Salt Lake City, 1987), 300.

[89] Whymper, *Scrambles amongst the Alps* (London, 1871), 352.

[90] Ibid., 76.

[91] Rabinbach, *The Human Motor* (cit. n. 6), 6.

had studied the body's response to balloon ascents, but they rarely stayed aloft long enough to determine either the causes or effects of altitude sickness, whether exhaustion, lack of nourishment, weather, or temperature mattered, and whether the effects were permanent or temporary. The question, so to speak, was still in the air.

It also had moved into the laboratory. French physiologists were perfecting the hyperbaric chamber, an instrument designed to mimic the high altitudes achieved by mountaineers. The French physiologist Paul Bert constructed numerous chambers in his laboratory in Paris, evacuating them to dangerously low pressures. His first instruments were big enough to house birds, dogs, cats, and rabbits, but he soon advanced to chambers large enough to accommodate humans, including himself, in 1874. Scientific mountaineers were interested in Bert's results but argued that humans could not remain in the chambers long enough to obtain meaningful conclusions. They believed that experiments in the air and laboratory were inconclusive at best and lethal at worst, resulting in more questions than answers.

Whymper believed that the only way to settle such questions was to spend a large amount of time at high altitudes, stationary, where he could perform experiments, obtain measurements, and confirm results. The Alps weren't high enough, maxing out below 16,000 feet. He needed a higher mountain, and he chose the one with the longest, most distinguished history in the sciences. On the bottom slopes of Chimborazo, Alexander von Humboldt had penned his famous "Essay on the Geography of Plants," which laid the foundation for studying the relationship between climate, elevation, and the distribution of species. The mountain had also never been summited. In 1879, along with his two Swiss guides, Jean Antoine Carrel and Louis Carrel, Whymper transformed the Andes into his third Switzerland, Chimborazo into his scientific laboratory, and his own body and those of his guides into his scientific instruments.

Whymper set up his entire *Travels amongst the Great Andes of the Equator* as a set of bodily scientific experiments to determine the physiological limits of the human frame. He began the narrative recounting the several weeks he and the Carrels took to climb the historic peak. It reads like a laboratory report: "I found that my residence upon Chimborazo had extended over seventeen days. One night passed at 14,370, ten more at 16,664, and six others at 17,285, and this is perhaps the greatest length of time that any one has remained continuously at such elevations."[92] At each stage and throughout each day, he performed physiological experiments on himself and his guides. He checked their pulses, measured their breathing, and jotted notes about their physiological responses. He counted his steps—657 without stopping between 15,000 and 16,000 feet, comparable to what he experienced on Mont Blanc—and he recorded their symptoms as they progressed upward, including his own inability to perform his work and his surprising lack of desire to smoke.[93] He also counted the steps required to travel specific distances, again comparing his results with earlier calculations he had made near sea level in London. These and similar experiments foreshadowed later comparative approaches to understanding the effects of high altitude on physical endurance.[94] He also administered to himself chlorate of potash, a supposed cure for alpine sickness, though Jean Antoine considered it an insult to his

[92] Whymper, *Travels* (cit. n. 88), 83.

[93] As a habitual smoker, Whymper noted, "when persons put aside their beloved pipes there is certainly something wrong." Ibid., 53.

[94] E. S. Williams, "Exercise and Altitude," *Postgraduate Med. J.* 55 (1979): 492–4, on 492.

intelligence and manliness.[95] After an exhausting two-day push, they finally made it to the top of the famed peak, "hungry, wet, numbed, and wretched, laden with instruments which could not be used."[96]

Whymper hoped to stay on the summit of Chimborazo for several days to take physiological measurements, but Jean Antoine had suffered severe frostbite to both of his feet. They were forced to retreat down to Quito and recuperate before attempting to climb the neighboring Cotopaxi, which had a large slope at its summit that Whymper hoped could serve as a laboratory. The narrative focused less on the climb itself up Cotopaxi and more on the twenty-six hours spent on the summit, where Whymper and the Carrels again took bodily measurements.

The narratives of all of the climbs focused heavily on the sciences: glaciers, botany, zoology, the correct use of aneroid barometers, and list after list of insects. The final chapter, however, focused exclusively on mountain sickness: its symptoms, when it hit, at what pressure, how long it lasted, when it died away, how to cure it—by going down!—and its relationship to temperature, breathing, food consumption, sleep, anxiety, and the deterioration of muscles. The mountains were not the object. Whymper's conclusions were all physiological: "There are strong grounds for believing," he concluded, "that they [symptoms of mountain sickness] are due to the expansion, under diminution of external pressure, of gaseous matter within the body; which seeks to be liberated, and causes an internal pressure that strongly affects the blood vessels."[97] His approach was a direct prefiguration of what historian John M. Hoberman has referred to as the search for an "experimental quantifiable biological phenomenon for otherwise invisible metabolic processes" that became so popular at the end of the nineteenth century.[98]

While the causes of mountain sickness were physiological, so too were its symptoms. And the ailments turned out to be the exact ones the Victorians used to define the absence of manliness: weakness, lack of will, timidity, lassitude, muscle deterioration, slow decision-making ability, and the like. As historian Robert Nye demonstrated in his study of fin de siècle France, questions of masculinity led to interests in its biological sources that were "inseparable from concerns about the strength and endurance of the human body."[99] Whymper always had this focus on the body in mind: What can it achieve? What are its limits? He attempted to reduce the frailty of the human body and how and when it succumbed to mountain sickness to quantitative measurement.[100]

Whymper's studies were part of the burgeoning field of high-altitude physiology, a new area of study that relied extensively on self-experimentation, whether in balloons, in hyperbaric chambers, or on the summits of mountains. In both the field and the laboratory, the experimental subjects were always male. High-altitude physiologists beginning with Whymper mapped the vertical gendered relations found in the standard mountaineering narratives directly onto their science.

This linking of the biological basis of human performance with scientific

[95] As Whymper said of his guide: "For all human ills, for every complaint, from dysentery to want of air, there was, in his opinion, but one remedy; and that was Wine; most efficacious always if taken hot, more especially if a little spice and sugar were added to it." Whymper, *Travels* (cit. n. 88), 50.

[96] Ibid., 69.

[97] Ibid., 374.

[98] Hoberman, *Mortal Engines* (cit. n. 7), 31.

[99] Nye, *Masculinity* (cit. n. 18), 222.

[100] Bruce Haley, *The Healthy Body and Victorian Culture* (Cambridge, Mass., 1978); Nye, *Masculinity* (cit. n. 18), 224.

experimentation was not the sole preserve of human physiology. It also found expression in broader cultural trends taking hold within Europe and America in the last quarter of the century, including the rise in popularity of sports more generally. As Hoberman has pointed out, "The cult of the athlete developed along with scientific interests in extreme physiological states."[101] A similar focus on human potential, for example, was part of the revival of the Olympic Games from the very beginning. Baron Pierre de Coubertin, the founder of the modern Olympics, was obsessed with the physical limits of the human body, as was William Henry Grenfell, 1st Baron Desborough, the president of the British Olympic Association, and the person most responsible for bringing the games to London in 1912. Desborough seemed to epitomize the model athlete, uniting "the strength of a Porthos, the heart of an Athos, and the body of an Englishman."[102] He also summited the Matterhorn three times, the last in conjunction with the Monte Rosa and the Rothorn within eight days.[103]

As one modern critic pointedly noted, the revival of the Olympic Games was nothing more than "a gigantic biological experiment carried out on the human organism."[104] It should come as no surprise that the gendered notions that constituted high-altitude physiology also found expression in the early Olympics. Just as women and men adapted to different regions, the women to lower environments, the men to the heights, so too were women justifiably confined to specific sports—namely, figure skating and equestrian events.[105] According to reports in the British press, women had never been allowed to participate, and those who did were "hurled down from the neighbouring rock of Typaeum."[106] Tellingly, Pierre de Coubertin, the "father" of the modern Olympics, reported in 1902 that athletics for women was "against the laws of nature."[107]

CONCLUSION

> Travellers, like plants, may be divided according to the zones which they reach. In the highest region, the English climber—an animal whose instincts and peculiarities are pretty well-known—is by far the most abundant genus. Lower down comes a region where he is mixed with a crowd of industrious Germans, and a few sporadic examples of adventurous ladies and determined sight-seers. Below is the luxuriant growth of the domestic tourist in all his amazing and intricate varieties.
> —Leslie Stephen, *Playground of Europe*, 1871[108]

With the initial founding of the Alpine Club of London, mountaineering seemed poised to become a safe haven for an all-male fraternity where the qualities of masculinity could be safely defined. The alpinists considered themselves special because they had risen above others to blaze new trails. "For a long time the tourists who annually visited the Alps were content to follow in certain beaten tracks," noted a founding

[101] Hoberman, *Mortal Engines* (cit. n. 7), 61.
[102] "The Olympic Games at Shepherd's Bush: Lord Desborough and His Exploits," *Review of Reviews* 37 (1908): 375.
[103] Ibid.
[104] As quoted in Hoberman, *Mortal Engines* (cit. n. 7), 4.
[105] Holt, *Sport* (cit. n. 7), 129.
[106] "The Revival of the Olympian Games," *Cosmopolis* 2 (1896): 59–74, on 66.
[107] As quoted in Holt, *Sport* (cit. n. 7), 129.
[108] Stephen, *Playground* (cit. n. 20), 184.

member of the Alpine Club in 1859, "each one copying with almost servile fidelity the route followed by his predecessors."[109] The new class of men was proud to break new ground in order to view the world from a unique vantage point. Stephen was the master of this line of argument. He always differentiated himself through verticality from "that offensive variety of the genus of primates, the common tourist." He stopped short of suggesting their death "by leaving arsenic about," but felt "perfectly satisfied if they be confined to a few penal settlements in the less beautiful valleys."[110] His solution was simple: he climbed into "regions still in all the freshness of their primitive innocence; regions where the 'Times' is never seen."[111]

But, alas, the mountains appealed to women as well. They successfully domesticated the passes and glaciers while the middle-class male mountaineers scurried to reserve the peaks for themselves. Men wrote and published most of the mountaineering narratives, and they are similar to the colonial adventure narratives of the time, like *Heart of Darkness*—in which the women aren't even named—and *King Solomon's Mines*—dedicated to "all the big boys and little boys who read it."[112] The absence of women was a means of fortifying an all-male preserve. Yet, just like their colonial counterparts, or Michael Robinson's manly Arctic narratives and Nathan Ensmenger's computer programmers (both in this volume), mountaineers aggressively affirmed an all-male zone that did not really exist.

By the last decade of the century, most of the high mountains were being attempted, if not summited, by women climbers, often in all-female climbing parties. When American Annie Pec climbed the Matterhorn in 1896, several women had already made the climb. On her way up, she ran into two other women who had just traversed the mountain, up the Lion's ridge from Italy and down the Hornli ridge into Switzerland, the route first climbed by Tyndall in 1868. By the last decades of the century, Gertrude Bell, Nelly Bly, Fanny W. Bullock, and many others were moving ever onward and upward, not only in the Alps, but also abroad into Africa, South America, and the Himalayas.

While Dickens could pejoratively zone the Alps according to class and stupidity, Stephen got the metaphor right. Humans, like plants, had certain innate limits. In the mid-Victorian era, these limits were regularly gendered by elevation. It was acceptable for women to domesticate the lowlands, even the temperate zone. They could have the valleys and glaciers, which were no longer considered places of manliness. "Those who keep to the low levels, who never carry an ice axe and never need a rope," the *Saturday Review* noted, "do not know the meaning of hardship or the virtue of courage."[113] Those who made "the ordinary tours in Switzerland," another scoffed, had no notion of the "higher altitudes of the Alps" where "eye, ear, and brain are constantly called into play."[114] Through extensive bodily privation, exertion, and self-experimentation, male alpinists attempted to codify the high Alps, the third and uppermost Switzerland, as an all-male preserve.

[109] "Alpine Travellers," *Bentley's Quarterly Review* 2 (1859): 214–43.
[110] Stephen, *Playground* (cit. n. 20), 150.
[111] Ibid., 168.
[112] I owe the connection to colonial adventure narratives to David Agruss. See also Kathryn Marie Smith, "Revis(it)ing Joseph Conrad's *Heart of Darkness*: Women, Symbolism, and Resistance" (MA thesis, Florida Atlantic Univ., 2009).
[113] "Hours of Scrambling Exercise" (cit. n. 15), 59.
[114] "Mountaineering" (cit. n. 21), 288.

Increasingly, Victorian mountaineers turned to their bodies, and the bodies of their guides, to zone the high Alps as masculine, pioneering a new approach to high-altitude physiology, especially within the British context. While British physiology stagnated between 1840 and 1870, the field flourished on the Continent, becoming according to historian Gerald Geison "an increasingly rigorous and more broadly based experimental science," represented by the laboratory work of Claude Bernard.[115] Bernard's student, Paul Bert, turned his attention to questions of endurance, and the physiological responses of humans to extreme environments, by building expensive evacuation chambers. Ironically, the supposed "father" of high-altitude physiology never left his laboratory near sea level in Paris.

In England, the rise of the physiology laboratory came later, in the 1870s and 1880s, spearheaded by Michael Foster in Cambridge and T. H. Huxley at the Royal School of Mines. They focused little on questions of endurance or the physiological responses of humans to extreme environments.[116] Yet a parallel approach, which has received far less attention, was also taking off at this exact time. British mountaineers, represented here by Edward Whymper, were moving not into the laboratory but out into the field—or, rather, they were transforming the field into a laboratory. Evacuation chambers and balloon ascents did not allow for long-term incursions into high-altitudinal zones. Only mountaineers, spending weeks at a time at extreme heights, could gain useful knowledge of how the human body responded to lack of oxygen for extended periods.[117] Whymper's treks to the snows of Chimborazo represent one of the earliest examples of a British alpinist mounting an expedition to high altitudes specifically to analyze the effects of low pressures on human physiology.[118]

This transformation of geographic space, turning a mountain into a laboratory, became, as Sarah Tracy has noted, "an essential element of physiological research during the first half of the twentieth century."[119] Whymper's vertical approach—his extended stays at high elevations, his method of systematic, bodily examination, and his comparative observations at lower elevations—all became accepted methods later in the nineteenth century and into the twentieth. They are seen, for instance, in Angelo Mosso's work on Monte Rosa in the 1880s, John Haldane's and C. G. Douglas's work on Pike's Peak beginning in the 1910s, and Ancel Keys's work on the Chilean volcano Cerro Aucanquilcha in the 1930s, among many others.[120] Although historians and physiologists alike point to these examples of high-altitude, field-based research as a new approach to general physiology,[121] they had their roots in mid-Victorian questions

[115] Gerald L. Geison, *Michael Foster and the Cambridge School of Physiology: The Scientific Enterprise in Late Victorian Society* (Princeton, N.J., 1978).

[116] Ibid. See also Michael Foster, *A Textbook of Physiology* (London, 1877); and T. H. Huxley, *Lessons in Elementary Physiology* (London, 1869).

[117] Charles S. Houston, "Lessons to Be Learned from High Altitude," *Postgraduate Med. J.* 55 (1979): 447–53.

[118] Michael P. Ward, James S. Milledge, and John B. West, *High Altitude Medicine and Physiology* (Philadelphia, 1998), 11. As Philip Felsch has argued, prior to this there seemed to be a "Victorian animosity toward fatigue studies." See Felsch, "Mountains of Sublimity, Mountains of Fatigue: Towards a History of Speechlessness in the Alps," *Sci. Context* 22 (2009): 341–64, on 358.

[119] Sarah W. Tracy, "The Physiology of Extremes: Ancel Keys and the International High Altitude Expedition of 1935," *Bull. Hist. Med.* 86 (2012): 627–60, on 629.

[120] Ibid; Felsch, "Mountains" (cit. n. 118).

[121] Tracy, "The Physiology of Extremes" (cit. n. 119); W. Brendel and R. A. Zink, *High Altitude Physiology and Medicine* (New York, 1982), xii–xiii; Ward, Milledge, and West, *High Altitude Medicine* (cit. n. 118), 11.

of manly endurance and the physical limits of the human frame.[122] Whymper provided an example of how to turn a mountain into a physiological laboratory and people into scientific objects.

By defining a high, vertical zone based on the inner workings of the human body, Whymper and other mountaineers incorporated gendered constructions into their science, concepts that could be translated seamlessly into cultural discourse. Similar constructions can be seen, for instance, in the rise of the popularity of sports, the cult of the athlete, and in the revival of the Olympic Games. They appear in what became "sports medicine" and "nutritional physiology" later in the twentieth century.[123] And they formed one of the foundations of high-altitude physiology, a science that is still primarily a male preserve.[124] The connection between human physiology and definitions of masculinity, first formulated on the sides of mountains, appeared natural, based on physical abilities. But it took quite a bit of endurance on the part of the mid-Victorian mountaineers to make it that way.

[122] Hoberman, *Mortal Engines* (cit. n. 7), 65.
[123] Tracy, "The Physiology of Extremes" (cit. n. 119).
[124] As an example, the editor-in-chief, both associate editors, and thirty-six of the forty-two members of the editorial board of *High Altitude Medicine and Biology* are men.

Masculine Knowledge, the Public Good, and the Scientific Household of Réaumur

by Mary Terrall*

ABSTRACT

In the Royal Academy of Sciences of Paris (founded 1666), expressions of a masculine culture of science echoed contemporary language used to articulate the aristocracy's value to crown and state—even though the academy was not an aristocratic institution as such. In the eighteenth century, the pursuit of science became a new form of manly service to the crown, often described in terms of useful knowledge and benefit to the public good [*le bien public*]. This article explores the connection of academic scientific knowledge to the domestic spaces where it was made and, in particular, to the household of R.-A. Ferchault de Réaumur, an exemplary academician. Although Réaumur had neither wife nor children, a complex net of affective ties, some of them familial, linked the members of the household, which accommodated women (the artist Hélène Dumoustier and her female relatives) as well as men (a series of assistants, many of whom eventually entered the academy). As head of this dynamic household, Réaumur produced not only scientific results but also future academicians.

Like virtually all Old Regime institutions, the Paris Royal Academy of Sciences was a bastion of male privilege, organized hierarchically to mirror the social and political structures in which it was embedded. As we would expect, the exclusion of women from the membership rolls did not have to be formally stipulated. As the academy and its governing ministers established modes of customary practices and intellectual style, prevailing notions of masculinity reinforced assumptions about who would be doing scientific work. Perhaps more to the point, these ideals worked to enhance the status of academicians living in a society based on rank and privilege. The academy operated as a small cog in the mushrooming apparatus of the absolutist French state, designed by the royal minister Jean-Baptiste Colbert in the 1660s. The early history of the institution, and the scientific work done under its auspices, unspooled in tandem with the consolidation of Louis XIV's power and the concomitant changes in the values and functions associated with nobility. This larger story about the fluctuating fortunes of the aristocracy, continuing beyond the reign of the Sun King and through the eighteenth century, lies well beyond the scope of this essay.[1] It is worth noting,

*Department of History, University of California, Los Angeles, Box 951473, Los Angeles, CA 90095-1473; terrall@history.ucla.edu.

[1] The literature is vast, but see esp. Jay Smith, *The Culture of Merit* (Ann Arbor, Mich., 1995); and the essays in Smith, ed., *The French Nobility in the Eighteenth Century: Reassessments and New Approaches* (University Park, Pa., 2006).

however, that expressions of the masculine culture of science in the academy often echoed language being used contemporaneously to define the nature and shifting valences of the aristocracy's relations to crown and state—especially the language of zealous service, social utility, the cultivation of talents, and reward for merit. When they argued for the economic and patriotic value of scientific knowledge, academicians were deploying a rhetoric that evoked aristocratic values of merit and service, even though their institution was not aristocratic as such. The pursuit of the sciences in this context became a new form of service to the crown, explicitly formulated to further "the public good" [*le bien public*].

In 1699, three decades after its founding, Louis XIV invited the academy to exchange its rooms in the royal library for "incomparably more commodious and more magnificent" quarters in the Louvre. Every Wednesday and Saturday afternoon, the anointed men of science gathered in these luxurious surroundings, seated according to academic status and seniority, to listen to their colleagues read papers on recent work and reports from distant correspondents. And twice a year they opened their sanctum to visitors who filled the galleries around the meeting room, "where the public could judge with its own eyes the form and utility of these meetings."[2]

Unlike nearby scientific institutions, such as the Royal Botanical Garden and the Paris Observatory, where specialists and technicians worked (and often resided) on site, the Royal Academy of Sciences was a venue for the presentation, discussion, certification, and publication of knowledge that had been made elsewhere. In particular, the results read to the assembled group in the Louvre could not have been produced without facilities and other resources located in and around the homes of academicians.[3] In these studies, laboratories, collections, and gardens, people who never set foot in the meeting room worked alongside academicians. Some of these people were themselves future academicians; others were artists, artisans, gardeners, medical students, or family members. The exclusive, and de facto masculine, academy thus depended on many kinds of personal, patronage, and intellectual relations that tied the institution to the places around the city (and sometimes beyond) where men and women with a variety of skills and aspirations worked alongside each other.

A great deal of scholarly attention has been devoted to Parisian salons as sites of philosophical and literary production frequented by both men and women. Originally the room where a hostess received her guests, the salon has come to designate quasi-institutionalized gatherings, each with its own local culture and rules, where intellectual and artistic matters were discussed alongside gossip and news. Some years ago, in my work on science in and around the Paris Royal Academy of Sciences, I explored the gendered distinction between participants and spectators. At times,

---

[2] *Le mercure galant* (1699), quoted in Ernest Maindron, *L'Académie des Sciences* (Paris, 1888), 28–9. For anecdotal evidence about everyday behavior of academicians at meetings, see Irène Passeron, "Une séance à l'Académie au XVIIIe siècle," in *Histoire et mémoire de l'Académie des sciences: Guide de recherches* (Paris, 1996), 339–50. All translations from the original French are my own.

[3] For a similar point about the early Royal Society, see Steven Shapin, "The House of Experiment in Seventeenth-Century England," *Isis* 79 (1988): 373–404. Alix Cooper argues that domestic venues for science in the early modern period were superseded by the new scientific institutions starting in the late seventeenth century; my point here is that these institutions depended crucially on all kinds of work done in the home. See Cooper, "Homes and Households," in *Cambridge History of Science*, vol. 3, *Early Modern Science*, ed. Katharine Park and Lorraine Daston (Cambridge, 2006), 224–37. For another example, from a very different cultural setting, of the connection between technical know-how, knowledge production, and the home, see Eugenia Lean, "Recipes for Men: Manufacturing Makeup and the Politics of Production in 1910s China," in this volume.

feminized (though not by any means exclusively female) audiences played a crucial role in validating masculine science, and in enhancing the visibility and reputation of men of science.[4] The interest and attention of elite female readers, interlocutors, and spectators enhanced the status of science, at a time when its practitioners were carving out a privileged niche for themselves by pushing the masculine values of utility and productivity.[5] I associated the making of scientific knowledge with the homosocial space of the academy and situated important aspects of its reception in the less formal and mixed-gender salons, where women played a key role in managing sociability and intellectual exchange among their guests. Revisiting some of these issues in this essay, I complicate the salon-academy binary by paying attention to the making, rather than the representing and consuming, of scientific knowledge, and by recuperating the venues where experiments and calculations were performed, texts were written, and drawings were made. This means looking into the homes of academicians, where they worked when they were not attending meetings of the institution that gave them their scientific identity. By extending our understanding of the domestic to include studies, cabinets, and laboratories as well as drawing rooms, we may also be able to uncover ties that bound the people in these spaces to each other, as well as to the academy. Household dynamics contrasted with the homosociality and ritualized formality of the academy's proceedings, and this contrast will illuminate some nuances in contemporary understandings of the masculinity of science. In the more formal institutional setting—whether in the semiweekly meetings or in the pages of official publications—the aristocratically inflected values of service and utility colored the public, and masculine, face of science. At home, the organization of scientific work varied enormously depending on financial resources and family situation, as well as on personal proclivities and on subject matter.

Here I investigate the household of a consummate academician, René-Antoine Ferchault de Réaumur, whose career spanned the better part of half a century.[6] A closer look at how he organized his home life around scientific work will point up some of the ways that masculine academic ideals played out in a domestic setting and will show in turn how the intellectual and interpersonal dynamics of the home fed back into the life of the institution. Réaumur worked on an impressive range of subjects over the years, from the locomotion of shellfish to regeneration in crayfish; from the strength of rope and spider silk to the management of forests; from thermometer design to processes for manufacturing steel, paper, and porcelain; from bees and their eminently useful wax to aphids, mayflies, and any number of other insects; and on to the artificial incubation of birds' eggs and taxidermy techniques. He hailed from the lower nobility in the Vendée region; though not enormously wealthy, he could nevertheless rely on his lands and his investments to provide a comfortable income.

---

[4] Mary Terrall, "Heroic Narratives of Quest and Discovery," *Configurations* 6 (1998): 223–42; Terrall, "Gendered Spaces, Gendered Audiences: Inside and Outside the Paris Academy of Sciences," *Configurations* 3 (1995): 207–32; Terrall, "Salon, Academy, Boudoir: Generation and Desire in Maupertuis's Science of Life," *Isis* 87 (1996): 217–29. For female readers as spectators, see Terrall, "Fashionable Readers of Natural Philosophy," in *Books and the Sciences in History*, ed. Nick Jardine and Marina Frasca-Spada (Cambridge, 2000), 239–54.

[5] Terrall, "Gendered Spaces" (cit. n. 4), 214.

[6] For Réaumur's biography, see Jean Torlais, *Réaumur: Un esprit encyclopédique en dehors de l'Encyclopédie* (Paris, 1936); on his natural history, see Mary Terrall, *Catching Nature in the Act: Réaumur and the Practice of Natural History in the Eighteenth Century* (Chicago, 2014).

He maintained experimental facilities at home and gradually expanded his household to accommodate collections, equipment, and people (residents and visitors). Most of the scholarship on making natural knowledge in the home focuses on family groupings. Wives, daughters, and sisters assist in astronomical observations; women distill and compound medicines in their kitchens to treat their families and their tenants; sons might be trained to follow in their fathers' footsteps.[7] In eighteenth-century Paris, well-established scientific families like the Cassinis and the Geoffroys and the Jussieus filled the roster of the academy with generations of astronomers, chemist-apothecaries, and botanists. Réaumur presents an interesting case because he explicitly chose a life of science as an alternative to fulfilling his default role as noble paterfamilias. When he moved to the capital as a young man, first to pursue his studies and then to seek election to the Academy of Sciences, he ceded his place as head of the family to his younger brother, who married and lived on the ancestral property in Poitou. Even after his brother died prematurely, leaving no children, Réaumur remained a bachelor—though he took over the management of his estates and spent six weeks in the manor house every year during academic vacations. Scientific pursuits took the place of marriage and the production of progeny, so that the family line was extinguished with his death. Before turning to the evolution of his household, with its constantly shifting cast of characters performing a variety of scientific work, let us look more closely at the gendered resonances of science in the academy, where Réaumur was a central figure.

USEFUL KNOWLEDGE

The rhetoric of useful knowledge, a central trope of academic discourse, marked science as masculine and justified the elite standing of the institution in the political landscape of the absolutist administrative state. As an arm of the state, and in the service of a mercantilist economic ideology, the academy functioned as both arbiter and creator of knowledge that was supposed to disseminate outward (and downward) and into the practices of entrepreneurs, navigators, doctors, colonists, and eventually, mediated by local gentry or parish priests, even artisans and farmers. This diffusionist ideal reflected the paternalist hierarchy inherent in the privileged institution and represented a particular notion of masculine knowledge. The academy's role as a reservoir of technological expertise available to the crown is a familiar story.[8] Without rehearsing the specifics of technical consulting by academicians (in such areas as waterworks, navigation and shipbuilding, dyeing, steel production, evaluation of inventions, and forest management), for present purposes I simply point out that the language of utility marked science as a patriotic and masculine endeavor, to

[7] Lynette Hunter and Sarah Hutton, eds., *Women, Science, and Medicine, 1500–1700: Mothers and Sisters of the Royal Society* (Stroud, 1997); Londa Schiebinger, "Maria Winkelmann at the Berlin Academy of Sciences: A Turning Point for Women in Science," *Isis* 78 (1987): 174–200; Deborah Harkness, "Managing an Experimental Household: The Dees of Mortlake and the Practice of Natural Philosophy," *Isis* 88 (1997): 247–62; Monika Mommertz, "The Invisible Economy of Science: A New Approach to the History of Gender and Astronomy at the Eighteenth-Century Berlin Academy of Sciences," in *Men, Women, and the Birthing of Modern Science*, ed. Judith Zinsser (DeKalb, Ill., 2005), 159–78.

[8] Roger Hahn, *The Anatomy of a Scientific Institution: The Paris Academy of Sciences, 1666–1803* (Berkeley and Los Angeles, 1971); Robin Briggs, "The Académie royale des sciences and the Pursuit of Utility," *Past & Present* 131 (1991): 38–88.

be distinguished from the "frivolous" or "amusing" pastimes of the effeminate and unproductive rich. At the academy, the rhetoric of utility was especially in evidence on public occasions and in print, central to the institution's raison d'être and the self-presentation of many members—even though the papers that filled the pages of the annual *Mémoires* did not necessarily translate directly into practical applications.[9] At the inaugural public assembly, in 1699, the meeting room in the Louvre was packed with spectators, including "a small number of ladies" who chose to watch the proceedings from discreetly screened balconies. The Abbé Jean-Paul Bignon, scion of a powerful noble family and the architect of the renewed academy, presided.[10] In his speech to the gallery, Bignon used unmistakably gendered language to contrast the sciences to the domain of the royal literary academy, the Académie française. In this equally exclusive learned body, poets and rhetoricians dealt in "the art of speech with all its pleasing ornaments," while the sciences "aspired only to the truth." Bignon warned the spectators that they might be in for some dry and unadorned discourse, since "it sufficed for the academy [of sciences] that the truth be useful and that it did not need to be pleasing."[11] His audience would have recognized the gendering of artful speech and pleasure as feminine—the Académie française was generally regarded as intimately connected to the feminized world of Parisian salons. The contrast to the sciences pointed up the masculinity of truth and utility.[12]

Bignon imagined a revitalized academy that would serve the state while engaging the interest of an elite public and worked to realize his vision of the kind of knowledge appropriate for the academy, and the kind of man an academician should be.[13] After about 1710, Réaumur served as his right-hand man, as a new recruit to the mathematics class of the academy. Over the course of a long career, Réaumur exemplified, in his research and in the conduct of his scientific life, Bignon's ideal academician, eschewing theory for a profusion of experimental and observational projects, many of them geared toward economic or technological utility. This contemporary assessment of his character and his devotion to solving problems frames the scientific work in terms of aristocratic service: "Zealous for the public good, [M. de Réaumur] has willingly devoted his talents to objects, small in appearance, but which are directed to perfecting the mechanical arts, or to anticipating social needs. The means of making a new dye, of increasing the fertility of fields, of preserving woolens from moth infestations, of keeping eggs fresh for three or four months—such are the objects of

[9] For a contemporary panegyric on the usefulness of the sciences, including the "abstract sciences," see Bernard le Bovier de Fontenelle, "Préface sur l'utilité des mathématiques et de la physique et sur les travaux de l'Académie des Sciences," in Fontenelle, *Oeuvres complètes*, ed. A. Niderst, 9 vols. (Paris, 1994), 6:35–50.

[10] On Bignon's family background, and his engagement with the administration of the academy over many decades, see David Sturdy, *Science and Social Status: The Members of the Académie des Sciences, 1666–1750* (Woodbridge, Conn., 1995), 222–6 and 367–74.

[11] *Mercure galant* (cit. n. 2).

[12] On the female audience for science, see Erica Harth, *Cartesian Women: Versions and Subversions of Rational Discourse in the Old Regime* (Ithaca, N.Y., 1992). Note that Bignon did not intend the gendered contrast to disparage the work of the Académie française, whose members, of course, were all men.

[13] Bignon did not conceive or implement this vision single-handedly. He worked closely with the minister Maurepas, and the perpetual secretary Fontenelle, as well as with Réaumur and other academicians. For a similarly gendered vision of scientific identity in twentieth-century psychology, see Alexandra Rutherford, "Maintaining Masculinity in Mid-Twentieth-Century American Psychology: Edwin Boring, Scientific Eminence, and the 'Woman Problem,'" in this volume.

his curiosity and of his work [*travail*]."[14] This zeal to benefit the public, an idealized and selfless devotion to useful work, implied nobility as well as manliness.

Réaumur's academic coming-of-age coincided with the revival of the academy's encyclopedic project to describe and improve "the arts and trades," and Bignon assigned him responsibility for this long-delayed work. A few years later he was asked to supervise a survey of the kingdom's natural resources and industries at the behest of the Duc d'Orléans, regent of France during Louis XV's minority.[15] Although the *Description des Arts et Métiers* was not completed in Réaumur's lifetime, he produced a steady stream of scientific papers based on material collected for that project and for the regent's survey. He rose rapidly in the ranks at the academy, promoted to a pensioned position after only three years in an entry-level slot, and soon thereafter served as director for the first of many terms.[16] In 1722, a generous supplementary pension, earmarked for expenses associated with his scientific work, singled out his work on forest management and metallurgy, as well as his contributions to the collective *Description des arts*.[17] Frequently presenting his work in the public assemblies, he was very much in the public eye. In this context he articulated his own vision of the utility of the sciences, and the special connection between what he called "the curious and the useful." "People often rush too easily to divide knowledge into the curious and the useful," Réaumur argued in his treatise on steel. "This distinction is not as easy or as certain as we might think, especially in this matter [the properties of iron and steel]. The useful, when considered properly, always has something of the curious about it, and it is rare that the curious when followed carefully, does not lead to the useful."[18] He maintained this position in all areas of natural history and experimental physics, where applications might be less obvious than for methods of steel production. "Those who see the natural history of insects as nothing but curious diversions, and who would gladly put it into the ranks of frivolous amusements, do not know enough about its scope; we have sufficiently proved that there are few areas of research from which we can anticipate so many actual uses [*utilités réelles*] as research on insects."[19] Further, he pointed out that a narrow search for utility could be counterproductive: "We would like first of all something useful, and we do not recognize that we must be brought to it by degrees; is it not fortunate that curious observations can lead us there?"[20] In these methodological ruminations, Réaumur was deflecting curiosity toward the useful and away from frivolity. Although he did not explicitly label this curious-useful nexus as masculine, I would argue that we can read it as such, since he

[14] "Idée des progrès de la philosophie en France," *Mercure de France*, December 1754, 7–30, on 23–4.

[15] On the *Description des arts et métiers*, see Charles Gillispie, *Science and Polity in France at the End of the Old Regime* (Princeton, N.J., 1980), 344–57; Geraldine Sheridan, "Recording Technology in France: The *Description des arts*, Methodological Innovation and Lost Opportunities at the Turn of the Eighteenth Century," *Cult. Soc. Hist.* 5 (2008): 329–54. The documents generated by the "regent's survey" have been published in full: C. Demeulenaere-Douyère and D. Sturdy, eds., *L'Enquête du Régent, 1716–1718: Sciences, techniques et politique dans la France pré-industrielle* (Turnhout, 2008).

[16] Réaumur entered the academy as Varignon's "student" in 1708; his accelerated election to a pension three years later skipped over the intermediate "associate" step.

[17] The official decree establishing the pension and its justification is reproduced in "Brevet de pension de Réaumur," *Archives historiques de la Saintonge et de l'Aunis* 15 (1887): 23–4.

[18] Réaumur, *L'Art de convertir le fer forgé en acier, et l'art de faire d'adoucir le fer fondu* (Paris, 1722), iii–iv.

[19] Réaumur, *Mémoires pour servir à l'histoire des insectes*, 6 vols. (Paris, 1734–42), 4:vii–viii. This comes in a discussion of the cochineal and kermes insects, used in medical remedies and in textile dyes.

[20] Ibid., ix.

laid it out in opposition to the "frivolous amusements" and "diversions" in the passage just quoted. These value-laden terms can plausibly be interpreted as feminized, in the context of contemporary usage.

IDEAL AND REALITY

If the academy were to be useful to the crown and to society, its members would have to be productive. Ensuring that academicians did actually work was a challenge from the early days of the institution. Sometime in the 1720s, Réaumur drafted a document justifying the academy's usefulness to the crown and arguing that lack of material resources was constricting scientific effort and output.[21] Though the recommendations were not put into practice, the document articulated a vision of what men of science should do and how they should be rewarded. For our purposes, it is particularly revealing for the specific ways that Réaumur used gendered categories to express his ideal. As encouragement to take their obligations seriously, he argued, academicians needed sufficient financial means to devote themselves fully to scientific work. The crown could optimize its interests by rewarding academicians so that they would not need to find additional remunerative work. (The first two levels on the academic ladder came with no stipend [*pension*] at all.) "Is it fair that someone who applies himself to research important to the welfare of the state cannot hope to achieve some degree of fortune? The soldier, the magistrate, the merchant can expect recompense for their work; only the savant has no expectations [for reward] from his work."[22] Why not model the pursuit of scientific knowledge on those other masculine occupations that contribute to the kingdom's security and prosperity, while bestowing honorable status on the loyal subjects who sit in the academy? In actuality, regardless of the honor and status conferred by membership, most academicians had other appointments or occupations: if they did not have personal fortunes, they were physicians, apothecaries, lecturers, hydrographers, engineers, or administrators.

Réaumur wanted his colleagues to be able to throw themselves into their scientific work as single-mindedly as he did himself, but he recognized that this would be impracticable without a change in the reward system. More than half his *confrères* could not afford to treat the sciences as more than "amusements"—and amusements or diversions, as we have seen, could easily devolve into effeminacy. "What works can we expect from savants constrained to pass their days on the streets of Paris when they should be [at home] working in their studies? Is a man who arrives home tired and distracted in any state to work at something that demands his full effort? Will he spend his nights doing experiments?" In short, under the existing system, very few had the option of pursuing their experiments or calculations while "living at that level of material comfort that puts the mind at rest and in a condition to give itself over to useful research."[23] This plea for more substantial support suggested that the stipend structure should be designed to free academicians to be more manly by using their relative

---

[21] Réaumur, "Reflexions sur l'utilité dont l'Académie des sciences pourroit être au Royaume, si le Royaume luy donnoit les Secours dont elle a besoin," in Ernest Maindron, *L'Académie des sciences* (Paris, 1888), 103–10. Original in Archives de l'Académie des sciences (hereafter AAS), dossier Réaumur.

[22] Réaumur, "Réflexions" (cit. n. 21), 107.

[23] Ibid., 106. The document argued that only a "small number" of the forty-eight regular members were actually "workers [*travailleurs*]"; the institution was not working up to its potential.

freedom not for leisure or amusements but for useful work, to be accomplished in their comfortable homes. They would be able to live not like the idle aristocrats of old, but like "soldiers, magistrates, merchants"—men of different estates making themselves useful to the crown and its subjects.

In this improved academy, Réaumur envisioned a kind of paternalistic oversight imposed on scientific workers by the state, with the institution as the mediating authority. "In order that the entire academy should work for the public good, it would be appropriate to give every academician a task every year, relative to his subject and his level of support."[24] Botanists could experiment with forest preservation; chemists could work on improving domestic production of saltpeter or finding new methods for refining metals; and so on. This vision took for granted the hierarchical patronage and status economy, with pensions dispensed by royal largesse to reward service. At the same time, he imagined the increase in scientific productivity that might follow from a more explicit link between tasks and remuneration. What kind of man would the academician be in such a system? Or, shifting the question slightly: what kind of manly occupation was science in the Old Regime? For our purposes—exploring the gendered conceptions and values attached to the sciences in this period—the analogies used by Réaumur in this document are illuminating. He compared the ideal academician to soldiers (probably meaning officers) and magistrates—because these eminently masculine professions served not only the crown, but the "public good."

A SCIENTIFIC HOUSEHOLD

Réaumur, unlike those of his colleagues who had to take up multiple posts to make ends meet, had sufficient financial resources to devote himself fully to his scientific work. Over the course of the roughly fifty years of his working life, he lived in at least five different houses in and around Paris, in addition to the estate in Poitou where he spent six weeks every autumn. The size and composition of the household fluctuated, and available sources do not allow a precise accounting of these fluctuations over his whole career. Nevertheless, a good deal about the residents and the places where they lived and worked can be pieced together. Although Réaumur had neither wife nor children, a complex net of affective ties, some of them familial, linked the members of the household; at any given time some of these people were engaged in various kinds of scientific work. For most of his first two decades in Paris, Réaumur shared a home with his close friend and fellow countryman Pierre Jarosson, an upwardly mobile barrister whose growing wealth allowed him to purchase the honorific office of *secrétaire du roi*.[25] Not a man of science himself, Jarosson was nevertheless involved in Réaumur's experimental trials on iron smelting and steel production; he invested in a manufacturing venture that put these methods into practice, briefly producing ornamental cast-iron objects for the Paris luxury market.[26] Jarosson's widowed mother and

[24] Ibid., 109.

[25] Jarosson owned property in Paris and shares in the Compagnie des Indes, and he eventually bought a large estate in the Maine region. In middle age, he married the wealthy daughter of an academician, Marie Madeleine Fantel de Lagny. Jarosson's financial holdings are listed in his marriage contract from 1742: Archives nationales, Minutier central, ET/CXV/533. Réaumur and Jarosson named each other legal executors in their wills; near the end of his life, Réaumur inherited his friend's provincial property.

[26] On the iron enterprise, see Daniel Bontemps and Catherine Prade, "Un magasin parisien d'ouvrages en fonte de fer ornée au XVIIIe siècle: Une réussite méconnue de Réaumur," *Bull. Soc. Hist. Paris l'Ile-de-France* 118 (1991): 215–61.

a family of her cousins became intimate members of Réaumur's circle as well. One of these distant cousins was Hélène Dumoustier, the artist who illustrated nearly all of Réaumur's voluminous writings on the natural history of insects.[27] She, her mother, and two of her sisters visited frequently and ultimately lived as part of the extended household for many years.

The Paris home he shared with Jarosson could not accommodate laboratory equipment or extra people, so Réaumur rented a country house in the village of Charenton, not far upriver from the city and easily accessible on horseback, or even on foot. He used this property as a retreat from urban life, but also as a workplace. Outbuildings housed furnaces and other equipment for chemical and metallurgical experiments; the garden, pond, riverbank, and nearby forests supplied material for natural history observations. Guests visited the glass-fronted beehives and collected insects on promenades in the neighboring forest at Vincennes. Until Réaumur moved to a larger house in the city, most of his scientific investigations were pursued in and around the Charenton establishment, and even after the household expanded, its members traveled regularly back and forth between city and country houses.[28] In 1728, intensifying the domestic resources devoted to natural history and other pursuits, he leased a large aristocratic townhouse, the Hôtel d'Uzès, located only a few minutes' walk from the academy's meeting room in the Louvre palace.[29] The several buildings of the Hôtel d'Uzès were arranged around two courtyards, with enough space to accommodate scientific instruments and equipment, an expanding natural history collection, a library, and a "menagerie" of living insects. The added space meant that Réaumur could lodge not only servants and occasional guests but a shifting set of assistants, artists, and companions as well.

A vignette drawn to illustrate Réaumur's natural history of insects by Philippe Simonneau, one of the academy's most prolific artists, depicts an idealized view of this work space, packed with observations and experiments in progress (see fig. 1). Simonneau had no doubt spent time at both town and country residences while making his drawings; this image combines elements of country garden and townhouse workroom, giving a kind of visual extract or overview of investigations that must have taken place in different seasons, and in different locations. A few butterfly specimens lie flattened on the table, but the rest of the insects portrayed here are alive. Chrysalises hang from a frame; caterpillars crawl on the table and hang on the twigs of a branch, ready to pupate; on the right-hand wall, rows of glass jars of eggs and caterpillars fill the shelves. The interior space of the room merges seamlessly with the highly engineered garden, where we can spot a small tank set into the ground, with dragonflies hovering over it, a glass-fronted beehive, a hothouse, a large spiderweb, and a geometric array of some sort of nest or trap in the left foreground. The butterflies crossing freely from inside to outside suggest the dynamic interplay between nature and the controlled spaces of workroom and garden.

[27] Hélène Dumoustier's paternal aunt was married to Pierre Jarosson's maternal uncle.

[28] On experiments and observations pursued at Charenton by members of the household, see Mary Terrall, "Frogs on the Mantelpiece: The Practice of Observation in Daily Life," in *Histories of Scientific Observation*, ed. Lorraine Daston and Elizabeth Lunbeck (Chicago, 2011), 185–205.

[29] The Hôtel d'Uzès had been built for the Marquise de Rambouillet, the famous seventeenth-century *précieuse* hostess. On this house and its occupants, see Terrall, *Catching Nature in the Act* (cit. n. 6), chap. 3. On the architecture, see Jean-Pierre Babelon, "L'Hôtel de Rambouillet," *Paris et Ile-de-France: Mémoires publiés par la Féderation des sociétés historiques et archéologiques de Paris et de L'Ile-de-France* 11 (1960): 313–49.

Figure 1. *Philippe Simonneau, title vignette for Réaumur,* Mémoires . . . insectes *(cit. n. 19),* vol. 1.

The representation of all this activity has been cleansed of people, masking the human activity that made possible the intensive examination of insect lives. In putting the people back into the picture, I am sketching out a different (if imaginary) composite image that will open up the life of the household to our retrospective view. My goal here is not so much to retrieve the contributions of invisible technicians as to situate the scientific work, destined ultimately for presentation to the academy, to visitors, and to readers around the world, in the highly articulated social and physical space that was Réaumur's household.[30] I use this term to include the occupants of the residence and others who came to work there as well as the architectural space of the houses, surrounding gardens, and outbuildings. The boundaries of this unit were elastic and accommodating, as residents came and went between city and country houses, and as visitors and assistants moved in and out of the laboratory, the library, and the collections. Réaumur himself functioned as the patriarch of the whole operation, but he was a patriarch without any immediate family—no wife, no children, no siblings. As he opened his home to a series of assistants—mostly young men in the early stages of their careers—he built, intentionally or not, a sort of scientific family, whose offspring recognized the formative role played by his tutelage and patronage. Emotional connections forged in this setting, like family ties, lasted long after individuals had moved on, sometimes forming the basis of lifelong friendships and collaborations.

The daily details of the roles played by the characters in my story remain elusive, but clues emerge from a variety of sources: passing references in correspondence and published scientific papers, eulogies delivered years later by the academy's secretary, manuscript laboratory notes, illustrations, and inventories. The density of interpersonal interaction, conversation, and collaboration can only be inferred from the faint traces left in these documents. The first of Réaumur's assistants was Henri Pitot, who came to Paris from Languedoc as a youth of twenty-three, eager to study mathematics.

---

[30] Steven Shapin develops the useful notion of "invisible technicians" in Shapin, *A Social History of Truth: Civility and Science in Seventeenth-Century England* (Chicago, 1994), chap. 8. On the invisibility of women's work in astronomy, see Mommertz, "Invisible Economy of Science" (cit. n. 7).

Introduced to Réaumur by a female relative, Pitot found himself with a patron and a mentor; the older man "took pleasure in encouraging young people whose talents he recognized" and supervised Pitot's scientific and philosophical reading for the next few years.[31] At this point, in the early 1720s, Réaumur's household was still small; his research questions derived primarily from the academic projects promoted by Bignon, especially the *Description des arts et métiers*. He was also responsible for the chemistry laboratory maintained by the academy, which was probably identical to the one at his house in Charenton. Pitot's first "useful" assignment—as his eulogist called it—was as assistant in this laboratory, with a modest stipend from the academy; this served as the next phase of his induction into experimental science. In practice this meant working with his mentor on "a great many investigations and experiments on cast iron, on porcelain, on different kinds of varnish [for metals], and in preparing a large collection of material for the *Histoire des arts*."[32] Soon Pitot was appointed to an opening as adjunct at the academy, again through Réaumur's patronage. While assiduously fulfilling a multitude of academic responsibilities, Pitot continued to work with Réaumur off and on for the next ten years, when he was finally elected to a pensioned slot at the academy.[33]

Pitot was the first of several future academicians who were, in varying degrees, supported by Réaumur while working in his laboratory. Jean-Antoine Nollet was the second. Already known as an adept enameler and scientific instrument maker, he had previously worked on electrical experiments with Réaumur's friend and fellow academician Charles François Dufay. Nollet took over Pitot's role in the Charenton laboratory, and, as the academy's secretary later recalled, "it was in this excellent school, which has supplied the academy with several of its most illustrious members, that [Nollet] completed his training."[34] In the 1730s, the main scientific preoccupations of the household were insects and thermometers. Though Pitot and Nollet did not live in the house, they were often in the laboratory in Charenton and in the various workrooms where the insects and the collections were kept in town.[35] Both men were intensely involved in Réaumur's early work on thermometry, and they each built numerous instruments following Réaumur's protocols.[36] As part of an extensive research

[31] Jean-Paul Grandjean de Fouchy, "Eloge de M. Pitot," *Histoire de l'Académie Royale des Sciences* (hereafter *HARS*), 1771, 145.

[32] Ibid., 147. Réaumur was working on making porcelain from glass in 1723, using "my large furnace at Charenton," probably with Pitot's assistance. Réaumur, "Notes on the manufacture of porcelain and glass, ca. 1722," Getty Research Institute; for other metallurgical experiments, manuscript notes in Biblioteca Laurenziana, Florence, MS Ashburnham 1804.

[33] Réaumur interceded on Pitot's behalf with the royal minister Maurepas, trying to get him a teaching post, noting that his protégé had been working for the academy for years without any monetary reward. Réaumur to Maurepas, 6 November 1732, Bibliothèque de Genève, MS Trembley 5.

[34] Grandjean de Fouchy, "Eloge de M. l'Abbé Nollet," *HARS*, 1770, 122.

[35] It is not always possible to determine exactly who was in residence at a given time. Pitot, who lived nearby, was making thermometers from Réaumur's design at least from 1730; Nollet took over soon thereafter, but they may well have been in the laboratory together for some experiments.

[36] C. Messier, "Mémoire sur le froid extraordinaire que l'on ressentit à Paris . . . au commencement de cette année 1776," *Mémoires de l'Académie Royale des Sciences* (hereafter *MARS*), 1776, 1–155 (transcription of label on a thermometer made by Pitot in 1730 on 140). On the early history of Réaumur's thermometer, see Jean-François Gauvin, "The Instrument That Never Was: Inventing, Manufacturing, and Branding Réaumur's Thermometer during the Enlightenment," *Ann. Sci.* 69 (2011): 514–49, on 22n80. Pitot's intervention in experiments mentioned in Réaumur, "Essais sur le volume qui résulte de ceux de deux liqueurs mêlées ensemble," *MARS*, 1733, 167–8. Nollet discusses working with Réaumur on thermometry in J.-A. Nollet, *Leçons de physique expérimentale*, 6 vols. (Paris, 1743–8), 4:407–11.

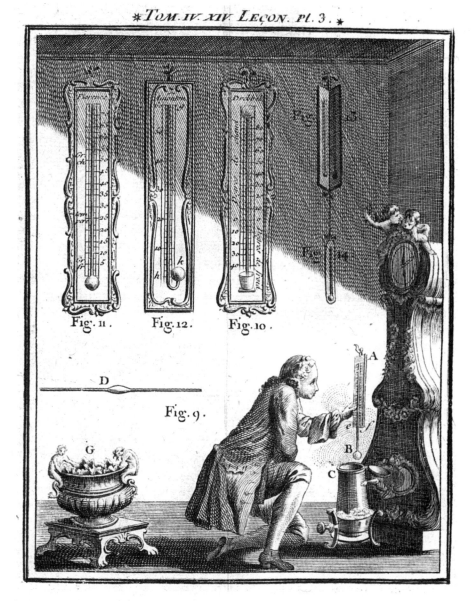

Figure 2. *Nollet observing boiling temperature of liquid with instruments and apparatus similar to what he would have used in Réaumur's laboratory. Nollet,* Leçons de physique expérimentale *(cit. n. 36), vol. 4, lesson 14, pl. 3.*

program on temperature and heat capacity, Pitot and Nollet helped with experiments on the expansibility of different kinds of alcohol and measurements on melting ice (see fig. 2). Nollet took charge of the jars of living insects, as well as the instruments in the laboratory. At times the insects became the subject of physical experiments, when he put caterpillars in the receiver of the air pump to investigate transpiration or measured the freezing temperature of the caterpillar's internal fluids.[37] He was also available to show visitors around the collections.[38]

Two other young men moved into the Hôtel d'Uzès while Pitot was calibrating thermometers in Charenton: a medical student by the name of Baron and a neophyte artist called Regnaudin. Baron had particular responsibility for feeding and keeping track of the denizens of the insect menagerie, replenishing the jars and boxes as necessary.[39] He stayed for a year or two, before moving to a provincial town to start his medical practice, after which he continued to send boxes of insects by post to Paris. Very little is known about Regnaudin, the illustrator, apart from the fact that he was trained to draw insects, learning to use a loupe and a microscope as well as pen and ink. Before he was in residence, Réaumur had employed the academy's regular illustrators but found it increasingly difficult, without an artist on call, to document the elusive habits and transformations of the many kinds of insects living on his shelves and in his garden. Simonneau, who drew the idealized work space represented in figure 1, had as much work as he could handle making the engravings for the academy's publications, and he could not be available at Réaumur's convenience. As benefactor and patron, the naturalist could be the master of Regnaudin's time, especially since he was living on the premises. Though inexperienced when he arrived, he soon became accomplished at his task, illustrating academic papers on thermometry and insects. But before his patron's ambitions for him could be fully realized, the young artist died unexpectedly.[40]

Meanwhile, Hélène Dumoustier had taken up drawing—she later recalled that her interest had been sparked by watching artists at work in the house—and she was soon recording the mechanisms and maneuvers of the insects that were to fill the pages of Réaumur's books for the next decade.[41] Dumoustier and her family had come into the Hôtel d'Uzès, where they kept their own apartment and servants, through their family connection with Jarosson and his mother, but once she took on the role of resident artist, Hélène was often working alongside the men, and she was well known to everyone who spent time in the collections, the laboratory, or the library. Her sisters do not seem to have participated in this scientific work, though all the women routinely traveled back and forth to Charenton and went along on the annual journey to Réaumur's estates, where the pursuit of natural history continued unabated. In later years, when asked in what capacity she had lived in Réaumur's house for so many years, Dumoustier replied "as a friend occupying an apartment that her mother rented." After

[37] "Experiences à faire faire par l'Abbe Nollet," AAS, Fonds Réaumur, dossier 47, fol. 46. For the freezing temperature of insects, see Réaumur, "Expériences sur les différents degrés de froid qu'on peut produire en mêlant de la glace avec différents sels," MARS, 1734, 187. See also Réaumur, Mémoires . . . insectes, (cit. n. 19), 3:178, for Nollet's participation in experiments with clothes moths.

[38] Jean-François Séguier was shown around the collections in the Hôtel d'Uzès and the laboratory in Charenton by Nollet in 1733: Séguier, "Fragments de quelques notes que je fis en voyageant en France," Bibliothèque municipale de Nîmes, MS 129.

[39] "M. Baron lived with me in Paris, and even took care of my insect menageries." Réaumur, Mémoires . . . insectes (cit. n. 19), 1:51. Regnaudin may have been related to Réaumur, based on his surname, but I have found no solid evidence for a blood relationship, nor have I identified first names for either of these men.

[40] Réaumur mentions the death of Regnaudin, without naming him, in Mémoires . . . insectes (cit. n. 19), 1:54. Regnaudin's drawings were engraved to illustrate two of Réaumur's papers printed in 1732 for the 1730 volume of the academy's journal. "Memoire des desseins faits par le Sr. Regnaudin," 22 October 1731, AAS, Fonds Lavoisier, 1065 ac 1731.

[41] Dumoustier's recollection in "Interrogatoire," 22 November 1759, Archives nationales, Y 13951.

the death of her mother in 1743, she and her sisters continued to occupy their rooms "as friends and companions keeping their home [*ménage*] distinct and separate . . . at their own expense, just as their mother had done."[42] She very likely was exaggerating the family's independence; they seem to have lived modestly on income from small investments. Although she downplayed this fact when asked about it later, records in the academy's archives show that Dumoustier was paid for her drawings from 1736 to 1747, at the same rate commanded by the professional artist Simonneau.[43]

Hélène Dumoustier's engagement with natural history did not stop with her drawings, nor was she simply carrying out instructions. Her constant presence around the house and on outings into the countryside and forest gave her the opportunity to see and collect things for herself. Like the others, she came and went, observing and manipulating her tiny subjects as well as drawing them. This should not surprise us, given that both natural history and physics investigations were materially integrated into the life of the household. Réaumur attributed numerous observations to her, in his private notes and in print; he made a point of acknowledging her crucial role in many investigations, when she noticed details and spotted unusual phenomena no one else had seen.[44] In print, the artist-observer was never explicitly named, at her own insistence, though she appeared from time to time, rendered anonymous by asterisks, as in this example: "While Mlle. *** was drawing one of these caterpillars, very near to the time of its metamorphosis, she observed that several drops of water emerged from different places on its skin. The next day, I observed the same caterpillar, and I saw it make a little move that I have not yet seen made by any other."[45] The two observations on subsequent days, by different people, blend into the narrative description of the metamorphosis unfolding on the worktable. The artist was clearly working on her own, unsupervised, and reported her noteworthy observation later. Réaumur took her testimony about ephemeral phenomena like the liquid emitted by the caterpillar to be entirely creditable—it did not need confirmation from a more authoritative witness. When such explicit references make it into the final text, we can assume that they represent only the tip of the iceberg, so to speak, from which we can extrapolate to a much larger mass of daily observations of all kinds.

At other times, we catch sight of sustained collaborative projects, as when they devised techniques for paralyzing and counting and sorting bees while investigating the seasonal cycle of the beehive. In this case, Réaumur took the opportunity, still without naming his artist, to point out the intimate connection between drawing and observing, and to acknowledge the value of Dumoustier's observations over the long term: "I had with me a person who loves natural history, and who has supplied me with observations recorded in the preceding volumes, and in addition to observations, very perfect drawings; a person who knows as much as I do about bees of different sexes since she has made drawings of them. She and I, we set out to examine them, to sort them, so to speak, one by one, with more care than one gives to sorting coffee beans."[46] Eventually they found the queen and determined that she was the only one

---

[42] Ibid.

[43] The total amount paid over eleven years was 8,000 *livres*. AAS, Fonds Lavoisier, comptabilité, 1066, 1068, 1069 ac.

[44] On Dumoustier and her family, see Terrall, *Catching Nature in the Act* (cit. n. 6), chap. 3.

[45] Réaumur, *Mémoires . . . insectes* (cit. n. 19), 2:75; on Dumoustier's refusal to be named in print, 1:55.

[46] Ibid., 5:545.

in the hive. Note that Réaumur's judicious use of pronouns here makes Dumoustier's presence obvious, although still anonymous. Réaumur also trusted her with keeping the work going when his other (masculine) obligations interfered. The bees lived in the garden at Charenton, where the observers were trying to see the mechanics of copulation. After a certain amount of trouble, they finally got bees to mate in a glass jar: "After observing these proceedings, and having seen them repeated for more than two hours, I was obliged to leave my two bees and the country house to go to Paris, where one of our academy meetings called me. But several people whom I left at my house, and one in particular, whose eyes I trust as much as my own, did not cease observing what was happening for the rest of the afternoon, and upon my return they gave me an account of all they had seen."[47]

Hélène Dumoustier was a fixture in the house over many years, working primarily in and around the insect collection. She was well known to the string of assistants, collaborators, and visitors who stayed for shorter periods, and developed lasting friendships with many of them. The assistants made the same kind of cameo appearances in Réaumur's texts, though unlike the artist, they were often named.

In 1740, Réaumur gave up the Charenton house and consolidated his research and his collections into one site, another elegant house just outside the city walls in the Faubourg Saint-Antoine. The Dumoustier family moved with him; Pierre Jarosson, too, was part of the entourage and lived in the new house until his marriage in 1742. The move out of the city ushered in a new period in the life of the household, with the rapid expansion of the museum of preserved specimens and the laboratory associated with it.[48] Réaumur was still working on the natural history of insects when he moved, but his focus shifted more and more to birds. With large numbers of dried and pickled specimens arriving daily, much of the attention of various assistants was taken up by problems of preservation and presentation. At the same time, the flock of poultry and other birds kept in the yard provided material for a raft of other work, especially the eminently utilitarian study of artificial incubation of chicken eggs and the preservation of eggs. And every autumn most of the house's residents would pack up collecting jars, microscopes, and books for the ten-day carriage journey to the manor house in Poitou.

Of the men who worked in and around Réaumur's collection in the 1740s, three went on to pensioned positions at the academy. Jean-Etienne Guettard came to Réaumur's attention while pursuing his botanical and medical studies. He was in Charenton in the spring of 1740, when the household was occupied with observing the mating habits of frogs, and shortly thereafter he moved into the house in Faubourg Saint-Antoine, where he lived for four or five years. Guettard was brought in as an all-around naturalist, working indoors on the collection in the winter months, and outdoors in the field whenever possible. "The season of botanizing and gathering insects has arrived, and for several months M. Guettard . . . has done nothing but traverse the countryside in our vicinity," his patron wrote to a correspondent interested in the collection.[49] In 1741, Guettard came along to Poitou and was dispatched to the coast

[47] Ibid., 504.
[48] Terrall, *Catching Nature in the Act* (cit. n. 6), chap. 6.
[49] Réaumur to Louis Bourguet, 29 July 1741, Bibliothèque publique et universitaire, Neufchâtel, MS 1278.

where he spent several weeks on his own, cutting up sea anemones and sea stars to see if they could regenerate.[50] Other times, he accompanied Réaumur and Dumoustier to the shore. Réaumur documented one such excursion in notes appended to Dumoustier's drawing of a colony of tiny sea stars: "An observation that M. Guettard made yesterday, and that I repeated today, convinced us both that each of these little stars is an animal. Some stars have eight arms, others six or seven, others fewer, and Mlle. Dumoustier found one which only had three."[51] Such passing comments offer fleeting glimpses of the kind of intensely focused but sociable investigation that would have been routine for this little group, out on the rocky shore inspecting the tide pools for unusual specimens.

In 1743, Guettard was elected to the botanical class of the academy. He continued to live with Réaumur for two more years as he pursued his own botanical research, while providing occasional observations and specimens to his patron.[52] He was replaced in the "laboratory" and collections by François David Hérissant, a young physician recommended to Réaumur for his manual dexterity and his expertise at dissection.[53] Hérissant did not live at the house; there was probably no room for him anyway, as the collections had expanded to take up the entire second floor of the house, and everyone except Réaumur occupied apartments in other buildings on the property. He spent his time in the laboratory, working on methods of preservation for the natural history specimens, and in the poultry yard, where he studied digestion in birds. Hérissant got a place in the academy after five years of working for Réaumur. By this time he had also established a medical practice, serving as Réaumur's personal physician, and eventually he took up lodgings in a building owned by an elderly widowed cousin of his patron.[54]

For the last eight years of Réaumur's life, his natural history collection, by this time filled with hundreds of birds from all over the world, was managed by Mathurin Jacques Brisson, supported as Pitot had been years before by a modest stipend from the academy. Brisson was the nephew of Réaumur's sister-in-law, thus a relative by marriage; he grew up near the estate in Poitou and came looking for a post after abandoning his clerical career.[55] In 1749, he moved into Réaumur's household, where he became a crucial player in all the ongoing projects, indoors in the laboratory and collection and outdoors in the poultry yard. He continued Hérissant's work on taxidermy and used the collection as the raw material for two major taxonomic works, one on quadrupeds and one on birds. He also made thermometers, distributed them to Réaumur's correspondents, and lent his youthful eyes to microscopic observations. Brisson

[50] For Guettard's biography, see Condorcet, "Eloge de M. Guettard," *HARS*, 1786, 47–62. Guettard appears in Réaumur's observation notes on frogs, "Grenouilles," 8 April 1740, AAS, Fonds Réaumur, dossier 35. For his experiments on sea creatures on the Atlantic coast, see Guettard to Réaumur, 27 September 1741, AAS, dossier Guettard. His manuscript inventory of Réaumur's collection is in Bibliothèque centrale du Muséum d'histoire naturelle (Paris), MS 1929 (iii).

[51] Bibliothèque centrale du Muséum d'histoire naturelle (Paris), MS 972, fol. 141.

[52] Guettard wrote from one of Réaumur's properties near La Rochelle, with observations of zoophytes: Guettard to Réaumur, 12 July 1745, AAS, dossier Guettard.

[53] "Eloge de M. Hérissant," *HARS*, 1773, 118–34.

[54] For Hérissant's work on preparing specimens, see Réaumur, "Moyens d'empêcher l'évaporation des liquides spiritueuses, dans lesquelles on veut conserver des productions de la nature," *MARS*, 1746, 507.

[55] See Arthur Birembaut, "Les liens de famille entre Réaumur et Brisson," in *La vie et l'oeuvre de Réaumur*, ed. P. Grassé (Paris, 1962), 168–70.

was elected to the academy only after the death of his patron, on the strength of his systematic work on ornithology.[56]

In his eulogy of Guettard, three decades after Réaumur's death, Condorcet memorialized the nexus of intellectual, patronage, and affective elements at play in the scientific household where a string of young men found their scientific bearings.

> M. de Réaumur undertook immense projects in the sciences and the arts which he could not have done alone; he sought to attach to himself young men whose burgeoning talents still needed support. They helped him in his work, succeeded in learning under his gaze, found in his books, in his collections, in his laboratory the kind of aid that is still so often not available to hardworking, but poor and obscure, young people, even in the midst of so many institutions designed to favor the sciences. Then, released after several years, they came into the world with a name that was already known, and saved by useful connections from those dangers that often block entry into a career in the sciences. Most of these students subsequently entered the academy, and all retained for M. de Réaumur a tender and permanent appreciation that proves both that he chose them well and that he knew how to avoid with them the kind of superior attitude that his age, his long researches, and a confirmed reputation might have given him. M. Brisson is the only one of these students of Réaumur who is left to us. In learned societies, we like to remember these filiations that make our talents more dear to us by linking them to the memory of those we have lost.[57]

The nostalgic tone here is of course appropriate to the eulogy genre. For his own polemical purposes, Condorcet oversimplified the distinction between the aristocratic master and his "poor and obscure" students, but he did capture the formative experience for these men of working with the "books, collections, and laboratory" in Réaumur's various residences. And in reminding his colleagues of the "filiations" linking the academicians of the 1780s to previous generations, Condorcet was also memorializing a kind of family connection that linked those who came to the academy by this particular route to each other, suggesting that we might view them as a band of brothers, passing in turn under the tutelage of a benevolent paternal figure.

CONCLUSION

Réaumur's aristocratic status, and the resources he commanded, made possible the combination of hierarchical and companionate relations that structured his household and its ties to the academy (and other royal institutions). His various homes—with space for experiments, accommodations, collections, gardens, and so on—were sites for making knowledge and for making scientists. Overseeing and managing all that went on there, Réaumur played a number of overlapping masculine roles: mentor and father figure, patron, employer, landlord, and host. Assistants came and went; in many cases, as we have seen, these young men parlayed their work for Réaumur into other positions, especially in the academy. Affective and professional ties developed in the course of the day-to-day operation of taking care of insects, operating microscopes and air pumps, reading and calibrating thermometers, and preserving and arranging specimens. This work carried over into correspondence, collaboration, and friendship after the men had moved on to other domestic situations. The women in the household

---

[56] M. J. Brisson, *Le règne animal divisé en IX classes, ou méthode contenant la division générale des animaux* (Paris, 1756); Brisson, *Ornithologie, ou méthode contenant la division des oiseaux en ordres, sections, genres, especes & leurs variétés*, 6 vols. (Paris, 1760).

[57] Condorcet, "Eloge de M. Guettard" (cit. n. 50), 49–50.

maintained their independence to some degree, but they were also part of the social and scientific life that filled most rooms of the houses they occupied. Certainly the work of Hélène Dumoustier was essential to the whole operation, as Réaumur recognized, especially for her drawings but also for her observations. She occupied a peculiar position, as part of a family of women embedded in an otherwise masculine domain tightly linked to the Paris Academy of Sciences, while contributing (unlike her sisters or her relative Jarosson) to the scientific work permeating the house.

Much of the scientific activity of Réaumur's household fell into the category of the kind of useful knowledge that he promoted throughout his career in the academy. Usually, results generated in this amended version of an extended-family setting then passed through the academy on the way to a wider public, whether in the public sessions or in print. Ties formed in the house and laboratory persisted through several intellectual generations, as Réaumur became godfather to Pitot's son René, and Nollet, on his retirement from a prestigious position at the College de Navarre, arranged for Brisson (a generation younger) to take over his appointment as professor of experimental physics. Nollet kept in his possession a thermometer made with his patron in 1732 and passed it down in his will to Brisson, who claimed it was still accurate in the 1770s.[58]

As for Hélène Dumoustier, she was revealed as Réaumur's universal legatee when his will was discovered in a locked cupboard after his death. In this document, the richest single source for understanding their relations, he acknowledged his debt accrued over her long years of illustrating and observing. If his assistants had inherited his intellectual legacy, carrying on their experimental science in the academy and elsewhere, Dumoustier was supposed to inherit some real property. As it happened, though, the will was contested by distant cousins, who dragged the case through the courts for years. She spent her last years with her one surviving sister, living comfortably but modestly on her investments. At the end of her life, the one material memento of her time as a scientific illustrator remaining in her possession was a terracotta portrait bust of Réaumur that had greeted visitors to the house in Faubourg Saint-Antoine years earlier (see fig. 3). In her last will and testament, she left this bust to Hérissant, "as a mark of my friendship; . . . he deserves it for the attachment that he had for [M. de Réaumur], and that he continues to have for his memory."[59]

Nollet's thermometer and the portrait bust bequeathed by Dumoustier to Hérissant represent, in material form, some of the myriad ties permeating and enlivening the living and work space of the household. These were ties of affection, certainly, as well as intellectual debts and collaborations that linked the home to the academy, where innumerable observations, measurements, techniques, and instruments were presented in the spirit of serving the public good and contributing to the mission of the institution. That mission was a masculine endeavor, not only because members were men, but also inasmuch as the double-barreled ideal of utility and public service was itself gendered male. The ideal also drew on language associated with aristocratic merit, as we have seen. As a member of the lesser nobility who chose not to continue his hereditary lineage in the usual way, Réaumur acted out his aristocratic values in the context

[58] On this thermometer, see Jean-Antoine Nollet, *L'art des expériences*, 3 vols. (Paris, 1770), 3:182. See also Gauvin, "Instrument That Never Was" (cit. n. 36).

[59] "Testament de Mlle. Dumoustier de Marsilly," Archives Nationales, Minutier central, ET/CXV/774. Hérissant left the bust to the Paris Academy of Sciences in his will; it is now in the Louvre.

Figure 3. *Jean Baptiste Lemoyne, portrait bust of Réaumur. Inherited by Hélène Dumoustier, passed on to F. D. Hérissant, and bequeathed by him to the Paris Academy of Sciences. Musée du Louvre. Copyright RMN-Grand Palais/Art Resource, New York.*

of his scientific career, at the academy and in his own home. Most of his projects, from early work on the kingdom's natural resources and steel manufacturing to his final treatise on poultry cultivation, incorporated explicitly utilitarian elements. From the story of the elaborate household operation, another aspect of the ideal of meritorious service to the state comes into focus, as we recognize, with Condorcet, that Réaumur was producing not only scientific results but also future academicians. He reproduced himself, or versions of himself, in his intellectual offspring. The household—where people from different families with different aspirations lived and worked together—provided the conditions for this peculiar form of asexual reproduction.

The activities and interactions of the people who spent time working in Réaumur's household left only imperfect traces in the historical record, making the domestic scene rather more invisible to us than it would have been for contemporaries. The infrastructure of the home—laboratory, study, collection, but also kitchen, bedrooms, and rooms where guests were entertained—was inextricably linked to the academy, and indeed the institution depended on these links. In Réaumur's case, the household was not a family as such, though various sorts of family connections came into play. Because of the scale of the operation and the shifting personnel, this example is quite complex; other households would no doubt repay study and would enrich our picture of how science was made, and how it was gendered. Only some of these would resonate with the kind of aristocratic values I have explored here. If we are to appreciate the diverse meanings of masculinity for the sciences in the Old Regime, we need to think about family ties, social status, friendships, institutions, domestic arrangements, and state patronage and investigate the interlinking of these categories and contexts.

MEASURES AND METAPHORS
OF GENDER

Detecting and Teaching Desire:
Phallometry, Freund, and Behaviorist Sexology

*by Nathan Ha**

ABSTRACT

During the 1960s and 1970s, Kurt Freund and other researchers developed phallometry to demonstrate the effectiveness of behaviorism in the diagnosis and treatment of male homosexuality and pedophilia. Researchers used phallometers to segment different aspects of male arousal, to discern cryptic hierarchies of eroticism, and to monitor the effectiveness of treatments to change an individual's sexuality. Phallometry ended up challenging the expectations of behaviorist researchers by demonstrating that most men could not change their sexual preferences—no matter how hard they tried or how hard others tried to change them. This knowledge, combined with challenges mounted by gay political activists, eventually motivated Freund and other researchers to revise their ideas of what counted as therapy. Phallometric studies ultimately revealed the limitations of efforts to shape "abnormal" and "normal" masculinity and heralded the rise of biologically determinist theories of sexuality.

INTRODUCTION

In 1977, Charles Bonnell penned an article entitled "A Freund Indeed" for *Body Politic*, at the time Canada's leading homosexual magazine. "Penile plethysmography?" the subtitle playfully asked. What was this exactly, and what could it reveal about sexual arousal in gay and straight men? According to Bonnell, the plethysmograph was a device consisting of a cylinder, attached to the penis, that could measure

* Institute for Society and Genetics, University of California, Los Angeles, 1320 Rolfe Hall, Los Angeles, CA 90095; nathan.q.ha@gmail.com.

I thank Erika Lorraine Milam and Robert A. Nye for their incisive commentary and support at crucial moments in the development of this essay. I also thank Ray Blanchard for sharing his perspective on Freund's life and work with me during an interview at the Centre for Addiction and Mental Health, where I also had the opportunity to talk with Michael Kuban about the use of phallometry in current research and to examine biographical materials about Freund made available by archivist John Court. This essay benefited from the stimulating conversations I had with participants of the "Masculinities in Science/Sciences of Masculinity" conference at the Philadelphia Area Center for History of Science in May 2012, especially Mary Terrall and Zeb Tortorici. I am grateful to Ted Porter, Rachel Lee, Lowell Gallagher, Hannah Landecker, Martine Lappe, and members of the University of California, Los Angeles (UCLA) History of Science, Medicine, and Technology Colloquium and the UCLA Center for the Study of Women Life Un(Ltd) Research Colloquium Working Group for their thought-provoking questions and comments. I also thank my two anonymous reviewers for suggestions that have honed the argument of the essay. The UCLA Institute for Society and Genetics provided generous funds to research this project. I am grateful to Eric Vilain, Soraya de Chadarevian, Ute Dormann, and Anthony Petro for their friendship and conversations that have improved the essay. And finally, I'd like to thank my two undergraduate research assistants, Keaton Savage and Shian Hong, for their helpful and enthusiastic labor.

changes in the volume of a man's erection and record these changes on a pen-and-ink graph. Bonnell provided an illustration of two men attached to this "1984-ish" contraption (fig. 1) that produced a tracing of their arousal as they watched a slideshow of naked people, and he cheekily compared it to the "AccuJak cuff," a masturbation toy. The plethysmograph, however, was more than an instrument of surveillance and a titillating novelty. It also played a prominent role in the scientific work of Kurt Freund, a researcher at the Clarke Institute of Psychiatry in Toronto who had emigrated from Czechoslovakia in 1969.[1]

Since the late 1950s, Freund had developed the use of the penile plethysmograph as a device that could quantify male sexual desire and make taboo preferences legible. In situations where a man's professed sexual preference was doubted, Freund contended that the tool could discern true homosexuals from those merely pretending to be homosexuals to avoid service in the Czechoslovakian Army, and to ascertain cases of pedophilic attraction in both gay and straight men. Because Freund also had attempted to cure homosexuality, Bonnell labeled him "homophobic." Yet, surprisingly, Freund had produced data that challenged several negative stereotypes of homosexuals. "Although Freund is no friend of Gay liberation," Bonnell wrote, "many of the results he has obtained can be used for good ends."[2] In particular, Freund had shown that male homosexuality was not correlated with disgust for women, that aversive conditioning did not cure homosexuality, and that gay men may be "less dangerous around 6-to-8 year olds" than straight men because it was actually straight men who were more aroused by images of naked children.[3] Even though Bonnell, a Harvard PhD, remained a skeptical critic of the plethysmograph, he punned that perhaps its promulgator would prove to become "a Freund indeed."[4]

During the 1960s and 1970s, Freund was one of many psychiatric and psychological researchers who worked to categorize and diagnose male sexual disorders. Since World War II, psychiatry had gained increased legitimacy from governments in the United States, Canada, and Europe because of the role that psychiatrists had played in screening out soldiers who could not handle the stress of combat.[5] War, in other words, had enlisted psychiatrists as adjudicators of adequate masculinity. This role continued even in peacetime as legal experts, the media, politicians, and the general public turned to psychiatry for answers provoked by a series of mid-twentieth-century panics about male sexual deviance. In medical and public discourse, rapists, pedophiles, and homosexuals were conflated; they all held a common status as dangerous sex offenders requiring state surveillance and psychiatric treatment. These figures of problematic masculinity constituted Freund's patients and research subjects.[6]

Seen in this light, it is not surprising that Freund applied the plethysmograph to a

---

[1] Charles Bonnell, "A Freund Indeed," *Body Politic*, 1977, 13–4. "Charles Bonnell" was a pseudonym. He published articles in the *Harvard Crimson* as well as gay and mainstream periodicals during the 1970s.

[2] Charles Bonnell, "Penile Plethysmography and Gay Liberation—On the Work of Kurt Freund," abstract in the program of the Fourth Annual Conference of the Gay Academic Union, Columbia University, New York, 26–8 November 1976.

[3] Ibid.

[4] Bonnell, "Freund Indeed" (cit. n. 1).

[5] Gerald Grob, *From Asylum to Community: Mental Health Policy in Modern America* (Princeton, N.J., 1991); Vernon Rosario, *Homosexuality and Science: A Guide to the Debates* (Santa Barbara, Calif., 2002).

[6] Estelle B. Freedman, "'Uncontrolled Desires': The Response to the Sexual Psychopath, 1920–1960," *J. Amer. Hist.* 74 (1987): 83–106.

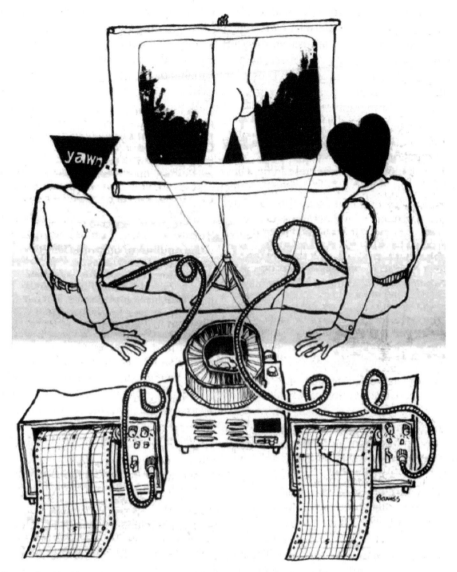

Figure 1. *Charles Bonnell's cartoon of the penile plethysmograph from "A Freund Indeed" (cit. n. 1).*

motley assortment of questions centered on male sexual deviance. He was initially concerned with homosexuality and then began working with pedophiles, rapists, and transsexuals. He also developed a gender identity scale and studied men with various paraphilias—sexual predilections for objects, bodily parts, or activities that were unusual or excessive. Eventually, he developed a theory called "courtship disorder" that explained why some of these paraphilias—exhibitionism, voyeurism, telephone scatologia, frotteurism, and biastophilia (rape)—often occurred in the same individual. Freund's research was and continues to be controversial. Indeed, ongoing debates about the validity of the plethysmograph only attest to the significant role that it has played in sexological research and forensic psychology. Scientists at many universities, including Northwestern University and Queen's University, currently use similar devices to study sexual arousal in men and women.[7] And a large proportion of sex offender programs in North America continue to use the plethysmograph to monitor and assess convicted offenders.[8] These deployments of the plethysmograph deserve further historical investigation, but this essay will focus on Freund's invention of the penile plethysmograph and the adoption of phallometry by other sexologists during the 1960s and 1970s.

In the middle decades of the twentieth century, homosexuality was a major topic of research for scientists who wanted to demonstrate the efficacy of behaviorism. I argue that Freund and other researchers developed phallometry as a critique of Freudian psychology and to demonstrate the effectiveness of learning theory and behavioral therapy in treating homosexuality. Through phallometry, Freund and his colleagues believed that they could obtain objective measurements of a patient's true sexual preferences, instead of relying upon that patient's own assertions. In other words, I argue that researchers used phallometry to *articulate* cryptic sexual desires. Phallometers enabled researchers to segment different aspects of arousal, to discern cryptic hierarchies of eroticism, and to ask the penis to speak on behalf of the mind.

Phallometric studies encompassed several underlying assumptions about masculinity shared by scientific researchers. First, they presumed that the possession of sexual desire was a key component of masculinity. Second, these studies were decidedly phallocentric and privileged the penis and its erections as indicative of male sexual desire. Third, they assumed that sexual preferences were akin to tastes that could be ranked hierarchically. Fourth, they took for granted the idea that normal masculinity necessitated the direction of sexual attention toward appropriate sex objects—namely, adult women. Fifth, and most significantly, they assumed that normal, masculine sexual preferences could be learned.

Ultimately, however, phallometry frustrated behaviorist researchers by showing that sexual preferences could be articulated but not inculcated. It demonstrated that most men could not change their sexual preferences—no matter how hard they tried and how hard others tried to change them. Here, the tracings left behind by the graph indicated the presence of intractable erotic preferences. This knowledge, combined with challenges mounted by gay political activists, eventually motivated Freund and other researchers to revise their ideas of what counted as legitimate therapy and what could be designated as deviance. Their work showed the limits of efforts to shape "normal" masculinity and presaged later biological determinist theories about sexuality.

---

[7] At the annual meeting of the International Academy of Sex Research held in Chicago 7–10 August 2013, researchers at Northwestern University presented a poster entitled "Sexual Arousal Patterns and Their Relationship to Autogynephilia," while researchers at Queen's University presented a poster on "Gender Specificity of Men's and Women's Attention to Sexual Stimuli."

[8] Jason R. Odeshoo, "Of Penology and Perversity: The Use of Penile Plethysmography on Convicted Child Sex Offenders," *Temple Polit. & Civil Rights Law Rev.* 14 (2004): 1–44; Lawrence Ellerby, Robert J. McGrath, Georgia F. Cumming, Brenda L. Burchard, and Stephen Zeoli, *Current Practices in Canadian Sexual Abuser Treatment Programs: The Safer Society 2009 Survey* (Ottawa, 2010), http://www.publicsafety.gc.ca/cnt/rsrcs/pblctns/2010-02-sss/index-eng.aspx (accessed 10 May 2015); Tom Waidzunas, "Measuring Desire," *Cabinet*, Summer 2009, http://www.cabinetmagazine.org/issues/34/waidzunas.php (accessed 10 May 2015).

THE LEGAL AND MEDICAL PROBLEM OF HOMOSEXUALITY

Until the middle decades of the twentieth century, doctors attempted to treat homosexuality as both a biological defect and a psychological disorder. Those who believed
that homosexuality was a biological defect attempted to cure it by administering hormones, following in the tradition of Viennese endocrinologist Eugen Steinach, who
first suggested in the 1910s that hormonal imbalances could lead to abnormal sexual
behavior.[9] By the 1950s, however, two schools of psychological explanation had displaced biological theories of homosexuality: psychoanalysis and behaviorism. In the
1930s and 1940s, psychoanalysts, inspired by Sigmund Freud, had argued that homosexuality was the outcome of improper psychodynamic development, originating in
pathological parental relationships and unresolved childhood traumas. Although
Freud doubted that he could cure homosexuality, psychoanalysts following him, such
as Edmund Bergler, Irving Bieber, and Charles Socarides, asserted that they could
guide their patients to heterosexuality through verbal explorations and resolutions of
unconscious traumas. Other psychiatrists, often critics of psychoanalysis, adhered to
behaviorist theories suggesting that human sexuality was acquired through learning.
According to this school, which traced its intellectual heritage through the work of
physiologist Ivan Pavlov and psychologist John Watson, desirable behaviors had to be
inculcated through a system of positive and negative reinforcements.[10]

Freund was one of these behaviorist practitioners even though his thinking on
homosexuality also reflected a deep engagement with biology and psychoanalysis.
In his 1962 book, *Homosexuality in Man*, Freund discussed various etiologies of
homosexuality spanning the gamut from chromosomes, hormones, and twin studies
to paternal alcoholism and childhood trauma. He also drew upon the work of anthropologists and animal behaviorists and speculated on the possible relationship between
homosexuality and intersexuality. His far-ranging treatise reviewed and connected
the work of early twentieth-century sexology, psychoanalysis, and endocrinology to
midcentury surveys of sexual behavior and theories of socially acquired gender.[11]
Like many other physicians at the time, Freund viewed homosexuality as a pathology. He sketched a "clinical picture" of it and drew upon the work of sexologists like
Richard von Krafft-Ebing to describe abnormal characteristics of male homosexuals,
including a higher incidence of "femininity" as scored by personality questionnaires.
In the end, Freund favored the idea that homosexuality and heterosexuality existed on
a continuum and that a predisposition toward homosexuality was likely mediated by

[9] Nelly Oudshoorn, *Beyond the Natural Body: An Archaeology of Sex Hormones* (New York, 1994);
Stephanie Hope Kenen, "Scientific Studies of Human Sexual Difference in Interwar America" (PhD
diss., Univ. of California, Berkeley, 1998); Anne Fausto-Sterling, *Sexing the Body: Gender Politics and
the Construction of Sexuality* (New York, 2000); Chandak Sengoopta, *The Most Secret Quintessence
of Life: Sex, Glands, and Hormones, 1850–1950* (Chicago, 2006).

[10] Ronald Bayer, *Homosexuality and American Psychiatry: The Politics of Diagnosis* (New York,
1981); Vernon Rosario, "Rise and Fall of the Medical Model," *Gay and Lesbian Rev.*, November 2012,
39–41; Thomas Waidzunas, "Drawing the Straight Line: Social Movements and Hierarchies of Evidence in Sexual Reorientation Therapy Debates" (PhD diss., Univ. of California, San Diego, 2010).

[11] Freund's bibliography lists the work of well-known sexologists such as Havelock Ellis, Magnus
Hirshfeld, and Richard von Krafft-Ebing; psychoanalysts from Sigmund Freud to Karen Horney; en-
docrinologists Eugen Steinach and Frank Lillie; cytogeneticists Murray Barr, Charles Ford, Patricia
Jacobs, and Paul Polani; biologists Konrad Lorenz and Alfred Kinsey; and psychologists John Money
and Albert Ellis. For more, see Freund, *Homosexualita u muze* [Homosexuality in man] (Prague, 1962).

a combination of biological and social factors. If so, then it may be possible to cure homosexuality. "The essence of the medical problem," he stated, "is the prevalence of the homoerotic motivation over the heteroerotic."[12] He then gave a lengthy discussion of hormonal, surgical, and psychotherapeutic treatments of homosexuality and compared them to his own attempts to treat homosexuality through the methodologies of behaviorism.

When Freund joined the Department of Psychiatry at Charles University in Prague, Czechoslovakia, in 1948, homosexuality was not only a medical problem but also a legal one. Sodomy, which included sex between men, was illegal in Czechoslovakia along with many other European countries. Many of Freund's homosexual patients were referred to him through the legal system. Of the sixty-seven patients he had treated for homosexuality between 1950 and 1953, twenty of them had "dealings with the police, the magistrates, or other official agencies."[13] In some cases, legal authorities delayed or dropped prosecution and punishments so that the individual could undergo medical treatment. Medically curing homosexuality was thus posited as a judicious alternative to legal punishment. But how could researchers identify effective cures?

Freund doubted the verbal reports that patients offered about their sexuality. As he put it: "This fact is particularly disturbing in the therapeutic situation where the patients are inclined to deny certain facts in order to appear as cured and therefore to avoid further treatment, which is often regarded as undesirable by them." In his analysis of the efficacy of his treatment program for homosexuality, he purposely excluded all the patients who had found their way to him through the legal system since their claims of being cured were obviously biased. Instead, he focused on the forty-seven patients who had come to his clinic because they had been sent by relatives, experienced "unrequited homosexual love," or had no "obvious external pressure" for seeking treatment.[14]

In 1960, Freund summarized his efforts to treat homosexuality in an edited volume dedicated to learning theory. In the foreword, German-British psychologist Hans Jurgen Eysenck explicitly argued that behavior therapy was "superior" to the methods of Freudian psychotherapists in the treatment of everything from childhood phobias to obsessive-compulsive behaviors to alcoholism and writer's cramp. Like Eysenck, Freund dismissed psychoanalysis and reformulated all attempts to treat homosexuality in behaviorist terms. If it were possible to cure homosexuality through psychotherapy, he reasoned, interventions had to systematically "devaluate homoerotic desires and

---

[12] Ibid. References in the remainder of this essay are to the German translation: Freund, *Die Homosexualität beim Mann* (Leipzig, 1965). The quote here is my translation of a sentence taken from *Die Homosexualität beim Mann*. The relevant passage is on page 33: "Es ist wohl anzunehmen, daß die gleichzeitige Disposition zu homo- und heterosexueller Appetenz weit verbreitet ist, daß aber die homoerotische Motivierung weitaus schwächer ist als die heteroerotische. Mit anderen Worten: vorzuherrschen scheint zwar Bisexualität, aber zwischen den Valenzen der beiden Einstellungen besteht ein großer Unterschied. . . . Der Kern der ärztlichen Problemstellung ist aber das Überwiegen der homoerotischen Motivierung über die heteroerotische."

[13] Kurt Freund, "Some Problems in the Treatment of Homosexuality," in *Behaviour Therapy and the Neuroses: Readings in Modern Methods of Treatment Derived from Learning Theory*, ed. H. J. Eysenck (Oxford, 1960), 312–26, on 317. The data and conclusions from this paper were later reproduced in Freund's 1962 book, *Homosexuality in Man*. The tension between legal, medical, and religious authorities over who has jurisdiction over sexuality is also discussed by Zeb Tortorici, "Sexual Violence, Predatory Masculinity, and Medical Testimony in New Spain," in this volume.

[14] Freund, "Some Problems" (cit. n. 13), 312, 318.

associations and to encourage and reward heteroerotic desires and associations."[15] Between 1950 and 1953, he set up an experiment that would punish homosexual urges and reward heterosexual ones. The punishment portion consisted of injecting his patients with nausea-inducing, emetic drugs and then showing them slides of dressed and undressed men. Seven hours later in the second, reward-oriented portion, he injected his patients with testosterone and showed them pictures of nude and semi-nude women. This procedure was repeated every day for at least five days. He then conducted follow-up interviews with his patients in 1956 and again in 1958 in order to ascertain the long-term outcomes of this treatment program. Overall, the results were not encouraging.

Immediately following treatment, over half of the patients (twenty-four out of forty-seven) reported no improvement. A few relapsed to having sex with men after only a few weeks or months, and only a quarter of them reported a heterosexual "adaptation lasting for several years." Even in 1956, Freund admitted the limitations of his program. Many of his patients had married women, had regular intercourse with their wives, and even fathered children, but when he probed them about their desires, Freund discovered: "In all of these heterosexually adapted homosexuals the intensity of homoerotic desires admittedly overbalances that of heteroerotic desires, although some patients claim that homoerotic desires only occur infrequently."[16] Freund arrived at an anticlimactic conclusion. He conceded that his treatment, like other psychotherapeutic treatments, had only limited effectiveness in curing homosexuality.

Freund's research made him sympathetic to the plight that homosexuals faced socially and legally, and by 1957 he began to argue that the criminalization of sodomy was unnecessary. Eventually, he and other psychiatrists submitted an official recommendation of this position to the Czechoslovakian government, which then passed a legal amendment effectively decriminalizing same-sex intercourse between consenting adults in 1961.[17] Hungary enacted a similar law the same year. The decriminalization of sodomy constituted part of a larger project undertaken by Eastern European Communist governments at the time, to reform and modernize laws regarding sexual deviance to fit a more scientific framework.[18] Even if its legal status had changed, however, homosexuality still figured as a medical pathology, one that Freund sought to understand and diagnose with more empirical precision. Freund's effort to master homosexuality then shifted from therapy to diagnosis, and he looked to physiology to create a mechanical device that could quantify psychological desire.

A MACHINE TO DETECT DESIRE

Curiously, Freund did not mention applying the penile plethysmograph to any of the homosexual patients in his 1960 chapter, even though he had begun developing the device around 1953. In a candid review many years later, he acknowledged that he was not the first researcher to attempt phallometry as several European scientists had ex-

---

[15] Ibid., 316.

[16] Ibid., 319–25, on 318, 319.

[17] Freund, *Die Homosexualität beim Mann* (cit. n. 12), VII. The relevant paragraphs penalized only same-sex acts in which one partner was under the age of eighteen or legally dependent, as well as acts of prostitution or ones that caused a public nuisance.

[18] Lynne Viola, *Contending with Stalinism: Soviet Power and Popular Resistance in the 1930s* (Ithaca, N.Y., 2002), 147.

perimented with the idea earlier in the century. In 1947, P. Ohlmeyer and H. Brilmayer reported that they had invented a ring that could monitor the presence of erections during sleep.[19] More immediately, Josef Hynie, founder of the medical sexology institute at Charles University and Freund's senior colleague, had made attempts in the 1930s to construct a device for the purpose of studying men with erectile difficulties. The Czechoslovakian government also sought a scientific means to determine whether potential soldiers were falsely claiming to be homosexual to avoid conscription.[20] This pressing concern motivated Freund to create a device that would allow him to quantify his patients' unarticulated erotic preferences. He tinkered with setups that measured breathing patterns, galvanic skin responses, and heart rate before publishing a paper in a Czechoslovakian journal of psychiatry about his invention of the volumetric penile plethysmograph in 1957. Publications in more widely distributed English-language journals soon followed in the 1960s, drawing the attention of sexologists worldwide.[21]

During the 1960s, the plethysmograph constituted part of a larger effort by sexologists to construct devices that could quantify the physiological responses of sexual arousal. Historian Donna J. Drucker has chronicled how sexologists produced a dizzying array of machines for sex research around this time.[22] In the United States, for example, Virginia Johnson and William Masters invented a "penis-camera," a camera mounted inside a Plexiglas dildo that filmed inside a woman's vagina as she masturbated. This device produced the empirical evidence supporting Masters and Johnson's famous four-stage sexual response theory: excitement, plateau, orgasm, and resolution.[23] The heydays of the 1960s sexual revolution seemed to have prompted sexual experimentations both within and outside the laboratory. By the early 1970s, researchers had invented dozens of instruments to quantify blood pressure and heart rate changes, electrodermal conduction in various body parts, respiration, scrotal tightening, vaginal blood flow, uterine contractions, temperature, pupillary responses, biochemical hormones, cortical responses, and penile erections.

Among the myriad devices invented during this period, the penile plethysmograph held a privileged place for researchers who believed that it was the best gauge of male sexual preferences. Sexologist Marvin Zuckerman reviewed the various instruments used to measure the physiological aspects of human sexuality and argued that "only tumescence, vasodilation, genital secretions, and rhythmic muscular movements are characteristic of sexual arousal alone." Thus, he asserted that "in the presence of sexual stimulation, penile erection would seem to have some 'face validity' as a specific measure of sexual arousal in the male." For sexologists of the mid-twentieth century, the erect penis indicated masculine desire if it occurred in the presence of an erotic

[19] Kurt Freund, "Psychophysiological Assessment of Change in Erotic Preferences," *Behav. Res. Therapy* 15 (1977): 297–301.

[20] Kurt Freund, "Reflections on the Development of the Phallometric Method of Assessing Erotic Preferences," *Ann. Sex Res.* 4 (1991): 221–8.

[21] Glen Kercber, *Use of the Penile Plethysmograph in the Assessment and Treatment of Sex Offenders: Report of the Interagency Council on Sex Offender Treatment to the Senate Interim Committee on Health and Human Services and the Senate Committee on Criminal Justice*, January 1993; Kurt Freund, "Diagnostika homosexuality u muzu," *Ceskoslovenská psychiat.* 53 (1957): 382–94; Kurt Freund, J. Diamant, and V. Pinkava, "On the Validity and Reliability of the Phalloplethysmographic Diagnosis of Some Sexual Deviations," *Rev. Czech. Med.* 4 (1958): 145–51. Also, Ray Blanchard, personal communication with Nathan Ha, 10 May 2012.

[22] Donna J. Drucker, *The Machines of Sex Research: Technology and the Politics of Identity, 1945–1985* (Amsterdam, 2014), 1–18.

[23] William H. Masters and Virginia E. Johnson, *Human Sexual Response* (Boston, 1966).

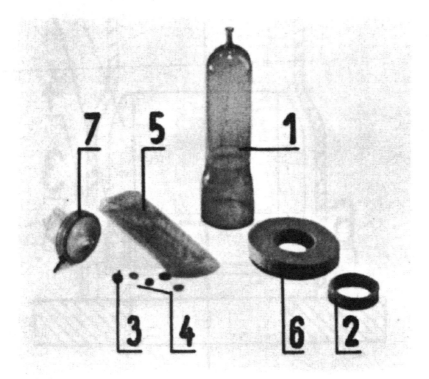

Fig. 1. Components of the transducer.
1 glass cylinder
2 plastic ring
3 metal tube with threads and perforated shield
4 locknut with washers and shield
5 thin rubber tube
6 flat soft sponge rubber ring
7 rubber cuff on the plastic ring

Figure 2. Diagram of apparatus designed to fit on the penis from Freund, Sedlacek, and Knob, "A Simple Transducer" (cit. n. 25).

stimulus. They therefore invested great efforts to quantify erections even though they knew that erections could occur unconsciously during sleep or because of friction, general excitement, and anxiety. Despite these confounding variables, Zuckerman believed that phallometers were "highly reliable and discriminating" and optimistically predicted that they would become the "methods of choice in future studies."[24]

Since Freund's device appealed greatly to sexologists, he published several papers over the course of the 1960s that taught other researchers how to build penile plethysmographs in their own labs. The physical apparatus consisted of a glass cylinder fitted around the penis with a cuff and attached to a volumetric recorder (fig. 2). Proper

[24] Marvin Zuckerman, "Physiological Measures of Sexual Arousal in the Human," *Psychol. Bull.* 75 (1971): 297–329, on 313.

assembly required tactful handling of bodies and equipment. As Freund and his co-authors instructed, "the sponge rubber ring acting as a pad for the glass cylinder, is fitted on the genital. Next apply the plastic ring with the elastic cuff and affix thereon the glass cylinder in such a manner that the piece of metal tube will protrude from the opening near its bottom. The glass cylinder is attached to the body of the patient with straps." After the cylinder was secured, the elastic cuff was inflated to form an airtight seal at the base of the penis. The cylinder's open tip allowed air to be displaced when the penis became erect. This tip was attached to a recorder that produced a line graph of the changing volume in the cylinder as the subject's erections waxed and waned.[25]

By examining the line graph produced from the recorder, a numerical "response" value was determined by averaging the peaks and valleys produced when the subject looked at pictures of naked males and females. This enabled researchers to define sexuality in purely operational terms—as quantified penile volume responses to the stimuli of nude male or female body forms. If the average for pictures of naked males was higher than the average for pictures of naked females, the subject was diagnosed as homosexual, and if the reverse, the subject was diagnosed as heterosexual. In some setups, Freund also repeated pictures and compared the "responses" to male and female pictures that occurred near the same time.[26] Most of his later refinements involved efforts to tinker with the stimuli given to his research subjects. He changed the number, order, and distribution of the photographs in the slideshow to ensure that his results were valid and reproducible. He also included photographs that focused on specific body regions. Adding pictures, however, increased the duration of the test to several hours, fatiguing some of his subjects. To ensure his subjects in later studies would pay attention, he compelled them to respond to a system of flickering lights and gave them caffeine and wine before each experiment. Finally, some of the men experienced inadequate erections, perhaps due to age or the anxiety created by the testing situation, so on the day before the examination, Freund began giving shots of testosterone to men over the age of twenty-five.[27]

Freund was thus well aware of the phallometer's limitations. According to him, the device could distinguish heterosexuals from homosexuals in the vast majority of cases, but it was possible to fool the machine. When Freund asked a group of self-identified heterosexual and homosexual men to fake their responses, he found that almost 17 percent of them (eleven out of sixty-six) could get the device to give a result contrary to their actual sexual preferences, especially if they had previous experience with it. Freund acknowledged this shortcoming but continued to advocate the general utility of the apparatus. It was useful for diagnosing not only homosexuality, he claimed, but pedophilia as well.[28]

[25] K. Freund, F. Sedlacek, and K. Knob, "A Simple Transducer for Mechanical Plethysmography of the Male Genital," *J. Exp. Anal. Behav.* 8 (1965): 169–70, on 170.

[26] Kurt Freund, "Laboratory Differential Diagnosis of Homo- and Heterosexuality: An Experiment with Faking," *Rev. Czech. Med.* 7 (1961): 20–31.

[27] Kurt Freund, "A Laboratory Method for Diagnosing Predominance of Homo- or Hetero- Erotic Interest in the Male," *Behav. Res. Therapy* 1 (1963): 85–93; Freund, Sedlacek, and Knob, "A Simple Transducer" (cit. n. 25); Freund, "Diagnosing Heterosexual Pedophilia by Means of a Test for Sexual Interest," *Behav. Res. Therapy* 3 (1965): 229–34.

[28] Freund, "A Laboratory Method" (cit. n. 27); Freund, Sedlacek, and Knob, "A Simple Transducer" (cit. n. 25).

ARTICULATING EROTIC PREFERENCES

Even though his initial goal had been the diagnosis of homosexuality, Freund's phallometric test already included pictures of adults and children in the 1950s. Since Freund worked at a state institution, many of his earliest experimental subjects were psychiatric patients who had been sent there for engaging in sexual activities with adolescents and children. Thus, Freund found it logical to expand the utility of his device from the diagnosis of homosexuality to the diagnosis of pedophilia, uniting both under the common rubric of sexual abnormalities. By also subjecting pedophiles to the phallometer and then overlaying preferences for age-specific body types across a primary axis of sex-specific preferences, Freund created a scale on which he could rank typologies of erotic deviance.[29]

The result was the intercalation of preferences for age with those for sex. To make the phallometric test sensitive for the detection of pedophilia, Freund added more pictures of nude children, between four and thirteen years old, and adolescents, between thirteen and seventeen years old. In previous setups, only a few pictures of children and adolescents appeared; now they constituted two-thirds of the images. Freund built symmetry into his research design, ensuring that two sexes (male and female) and three age groups (children, adolescents, and adults) were equally represented in the sixty images shown to subjects over the course of three hours. He believed that he could now simultaneously ascertain erotic preferences for age and sex.[30]

By experimentally conjoining sex and age preferences, Freund drew pedophilia and homosexuality closer together even as he attempted to draw distinctions between them. He used a group of alcoholics as nonpedophiliac "heterosexual controls" to validate his setup and then tested 130 subjects grouped into five categories: heterosexual and homosexual pedophiliacs, ephebophiliacs (men who preferred adolescent males), androphiliacs (men who preferred adult males), and normals (men who preferred adult women). The very naming of these groups made clear what was considered "normal" and "abnormal" masculinity. Furthermore, by comparing these different groups, Freund hoped to gain insights into the etiology of pedophilia. Did pedophiles prefer children because they were aroused by young bodies in general, or was it because they had an aversion to the full development of sex-specific characteristics in adults?

Freund thought that comparing the erectile responses of homosexuals and heterosexuals with different age-group preferences would yield answers by revealing minute differences in these cryptic preferences and thus shed light on the underlying nature of pedophilia. He discovered that heterosexual pedophiles were most aroused by pictures of female children and then, in order from most arousing to least, female adolescents, adult women, male children, male adolescents, adult men (see fig. 3). Although sexual preferences seemed to trump age preferences in general, the data also showed that preferences for specific age groups were conserved, even in the nonpreferred sex. Similar patterns of preference were repeated in all other groups, except for "normals," who found all males, regardless of age, equally unattractive. These results led Freund

---

[29] Freund, Diamant, and Pinkava, "Phalloplethysmographic Diagnosis of Some Sexual Deviations" (cit. n. 21).
[30] Freund, "Diagnosing Heterosexual Pedophilia" (cit. n. 27).

RESULTS

Figure 1 shows the mean ranks of picture types for each group,

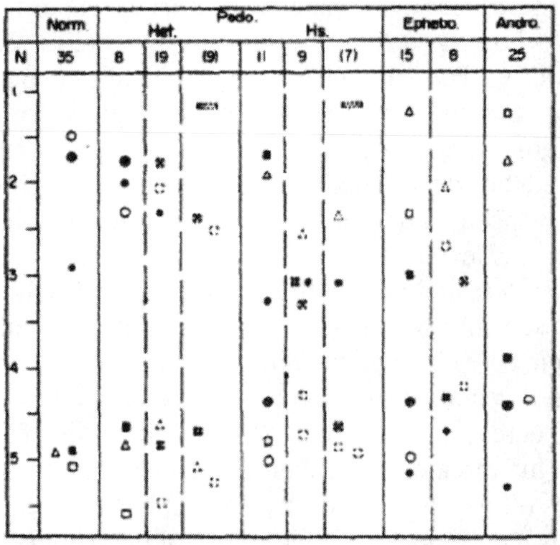

FIG. 1. Preference order of picture-types. Vertical axis, mean rank in terms of response. Column 1, normal controls; column 2, nonpretending heterosexual pedophiliacs; column 3, heterosexual pedophiliacs pretending erotic preference for adult women; column 4, heterosexual pedophiliacs with whom pictures of female children had rank 1 or shared first place with other type; column 5, nonpretending homosexual pedophiliacs; column 6, homosexual pedophiliacs pretending erotic preferences for adult women; column 7, homosexual pedophiliacs with whom pictures of female children had rank 1 or shared first place with other type; column 8, nonpretending ephebophiliacs; column 9, ephebophiliacs pretending erotic preference for adult women; column 10, androphiliacs. Row, number of subjects. Numbers in brackets indicate subset of preceding two columns.

■, male children; △, male adolescents; □, male adults; ●, female children; ⊗, female adolescents; ○, female adults.

Figure 3. *Hierarchy of erotic preferences of normal men, homosexual and heterosexual pedophiles, ephebophiles, and androphiles from Freund, "Erotic Preference" (cit. n. 31).*

to conclude that pedophiliac attraction was motivated by a preference for sexually immature characteristics and not an aversion to sex-specific adult characteristics.[31]

Ironically, Freund's paper ended up having more to say about hetero- and homosexuality than it did about pedophilia. "Normal" men found female children somewhat arousing, whereas androphiliacs did not find male children arousing at all. This led Freund to a series of striking conclusions regarding the etiology of homosexuality. Since androphiliacs found adult and adolescent women more attractive than female children, this refuted the psychoanalytic hypothesis that men turned toward homosexuality because they feared and loathed women, who symbolized the threat of castration embodied by the colorful phrase, "vagina dentata." Also, Freund's study showed that homosexual pedophiles demonstrated moderate arousal when exposed to pictures of female children, but heterosexual pedophiles found pictures of male

[31] Kurt Freund, "Erotic Preference in Pedophilia," *Behav. Res. Therapy* 5 (1967): 339–48.

children less arousing. "This result may be interpreted if," Freund wrote, "as a product of social adaptation, heterosexuals learn to inhibit as much as possible their erotic consideration of their own sex, and if homosexuals, in contrast, through a similar process come to enhance as much as possible their erotic feelings regarding the opposite sex."[32] Instead of revealing truths about the nature of pedophilia, the results had revealed more about the social factors involved in the development of hetero- and homosexuality.

By intercalating preferences for age and sex, Freund had generated a hierarchical scale upon which increasingly specific erotic tastes could be organized. This scale yielded data that challenged some Freudian theories and also suggested disturbing findings regarding "normal" masculinity. If the tracings were credible, then many "normal" men concealed an underlying sexual attraction to naked female children. What social, medical, and legal implications were there to this discovery? Freund did not ask this question, and the plethysmograph itself did not provide obvious answers. He believed that the device's tracings only represented a physiological response that could help him to divide and subdivide classes of normal and deviant sexuality. This articulation of taboo sexual preferences was ironic because it simultaneously made an analytical distinction between homosexuality and pedophilia while interweaving the two in a common rubric of sexual deviance. It also assumed that unspoken sexual desires could be articulated by asking the penis to testify on behalf of the mind. These ironies would multiply as other researchers adopted phallometry in their own labs.

MEASURING AND CONTROLLING DEVIANCE

During the 1960s, both Freund and phallometry would travel far beyond Czechoslovakia. In 1969, he emigrated to Canada after the Soviet Union and other Warsaw Pact countries invaded Czechoslovakia in 1968 and brought an end to the Prague Spring. As for the plethysmograph, it inspired several imitations. Researchers in the United States, the United Kingdom, and Australia created different renditions of phallometric devices that could be produced from locally available materials. Some, for example, measured changes in penile circumference instead of volume, which made these devices cheaper to produce and less cumbersome to wear. The erections of men could now be monitored so that an electric shock could be applied at the point of maximum arousal. The analogy between physiology and psychology grew increasingly narrow as the inscriptions of various phallometers became more than just shorthand signs of arousal. During a period when sexual deviance was a major medical, legal, and cultural concern, adopters of phallometry believed that they could use it to diagnose, shape, and control male sexual desire.[33]

By the 1960s, both Canada and the United States had enacted laws that shifted jurisdiction over the management of sex offenders from legal authorities to psychiatrists. Instead of confinement in penal institutions, sex offenders who were designated as "sexual psychopaths" could be sentenced to treatment in a psychiatric institution indefinitely. As historian Estelle Freedman has argued, the passage of these laws in the

---

[32] Ibid., 348. For more, see Kurt Freund, "Diagnosing Homo- or Heterosexuality and Erotic Age-Preference by Means of a Psychophysiological Test," *Behav. Res. Therapy* 5 (1967): 209–28.

[33] Concerns about adequate masculinity also figured prominently in the Soviet Union, even if desire was not always the main focus. See Frances Bernstein, "Prosthetic Manhood in the Soviet Union at the End of World War II," in this volume.

United States was closely tied to a series of sex crime panics that occurred between 1930 and 1960. Anxious about the economic uncertainties created by the Great Depression and then paranoid about the threat of Communism fomented by the Cold War, Americans turned away from a previous obsession with deviant female sexuality and became increasingly concerned about potentially dangerous men lurking on the fringes of society. Unemployed hobos, stealthy Communists, seductive homosexuals, predatory pedophiles, and murderous rapists all figured in the collective imagination as threats requiring containment and surveillance. During the 1930s and again in the late 1940s, the media, politicians, and private citizens' groups whipped up a frenzy about sex crimes committed against women and children. In response, many states passed laws that codified the "sexual psychopath" as a dangerous individual who could not restrain his sexual impulses. They authorized psychiatrists to treat sex offenders and transformed the sex criminal into a sexual patient.[34]

Canada experienced a similar storm of media coverage about violent sex crimes and imported the sexual psychopath model from the United States after World War II. In 1948, the Canadian Parliament passed legislation creating the category of the sexual psychopath and recommended that these individuals be treated in separate psychiatric institutions. Historian Elise Chenier has argued that the passage of this legislation culminated a century-long trend toward the psychiatrization of deviant sexuality motivated by progressive politics. The legislative process initiated years of public forums and committee hearings about sexual deviance that introduced many Canadians to the details of sexology. Combined with the media attention given to Alfred Kinsey's sexual behavior surveys, published in 1948 and 1953, sex became a legitimate topic of public discussion, and sexologists became recognized experts in formulating the legal, medical, and moral boundaries of sexual norms.[35]

Specifically, the Canadian government called upon sexologists to help them identify homosexuals during the Cold War. According to the logic of the period, homosexuals were national "security risks" because they engaged in illegal sexual activities (sodomy) that made them vulnerable to blackmail. In the United States, the State Department forced suspected homosexuals to submit to a polygraph test. During the 1950s, an estimated 1,000 men and women lost their jobs at the State Department and 5,000 in the federal government overall.[36] Similar purges occurred in Canada, where psychologist F. R. Wake advised the Royal Canadian Mounted Police (RCMP) on the development of a test that indicated sexual arousal by measuring pupil dilation when subjects viewed erotic pictures of male fitness models. This device was derisively referred to as the "fruit machine." However, the setup for the "fruit machine" proved difficult to standardize for different heights and pupil distances, and the project was discontinued in 1967.[37]

[34] Freedman, "'Uncontrolled Desires'" (cit. n. 6).

[35] Elise Chenier, "The Criminal Sexual Psychopath in Canada: Sex, Psychiatry and the Law at Mid-Century," *Can. Bull. Med. Hist.* 20 (2003): 75–101.

[36] Ken Alder, *The Lie Detectors: The History of an American Obsession* (New York, 2007), 215–28. For more on what has been called the "Lavender Scare," see David K. Johnson, *The Lavender Scare: The Cold War Persecution of Gays and Lesbians in the Federal Government* (Chicago, 2006); Robert D. Dean, *Imperial Brotherhood: Gender and the Making of Cold War Foreign Policy* (Amherst, Mass., 2001).

[37] Gary Kinsman, "'Character Weaknesses' and 'Fruit Machines': Towards an Analysis of the Anti-Homosexual Security Campaign in the Canadian Civil Service," *Labour/Le Travail* 35 (1995): 133–61; Gary William Kinsman and Patrizia Gentile, *The Canadian War on Queers: National Security as Sexual Regulation* (Vancouver, 2010).

By the time Freund arrived in Canada, the RCMP had closed its program and the government would soon decriminalize homosexuality, so it does not appear that the plethysmograph was used in Canada's surveillance program against homosexuals. Moreover, there is no evidence that Freund conducted aversive conditioning treatments for homosexuality after the 1950s. However, efforts to discern and cure homosexuals using science remained strong in other countries, where the penile plethysmograph inspired at least three emulations.

SHAPING NORMAL MASCULINITY

For many researchers during the mid-twentieth century, phallometry presented an opportunity to develop a new technology to ascertain and inculcate appropriate sexual preferences. A fundamental assumption of this paradigm was that sexual attraction to adult women was a crucial component of adequate masculinity. Yet the phallometer showed that most men could potentially respond to a hierarchy of preferred stimuli. Some researchers then came to the conclusion that they could intervene to rearrange this hierarchy and teach their subjects "normal" erotic preferences.[38] Extending Freund's work, many researchers believed that phallometry could be used not only to diagnose sexual deviance but also to monitor the effectiveness of treatments (e.g., positive and aversive conditioning) in the short and long terms.

In 1966, British psychiatrist John Bancroft, then a graduate student at London's St. George's Hospital, reported the design of a phallometric device consisting of an elastic ring filled with mercury.[39] This ring was fitted around a penis, and an attached transducer recorded changes in the electromagnetic field that occurred as erections stretched the ring (fig. 4). Bancroft claimed that his device was easier to use and less "cumbersome" than Freund's apparatus, but Bancroft's design was also open to the critique that it measured only penile circumference and not volume and hence offered a less complete assay of tumescence. A year later, Australian psychiatrist Nathaniel (Neil) McConaghy published a paper about his phallometer. McConaghy's device was a simplification of Freund's original volumetric design. Instead of a glass cylinder and a cuff, McConaghy used a cylindrical tin and stretched a finger stall over the open end to form an airtight seal around the penis. Finally, American psychologist David Barlow invented a circumferential phallometer that modified Bancroft's design. Barlow found the mercury gauge too "awkward" and replaced it with a strain gauge. This gauge contained an elastic piece of foil with an electrical resistance that could be monitored as the penis expanded or contracted. Each of these renditions was inspired by Freund's plethysmograph but varied according to their local contexts. Sometimes the variation was due to the availability of materials. It was expensive, for example, to get a glass blower to produce the same kind of penile cylinder that Freund used. Other times, researchers favored designs that increased the comfort and convenience of their experimental setups.[40]

---

[38] The expectation that bodies and minds should cohere with social norms of sex, gender, and sexuality is a common historical theme. For a much earlier rendition of medical interventions to "fix" abnormal sex/gender, see Leah DeVun, "Erecting Sex: Hermaphrodites and the Medieval Science of Surgery," in this volume.

[39] Bancroft had an illustrious career in sexology, rising eventually to become the director of the Kinsey Institute from 1995 to 2004.

[40] J. H. J. Bancroft, H. Gwynne Jones, and B. R. Pullan, "A Simple Transducer for Measuring Penile Erection, with Comments on Its Use in the Treatment of Sexual Disorders," *Behav. Res. Therapy* 4

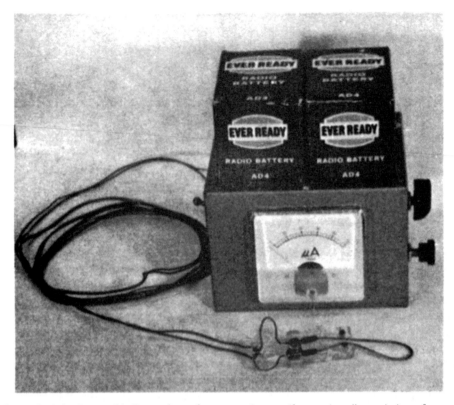

Figure 4. *John Bancroft's "transducer for measuring penile erections," consisting of an adjustable mercury-filled silicone tube. Bancroft, Jones, and Pullan, "A Simple Transducer" (cit. n. 40).*

All three of these researchers and their collaborators followed a similar path in deploying their phallometers to diagnose, treat, and assess changes in homosexuality. Phallometric adopters believed that the diagnosis of homosexuality had always been plagued by nonverifiability and nonspecificity. They applauded Freund for creating an innovation that could resolve these long-standing dilemmas. Bancroft wrote that the measurements provided by phallometers provided "objective data" that was better than Kinsey's "crude" scale that assigned a number between 0 (completely heterosexual) and 6 (completely homosexual) depending upon a man's self-reported sexual fantasies and behaviors. Likewise, Barlow reiterated that the advantage of measuring penile circumference change was its "specificity." Penile erections were a "necessary precursor to any consummatory behavior," and their measurement in response "to different stimuli should therefore be a valid indicator of the potential for further sexual activity involving that object." In this interpretation, Barlow justified the measurement

(1966): 239–41; N. McConaghy, "Penile Volume Change to Moving Pictures of Male and Female Nudes in Heterosexual and Homosexual Males," *Behav. Res. Therapy* 5 (1967): 43–8; D. H. Barlow, R. Becker, H. Leitenberg, W. S. Agras, "A Mechanical Strain Gauge for Recording Penile Circumference Change," *J. Appl. Behav. Anal.* 3 (1970): 73–6.

of erections as a proxy for sexual preferences since penile penetration of a sex object was the usual telos of masculine desire.[41]

After arguing that their phallometric devices were just as good as Freund's plethysmograph in diagnosing sexual preference, researchers quickly turned their attention to treatments for homosexuality. Here the attempted treatments could be divided into two classes: negative and positive conditioning. Negative or aversive conditioning treatments, of course, were not new since Freund himself had conducted such treatments with emetics during the 1950s. During the 1960s, however, other researchers transitioned to electroshock therapy because the effects of electric shocks were more immediate than the nausea induced by emetics. British psychologists M. P. Feldman and M. J. MacCulloch, for example, attempted to treat homosexuality by a technique known as "anticipatory avoidance learning." In this setup, the researchers projected pictures of men rated as attractive by subjects onto a screen and then gave them an electric shock. This painful shock would continue until the subject pushed a switch that removed the slide. Over time, it was hoped that the subjects would eradicate the sexual attraction they felt for men. Feldman and MacCulloch compared their experimental design to successful attempts made by other researchers to train dogs to develop an aversion to meat.[42]

Bancroft appears to have been impressed by Feldman and MacCulloch's results, especially their claim that 57 percent of their subjects had learned to diminish their homosexual attractions. He had the innovative idea, however, of combining electroshock treatments with phallometry. This would allow experimenters to monitor the erections of homosexuals and pedophiles and control the delivery of electric shocks. In 1966, he related the case history of a twenty-five-year-old pedophile who had been recommended for a lobotomy because of his sexual inclinations. Instead, he underwent Bancroft's treatment program. Five times a day for two months, Bancroft attached the phallometer to the man's penis and gave him pictures of children to fantasize about. Whenever the man's penis grew erect, Bancroft administered painful electrical shocks to the man's arm until the erection flagged. He was then given pictures of adult women to fantasize about. When the patient returned for a checkup eighteen months later, Bancroft declared success. He noted that the man had learned to desire adult women instead of children and had even married his former girlfriend. Proudly, Bancroft claimed that the man had so far avoided a lobotomy and "no longer runs the risk of being convicted."[43]

Bancroft also conducted similar aversive conditioning treatments for homosexuals, but his results here were equivocal. In 1969, he reported that he had applied shock treatments to ten men while monitoring their erections with a phallometer. Although

[41] John Bancroft, "Aversion Therapy of Homosexuality A Pilot Study of 10 Cases," *Brit. J. Psychiat.* 115 (1969): 1417–31; Barlow et al., "Mechanical Strain Gauge" (cit. n. 40).

[42] M. P. Feldman and M. J. MacCulloch, "The Application of Anticipatory Avoidance Learning to the Treatment of Homosexuality: 1. Theory, Technique and Preliminary Results," *Behav. Res. Therapy* 2 (1964): 165–83; M. P. Feldman, "Aversion Therapy for Sexual Deviations: A Critical Review," *Psychol. Bull.* 65 (1966): 65–79. Historian Vernon Rosario notes that electroshock and emetics had been used to treat schizophrenia in the 1930s and 1940s but that their application to treating homosexuality was uncommon at the time. See Rosario, ed., *Science and Homosexualities* (New York, 1997), 97.

[43] Bancroft, Jones, and Pullan, "A Simple Transducer" (cit. n. 40), 240. See also J. Bancroft and I. Marks, "Electric Aversion Therapy of Sexual Deviations," *Proc. Roy. Soc. Med.* 61 (1968): 796–9; Bancroft, "Aversion Therapy" (cit. n. 41).

the treatment decreased the homosexual attraction that seven of these men reported feeling, this effect was "long lasting" for only three individuals.[44] In a follow-up paper, Bancroft also acknowledged that reductions in homosexual erections in treatment settings did not necessarily lead to reduced homosexual behavior in real life. Even worse, when it came to the treatment of a homosexual masochist, the application of electric shocks actually increased the patient's erections. For other homosexuals, Bancroft found that for some inexplicable reason it was better to apply electric shocks before subjects developed a full erection in response to homosexual stimuli since this would increase the likelihood that they would develop erections in the presence of heterosexual stimuli. Ultimately, Bancroft remained optimistic about the utility of phallometry but became uncertain about the effectiveness of aversive conditioning for altering sexuality.[45]

As this skepticism grew, researchers designed new setups where phallometry could be used to assay the effectiveness of positive conditioning treatments. In positive conditioning, researchers tried to encourage and inculcate heterosexual desires instead of punishing homosexual ones. The intellectual justification for positive conditioning ironically came from interpretations of Freud's data indicating that men could control their erections and fool phallometers. American psychologist Richard Laws, for example, relied upon phallometry (using a self-built mercury strain gauge) to demonstrate that penile erections were not completely involuntary reflexes under the control of the autonomic system and were also subject to voluntary control. To prove this, he asked heterosexual men to inhibit their erections while being shown an erotic film and to develop erections in the absence of the film. With varying degrees of success, they were able to perform both activities. Laws thus concluded optimistically that since voluntary control of erections was possible, "patients with sexual problems might be able to learn to control their erections if they were trained to concentrate on specific thoughts at appropriate times."[46]

The rhetoric of control underpinned positive conditioning therapies that tried to teach homosexual men to develop heterosexual desires by two different techniques, "shaping" and "fading." "Shaping" itself could take different forms. In one form practiced by Bancroft, patients were taught to alter deviant fantasies incrementally toward normal ones. Researchers would monitor their erections and let them know if their erections flagged. Patients could then pause at graduated steps and ensure that they were fully aroused before modifying their fantasies further. Eventually, subjects were expected to achieve full erections with completely normal fantasies.[47] In the hands of other researchers, "shaping" resembled negative conditioning more than positive. British psychologist J. T. Quinn and his collaborators, for example, attempted to "shape" a homosexual man's preferences by creating a state of water deprivation. They denied him all fluids for eighteen hours and then required him to ingest salt and a diuretic. After affixing a phallometer and showing him pictures of an attractive

[44] Bancroft, "Aversion Therapy" (cit. n. 41), 1430.

[45] John Bancroft, "The Application of Psychophysiological Measures to the Assessment and Modification of Sexual Behaviour," *Behav. Res. Therapy* 9 (1971): 119–30.

[46] D. R. Laws and H. B. Rubin, "Instructional Control of an Autonomic Sexual Response," *J. Appl. Behav. Anal.* 2 (1969): 93–9. His paper mentioned, however, that he started with seven subjects and only four were able to produce any erection when shown the erotic film. So he excluded three subjects from the final paper.

[47] Bancroft, "Assessment and Modification of Sexual Behaviour" (cit. n. 45).

woman, they told him to fantasize about having sex with this woman. He was also told "that his phallic blood flow would be monitored and as a result of a change in this he would receive a reward (a drink of iced lime)." Over time, researchers increased the volume change required to earn a reward and claimed that they had "shaped" an overall augmentation of his heterosexual desire.[48]

Another class of positive conditioning treatments was known as "fading." Here, subjects were shown pictures of nude men and women superimposed upon one another. At first, homosexual subjects were allowed to achieve full erections while looking at clear pictures of a nude man. Researchers then slowly blurred the image of the naked man while resolving the picture of the naked woman. If the phallometer indicated a flagging erection, researchers paused the blurring process and allowed the subject to regain a full erection. The procedure continued until the subject could achieve adequate erections while looking at pictures of naked women. Barlow reported modest success with "fading" as a treatment for homosexuality and optimistically opined that it could show that "aversive techniques may not be necessary in the treatment of sexual deviation."[49] Similarly, Laws also applied a similar "fading" procedure to treat pedophilia but reported that he had encountered disappointing results; his subjects had demonstrated few alterations of their sexual preferences.[50]

Perhaps the most enthusiastic adopter of phallometry was Australian psychiatrist Nathaniel McConaghy, who used it to conduct multiple studies of aversive and positive conditioning treatments for homosexuality. Importantly, McConaghy also used phallometry to assess the long-term success of treatments. Starting in the late 1960s and continuing until the mid-1970s, McConaghy attempted a variety of methods (including many discussed above) to cure his patients. These included conditioning protocols in which the subjects were taught to associate nausea with male nudes or the cessation of electric shock with female nudes. He also tried pairing male and female nudes together and exposed his subjects to depictions of heterosexual sex to elicit positive learning experiences. Before and after the completion of each treatment program, McConaghy would use phallometry to see whether a change in sexual orientation had occurred. By 1977, however, McConaghy reported that his efforts had not been successful. Most of his patients claimed decreased homosexual desire and behavior, but McConaghy remained skeptical. The plethysmograph still scored most of them as homosexuals. Although they had learned to suppress the desire to engage in sex with other men, McConaghy admitted that he had merely created cases of "experimental neurosis." He had taught them new associations of pain and avoidance, but this had not changed who they were. Ultimately, he concluded that "present treatments may reduce or eliminate patients' homosexual behavior and awareness of homosexual feeling without altering their sexual orientation."[51]

Over the course of the 1960s and 1970s, researchers around the globe had adopted

[48] J. T. Quinn, J. J. M. Harbison, and H. McAllister, "An Attempt to Shape Human Penile Responses," *Behav. Res. Therapy* 8 (1970): 213–6.

[49] D. H. Barlow and W. S. Agras, "Fading to Increase Heterosexual Responsiveness in Homosexuals," *J. Appl. Behav. Anal.* 6 (1973): 355–66.

[50] D. R. Laws and A. V. Pawlowski, "An Automated Fading Procedure to Alter Sexual Responsiveness in Pedophiles," *J. Homosexuality* 1 (1976): 149–63.

[51] N. McConaghy, "Is a Homosexual Orientation Irreversible?" *Brit. J. Psychiat.* 129 (1976): 556–63, on 563. For more, see R. F. Barr and N. McConaghy, "Penile Volume Responses to Appetitive and Aversive Stimuli in Relation to Sexual Orientation and Conditioning Performance," *Brit. J. Psychiat.* 119 (1971): 377–83.

phallometry enthusiastically. Unlike Freund, they were less interested in using phallometry to organize hierarchies of sexual preference. Rather, Bancroft, Barlow, and McConaghy came up with creative experimental setups using phallometry to train men to decrease their homosexual and increase their heterosexual impulses. As practitioners of both negative and positive conditioning, these behaviorists believed that sexual preferences could be measured, and more importantly, taught. Phallometry was their tool of choice in the surveillance and control of abnormal desire. Ironically, this tool produced evidence that disappointed researchers. Instead of demonstrating the effectiveness of treatments, phallometers indicated the intractability of sexual preferences.

SUSTAINING HOMOSEXUALITY

Over time, phallometers left tracings that forced scientists to consider the existence of a sexuality that seemed potentially modifiable but was in fact immutable and "irreversible." During the 1950s and 1960s, for example, Freund wrote mostly of erotic preferences and rarely used the language of sexual orientation.[52] His choice of terms underscored a tension between the rigidity and plasticity that could exist within a presumed hierarchy of erotic preference. In the 1970s, he made this tension explicit when he formally defined homosexuality and heterosexuality.

> Homosexuality is the sustained erotic preference for same sexed persons when there is a virtually free choice of partner as to sex and to other attributes which may co-determine erotic attractiveness. In this definition, the term "sex" denotes male type or female type of externally visible gross somatic features ("body shape"), particularly the type of external genitalia. . . . The terms homo- and heterosexuality, according to this definition, denote only an erotic preference for body shape and not a preference for the type of sexual behavior of a potential partner or for the type of one's own preferred sexual behavior.[53]

What is remarkable about this passage are the implications of the word "sustained." During the 1960s, Freund had defined homosexuality in relative terms, as the preponderance of homoeroticism over heteroeroticism, implying that the balance between the two could be shifted in any given individual. His definition of homosexuality in the 1970s, on the other hand, implied that sexual behaviors could shift but that ideal sexual preferences for male or female body forms appeared to remain constant. What prompted this subtle shift? One hypothesis is that he already had some insights into the immutable quality of erotic preferences during the early 1960s but had not yet fully developed these ideas. Another hypothesis is that the collective work of researchers who had adopted phallometry led them to similar conclusions about the intractability of sexuality. A third hypothesis is that the late twentieth century witnessed a shift of intellectual currents away from behaviorism and toward biological determinism.[54]

---

[52] One example where he did use the term "orientation" was in 1963. See Freund, "A Laboratory Method" (cit. n. 27).

[53] Kurt Freund, "Male Homosexuality: An Analysis of the Pattern," in *Understanding Homosexuality: Its Biological and Psychological Bases*, ed. John A. Loraine (New York, 1974), 25–81, on 25.

[54] Garland E. Allen, "The Double-Edged Sword of Genetic Determinism: Social and Political Agendas in Genetic Studies of Homosexuality, 1940–1994," in Rosario, *Science and Homosexualities* (cit. n. 42), 242–70.

Indeed, the 1970s did mark a renewed interest in the moral lessons that could be drawn from ethology. As historian Erika Lorraine Milam demonstrates in this volume, social commentators during this period often relied upon studies of animal behavior to naturalize social concepts of male aggression.[55] Freund, who had long been interested in both ethology and endocrinology, paired these lessons with hormonal studies suggesting that androgens, like testosterone, affected early brain development and predisposed males to become more aggressive than females, in terms of both gender and sexual behavior. "Male type mating behaviour," he suggested, was "related to fight threat and dominance," whereas "female type mating behaviour" was related to "flight, retreat, [and] submission." This general pattern applied to many animals, including humans, and Freund remained optimistic that searching for the biological components underlying this pattern would yield a "wider and firmer base for the masculinity-femininity and gender identity concepts than is at present available."[56]

Simultaneously, Freund and his phallometric colleagues engaged with political and cultural movements that reordered medical categories of normal and abnormal sexuality. In 1973, the American Psychiatric Association decided to remove homosexuality from the *Diagnostic and Statistical Manual of Mental Disorders* (DSM). This event culminated years of effort by homosexual activists as well as sympathetic mental health professionals who critiqued the classification of homosexuality as an illness.[57] A couple of years after this controversial decision, the Association for the Advancement of Behavior Therapy held a symposium on "Homosexuality and the Ethics of Behavioral Treatment" that questioned the ethics of therapies aimed at curing homosexuality, especially if such therapies were requested by patients themselves. The *Journal of Homosexuality* published the proceedings and invited Kurt Freund, Neil McConaghy, and other researchers to comment. McConaghy held the position that it was unethical for physicians to deny patient requests for treatment based upon the physician's own moral judgments and drew an analogy to medical treatments for abortion.[58]

Freund issued a remarkable apology for his previous attempts to cure homosexuality. "I am not happy about my therapeutic experiment," he averred, "which, if it has 'helped' at all, has helped clients to enter into marriages that later became unbearable or almost unbearable." With great sensitivity, he argued that it was pointless to bicker about whether or not the removal of homosexuality from the DSM was motivated by political pressure or scientific evidence. The role of a physician, Freund suggested, was to help those who are in distress, and since homosexuals had to endure societal disfavor and oppression, Freund advised doctors to "first wait to see whether or not reasonable social changes may virtually abolish all specific distress of homosexual persons." In the meantime, he recommended counseling homosexuals toward self-acceptance because "there is no 'cure'" that would turn homosexuals into heterosexuals.[59]

[55] Erika Lorraine Milam, "Men in Groups: Anthropology and Aggression, 1965–75," in this volume.

[56] Freund, "Male Homosexuality," (cit. n. 53), 64.

[57] For details, see Bayer, *Homosexuality* (cit. n. 10).

[58] N. McConaghy, "Behavioral Intervention in Homosexuality," *J. Homosexuality* 2 (1977): 221–7; John Money, "Bisexual, Homosexual, and Heterosexual," *J. Homosexuality* 2 (1977): 229–33.

[59] Kurt Freund, "Should Homosexuality Arouse Therapeutic Concern?" *J. Homosexuality* 2 (1977): 235–40, on 239.

CONCLUSION

Over the ensuing decades, the aversive conditioning experiments of Freund and Mc-Conaghy would be reinscribed with ironic new meanings. Vilified by gay activists as forms of torture in the early 1970s, the studies would be cited by later experts to advance scientific and political positions that normalized homosexuality. In the 1980s, for example, psychologist Richard Green cited the studies as evidence that legal protections should be granted to homosexuals because sexual orientation was immutable.[60] A decade later, neuroscientist Simon LeVay would also refer to Freund's work, alongside a vast compendium of other studies, to argue that homosexuality has a biological basis.[61] And in 2009, the American Psychological Association would cite McConaghy's work in order to refute religious groups who claimed that they could help individuals to change their sexual orientation.[62]

This debate over the plasticity of sexual orientation continues into the present since California in 2012 and New Jersey in 2013 have passed laws that ban so-called gay reparative therapies for minors. Advocates of the laws have argued that efforts to change people's sexual orientations are scientifically unsound and psychologically damaging, and a federal appeals court has recently upheld their constitutionality.[63] The phallometer itself has also traversed a durable political history. Czechoslovakian physicians relied upon the device during the 1950s to detect army recruits falsely claiming to be homosexual to avoid conscription, and in 2010, the Czech Republic was criticized for requiring men to submit to a phallometric test if they sought political asylum to escape a country where homosexuals were persecuted.[64]

The longevity of phallometry as an instrument of sexological research and state surveillance should remind us of several ironies in its history. Freund had originally invented the device to operationalize the measurement of erotic preferences and hoped to demonstrate the effectiveness of behavioral over psychoanalytic therapies in changing those preferences. Instead, phallometers substantiated the intractability of homosexuality and exposed the limits of all therapies, including those of behaviorism, to inculcate normative desires. As the determinism of behavioral theories waned, however, the determinism of biological theories resurfaced. As early as the 1960s, Freund had favorably cited the 1952 work of Franz Kallmann, who had reported that

[60] Bayer, *Homosexuality* (cit. n. 10), 103; Richard Green, "Immutability of (Homo)sexual Orientation: Behavioral Science Implications for a Constitutional (Legal) Analysis," *J. Psychiat. Law* 16 (1988): 537–75.

[61] Simon LeVay, *Queer Science: The Use and Abuse of Research into Homosexuality* (Cambridge, Mass., 1996), 52–3.

[62] B. S. Anton, "Proceedings of the American Psychological Association for the Legislative Year 2009: Minutes of the Annual Meeting of the Council of Representatives and Minutes of the Board of Directors," *Amer. Psychol.* 65 (2009): 385–475.

[63] Erik Eckholm, "Gay 'Conversion Therapy' Faces Tests in Courts," *New York Times*, 27 November 2012, sec. U.S., http://www.nytimes.com/2012/11/28/us/gay-conversion-therapy-faces-tests-in-courts.html (accessed 10 May 2015); Susan Livio, "Gay Rights Advocates Hopeful after Christie Signs Bill Banning Conversion Therapy," *New Jersey Star-Ledger*, 19 August 2013, http://www.nj.com/politics/index.ssf/2013/08/christie_signs_bill_banning_licensed_therapists_from_using_gay-to-straight_conversion_therapy_on_kid.html (accessed 10 May 2015); Howard Mintz, "California Gay Conversion Therapy Ban Upheld," *San Jose Mercury News*, 29 August 2013, http://www.mercurynews.com/crime-courts/ci_23973557/ (accessed 10 May 2015).

[64] Organization for Refuge, Asylum, and Migration, *Testing Sexual Orientation: A Scientific and Legal Analysis of Plethysmography in Asylum and Refugee Status Proceedings* (San Francisco, 2010).

monozygotic twins were 100 percent concordant for homosexuality.[65] Likewise, McConaghy moved toward biological etiologies when he used the phallometer to confirm that a pair of identical male twins was "truly discordant for homosexuality." To explain this puzzle, he hypothesized that a hormonal difference during a critical period of development caused their divergent sexualities.[66] In 1994, Bancroft also wrote about the hormonal and genetic factors that contributed to differences of sexuality even though he was skeptical of biological determinism.[67] Specifically, he discussed the work of scientist Simon LeVay, who announced the discovery of a correlation between a region of the brain (INAH3) and male homosexuality in 1991, and Dean Hamer, who reported a linkage between male homosexuality and a section of the X chromosome (Xq28) in 1993. Though there are many reasons for this resurgent interest in the biological causes of homosexuality (including increased funding for AIDS research and the strategic deployment of a rights discourse based upon immutable differences), the story of phallometry helps us to understand how psychological researchers might also have become more amenable to biological arguments about homosexuality and heterosexuality.[68] By demonstrating that sexual preferences for masculine and feminine body shapes were relatively durable, phallometry represented a pivotal moment in the history of psychiatry, when behavioral psychologists helped to lead expert discourses away from Freudian psychoanalysis toward a rebiologization of the sexual minds of men.

[65] Franz J. Kallmann, "Comparative Twin Study on the Genetic Aspects of Male Homosexuality," *Journal of Nervous and Mental Disease* 115 (1952): 283–98; Kurt Freund and V. Pinkava, "Homosexuality in Man and Its Association with Parental Relationships," *Rev. Czech. Med.* 7 (1961): 32–9.

[66] N. McConaghy and A. Blaszczynski, "A Pair of Monozygotic Twins Discordant for Homosexuality: Sex-Dimorphic Behavior and Penile Volume Responses," *Arch. Sexual Behav.* 9 (1980): 123–31.

[67] John Bancroft, "Homosexual Orientation. The Search for a Biological Basis," *Brit. J. Psychiat.* 164 (1994): 437–40. Bancroft ultimately concluded that sexual orientation resulted from a multifactorial process that involved both biological and psychosocial factors.

[68] Jennifer Terry, *An American Obsession: Science, Medicine, and Homosexuality in Modern Society* (Chicago, 1999).

Half a Man:

The Symbolism and Science of Paraplegic Impotence in World War II America

by Beth Linker* and Whitney Laemmli†

ABSTRACT

At the conclusion of the Second World War, more than 600,000 men returned to the United States with long-term disabilities, profoundly destabilizing the definitions, representations, and experiences of male sexuality in America. By examining an oft-neglected 1950 film, *The Men*, along with medical, personal, and popular accounts of impotence in paralyzed World War II veterans, this essay excavates the contours of that change and its attendant anxieties. While previous scholarship on film and sexuality in the postwar period has focused on women's experiences, we broaden the analytical lens to provide a fuller picture of the various meanings of male sexuality, especially disabled heterosexuality. In postwar America, the paralyzed veteran created a temporary fissure in conventional discussions of the gendered body, a moment when the "normality" and performative features of the male body could not be assumed but rather had to be actively defined. To many veterans, and to the medical men who treated them, sexual reproduction—not function— became the ultimate signifier of remasculinization.

"One reason people aren't going to the movies anymore," famed actress Gloria Swanson complained in 1951, is that "Hollywood deals too much with the problems of paraplegics and the blind people."[1] The decade after World War II witnessed a flurry of reintegration dramas, films that featured permanently wounded soldiers overcoming their disabilities to fit back into a well-ordered society. The genre was by no means new: films depicting blinded and deaf World War I soldiers were also popular in the 1920s and 1930s. As film scholar Martin Norden points out, however, post–World War II films differed from their predecessors in one important way: a tendency to es-

* Department of the History and Sociology of Science, 303 Cohen Hall, University of Pennsylvania, Philadelphia, PA 19104; linker@sas.upenn.edu.

† Department of the History and Sociology of Science, 303 Cohen Hall, University of Pennsylvania, Philadelphia, PA 19104; laemmli@sas.upenn.edu.

We wish to thank Erika Lorraine Milam, Robert A. Nye, the participants of the 2012 "Masculinities in Science/Sciences of Masculinity" workshop, and the anonymous *Osiris* reviewers for their helpful comments and suggestions on earlier drafts of this essay. We would also like to thank Cynthia Connolly, Patricia D'Antonio, Rebecca L. Davis, Julie Fairman, David Gerber, Dorothy Porter, Dominique Tobbell, and Daniel J. Wilson for sharing their historical expertise and feedback at various stages of researching and writing this essay.

[1] "Gem of the Month," *Paraplegia News* 5 (August 1951): 7.

chew the "repaternalizing miracle cure" trope.[2] Post–World War II audiences no longer naively bought into the rosy picture of frictionless reintegration, and Hollywood responded with increased realism. World War II reintegration dramas, therefore, embodied lofty hopes that veterans might be seamlessly reincorporated into civil society, while also conceding that these men's feelings of anger, alienation, and frustration could easily undermine postwar harmony.[3]

In many of these World War II–era dramas, women appeared to shoulder the burden—practical, emotional, sexual—of reintegration. In the enormously popular *The Best Years of Our Lives,* double amputee Homer Parrish is rescued from self-pity and depression by the unconditional love of Wilma, his fiancée.[4] In a climactic scene, Homer allows Wilma into his bedroom and accepts her help in removing his prosthetic hooks. Wilma shows no horror at the sight of his stumps but rather kisses him passionately and gently tucks him into bed. Many scholars see this scene as the crucial element in Homer's reintegration, the instant when viewers become certain that Homer will indeed triumph over adversity. "If women played their prescribed role in the demobilization drama," historian David Gerber writes, it was assumed that "all would work out well for the nation, their loved ones, and themselves."[5] The assumption that women could and should play a central role in bringing about postwar order extended into advice literature of the day as well. Women's magazines and marriage manuals encouraged would-be wives to engage in "serious study of the veteran and his psychology"[6] in order to establish stable heterosexual unions with men coming home from the war.

Gender historians have found such postwar demands on—and portrayals of—women to be highly problematic.[7] Susan Hartmann criticizes post–World War II advice literature for its tendency to push newly independent women back into stereotypically gendered structures: "While women were assigned the crucial responsibility for solving this major postwar problem, they were to do so in terms of traditional female roles. Through self-abnegation, by putting the needs of the veteran first, women might successfully renew their war-broken relationships, but they would do so at the price of their own autonomy."[8] Historian Sonya Michel finds the denial of women's sexual agency most troubling. "Though the soldiers' sexual longing was a persistent theme in wartime popular culture," Michel claims, "women were instructed to temper expressions of their own sexual needs and behave submissively." Michel notes that, in

[2] Martin Norden, "Bitterness, Rage, and Redemption," in *Disabled Veterans in History,* 2nd ed., ed. David Gerber (Ann Arbor, Mich., 2003), 96–114, on 105.

[3] David Gerber, "Heroes and Misfits: The Troubled Social Reintegration of Disabled Veterans of World War II in *The Best Years of Our Lives,*" in Gerber, *Disabled Veterans* (cit. n. 2), 70–95; Gerber, "Anger and Affability: The Rise and Representation of a Repertory of Self-Presentation Skills in a World War II Disabled Veteran," *J. Soc. Hist.* 27 (1993): 5–27.

[4] *The Best Years of Our Lives,* directed by William Wyler (1946; Los Angeles, 2000), DVD.

[5] Gerber, "Heroes and Misfits" (cit. n. 3), 552.

[6] Willard Waller, "What You Can Do to Help the Returning Veteran," *Ladies' Home Journal,* February 1945, 94, as quoted in Rebecca Jo Plant, "The Veterans, His Wife and Their Mothers: Prescriptions for Psychological Rehabilitation after World War II," *Amer. Hist.* 85 (1999): 1468–78, on 1475.

[7] Jessamyn Neuhaus, "The Importance of Being Orgasmic: Sexuality, Gender, and Marital Sex Manuals in the United States, 1920–1963," *J. Hist. Sexual.* 9 (2000): 447–73; Susan Hartmann, "Prescriptions for Penelope: Literature on Women's Obligations to Returning WWII Veterans," *Women's Stud.* 5 (1978): 223–39; Plant, "The Veterans" (cit. n. 6).

[8] Hartmann, "Prescriptions for Penelope" (cit. n. 7), 236.

The Best Years of Our Lives, Wilma's sexual desire was expressed only "through her demure, quasi-maternal affection toward Homer."[9] Like other young wives and sweethearts of the era, Wilma's role as a restorer of lost masculinity required that she tailor her own gratification to Homer's physical and psychological needs and limitations.

Such critiques of postwar film and advice literature are valid, but these analyses almost uniformly sidestep the question of how male sexuality, and in particular disabled-male heteronormative sexuality, should be understood. At the conclusion of the Second World War, more than 600,000 men returned to the United States with long-term disabilities, contributing to a profound destabilization of the definitions, representations, and experiences of male sexuality in America. For many, the unprecedented number of paraplegic veterans was particularly disorienting. Prior to World War II, few soldiers with spinal cord injuries survived longer than a handful of days or weeks, frequently felled by infections of the urinary tract, respiratory system, and pressure sores; mortality rates during World War I hovered around 80 percent. By the end of World War II, however, blood transfusions, the mass production of penicillin and sulfa drugs, new techniques of catheterization, and innovations in hospital procedure had decreased mortality rates to around 10 percent.[10] But while, to some, such improvements in care made paraplegic veterans "living memorials to the skill of medical officers during World War II," the survival of this new patient group also raised potentially troubling questions.[11] What did it really mean to be paralyzed below the waist? Would these men be able to experience sexual pleasure, satisfy their partners, reproduce?

These questions seemed particularly pressing because, while sex experts of the early twentieth century measured "successful" heteronormative intercourse by the achievement of female orgasm, during the Second World War, attention shifted toward male performance, with men increasingly regarded as the "more fragile and sexually vulnerable" sex. As a man's sexuality became increasingly "entangled with his sense of self-worth," any inability to perform—reproductively or romantically—became a serious threat to his masculinity.[12] For disabled veterans, such anxieties were often particularly severe, threatening not only their manhood, but also their marriages, and thus the larger project of national rehabilitation.[13]

The specific link between erectile function and masculinity was not, of course, entirely unique to the postwar period. In this volume, Leah DeVun shows us that as early as the Middle Ages, physicians used "impotency tests" (in which a prostitute was used to test a man's capacity for arousal) to assess masculinity. When a patient failed this test, DeVun writes, he was deemed "less than a 'real man'" and dismissed as "lacking in hardness, substance, and maleness." Nathan Ha (also in this volume) demonstrates

---

[9] Sonya Michel, "Danger on the Home Front: Motherhood, Sexuality, and Disabled Veterans in American Postwar Films," *J. Hist. Sexual.* 3 (1992): 109–28, on 119.

[10] Mary Tremblay, "The Canadian Revolution in the Management of Spinal Cord Injury," *Can. Bull. Med. Hist.* 12 (1995): 125–55; J. J. Mattelaer and I. Billiet, "Catheters and Sounds: The History of Bladder Catheterization," *Paraplegia* 33 (1995): 429–33.

[11] Robert Kennedy, "The New Viewpoint toward Spinal Cord Injuries," *Ann. Surg.* 124 (1946): 1057–62, on 1061.

[12] Neuhaus, "Importance of Being Orgasmic" (cit. n. 7), 450, 465. See also Rebecca L. Davis, *More Perfect Unions: The American Search for Marital Bliss* (Cambridge, Mass., 2010).

[13] As Angus McLaren notes, "Though every era has employed discourses to represent and control sexuality, certain ages clearly manifested a heightened anxiety about the issue of male sexual dysfunction," and the postwar period clearly falls within this category. McLaren, *Impotence: A Cultural History* (Chicago, 2007), xiii.

that twentieth-century scientists made similar associations, using the rings and pressure cuffs of phallometry to gauge both masculinity and sexual orientation.[14]

For veterans with spinal cord injuries, however, fears about "impotence" extended far beyond simple measures of penile function. To more deeply explore the contested nature of masculinity, heteronormative sexuality, and disability during the World War II years, this essay focuses on an oft-neglected reintegration film of the postwar era: Fred Zinnemann's *The Men* (1950).[15] Starring Marlon Brando as Ken "Bud" Wilcheck, a World War II soldier who returns home with a sniper bullet wound to the thoracic spine, the film is one of the first to feature a paraplegic veteran bound to a wheelchair. By featuring paraplegic men on the silver screen, Zinnemann tapped into postwar fears of impotence as no other reintegration film had before.[16] In theater posters and advertisements, the film was billed as "a completely new experience between men and women!" (fig. 1).

Significantly, the film crew conducted weeks of research at the Birmingham VA Hospital in Los Angeles, California, where they observed (and in some cases lived with) dozens of paraplegic veterans and their treating physician, Dr. Ernst Bors.[17] At Birmingham, the film crew also encountered the Paraplegic Veterans of America (PVA), a group of spinal cord injured veterans who formed their own association to advocate for better medical care, pensions, and adaptive equipment that would meet their needs and rights as disabled citizens. Zinnemann became especially close to paraplegic Ted Anderson, the president of the PVA, on whom "Bud" Wilcheck is based.[18]

Toggling between filmic representation and published literature produced by the PVA and by medical researchers, this essay will demonstrate that impotence was not only a symbolic concern but an immediate physical and emotional issue for patients and for the medical scientists who treated them. Because the film closely tracked a historical situation that we have been able to reconstruct from published records, *The*

[14] See Leah DeVun, "Erecting Sex: Hermaphrodites and the Medieval Science of Surgery," 23, and Nathan Ha, "Detecting and Teaching Desire: Phallometry, Freund, and Behaviorist Sexology," both in this volume.

[15] *The Men,* directed by Fred Zinnemann (1950; Los Angeles, 2009), DVD. While scholars like Gerber, Michel, and Norden have all made reference to *The Men* in their work on postwar cinema, a sustained analysis of the film—its production, its reception, and the historical realities it sought to reflect—is still lacking. Similarly, Andrew Huebner's discussion of *The Men* in the context of postwar representations of the American soldier provides suggestive, albeit brief, analysis. Huebner, *The Warrior Image: Soldiers in American Culture from the Second World War to the Vietnam Era* (Chapel Hill, N.C., 2008).

[16] Oliver Stone's *Born on the Fourth of July* is often seen as the film that deploys the paraplegic veteran as the symbol for US national impotence in a war gone wrong (*Born on the Fourth of July,* directed by Oliver Stone [1989; Los Angeles, 2004], DVD). Zinnemann, however, delved into similar issues, albeit somewhat indirectly, as early as 1950.

[17] Brando reportedly lived three weeks with paraplegic veterans recovering at Birmingham before *The Men* was shot. For more on this and the filmography of *The Men,* see Fred Zinnemann, *A Life in the Movies: An Autobiography* (New York, 1992); *Fred Zinnemann: Interviews* (Jackson, Miss., 2005); Arthur Nolletti, ed., *The Films of Fred Zinnemann: Critical Perspectives* (Albany, N.Y., 1999).

[18] For more on Ted Anderson, see Anderson, "Paraplegic GI Relives War in Searing Scene," *Los Angeles Times,* 26 February 1950, D3; "13-Year Struggle Ended by Paralyzed Veteran," *Los Angeles Times,* 5 October 1958, A; "Paralyzed Vets Elect Officers," *Los Angeles Times,* 12 June 1948, A1; "Meet Ted Anderson," *Paraplegia News* 9 (June 1955): 4. Anderson also authored a column, "Paraplegia in Review," in *Paraplegia News* for several years. See, e.g., Ted Anderson, "Paraplegia in Review," *Paraplegia News* 9 (February 1955): 2.

Figure 1. *Original theatrical poster advertising* The Men.

Men offers a unique opportunity to look at both the symbolism and the science of impotence during the immediate post–World War II years in America.

By looking at paraplegic impotence, we hope to deepen existing discussions of
gender representation and heteronormative male sexuality in midcentury America.
Our focus on disabled-male sexuality is a particularly productive way of excavating
this history, in part because the disabled body—for better or for worse—has frequently served as a site of gender fluidity and complication.[19] In postwar America,
the paralyzed veteran created a temporary fissure in conventional discussions of the
gendered body, a moment when the standard measures of sexual performance could
not be assumed but rather had to be actively defined.[20] While this essay will demonstrate that there was a chorus of conflicting views on what constituted paraplegic

[19] See, e.g., Abby Wilkerson, "Disability, Sex Radicalism, and Political Agency," *Nat. Women's Stud.
Assoc. J.* 14 (2002): 33–57; Tobin Siebers, *Disability Theory* (Ann Arbor, Mich., 2008).

[20] It is also interesting to note that this was a historical moment when the concept of gender as
performance—and as potentially biologically modifiable—began to enter the American consciousness. See David Serlin, *Replaceable You: Engineering the Body in Postwar America* (Chicago, 2004);
Bernice Hausman, *Changing Sex: Transsexualism, Technology, and the Idea of Gender* (Durham, N.C.,
1995); Judith Butler, *Gender Trouble: Feminism and the Subversion of Identity* (New York, 1990).

masculinity and sexuality, many paraplegic veterans—and especially their treating physicians—wished to downplay the centuries-long emphasis on sexual performance and instead define masculinity in terms of the ability to sire children.

SYMBOLISM OF IMPOTENCE IN *THE MEN*

In many ways, *The Men* is a standard American post–World War II rehabilitation film. Bud goes off to war and, in the course of an act of military heroism, is felled by a German sniper's bullet. Paralyzed below the waist, Bud is consigned to a Veterans Administration (VA) hospital, where he struggles to regain his former self through a slow and painful course of physical and psychological rehabilitation. As Beth Linker has argued elsewhere, inherent to the ethic of rehabilitation is the demand for remasculinization, a process whereby a man is viewed as a hero for having struggled to recover from a debilitating wound to become "whole" again.[21] This process was by no means unique to the United States. In this volume, Frances Bernstein shows the extremes to which the Soviet Union went in order to erase the potentially emasculating "problem" of World War II disabled veterans.[22] Similar to earlier US rehabilitation efforts, the USSR focused its efforts on remaking—or, as Bernstein puts it, "unmaking"— amputee veterans, a group that became a model for how a war-torn nation could overcome the physical and economic realities wrought by years of battle. Just as amputees could "fix" themselves through the simple flip of a prosthetic strap, so too could a nation paper over the past with a new political image of normalization and vitality.[23]

The director and producers of *The Men* took a risk by filming paraplegic veterans who, by virtue of their wheelchair dependency, could not be made to appear as "normal" again. But Zinnemann knew that the severity of the disability would add to the film's dramatic effect and, if anything, allow him to embellish the struggles inherent in reintegration stories. The concern of *The Men*—as with most films of the genre from this time period—was the problem of reconciling Bud's wartime valor with his postinjury paralysis and symbolic emasculation. Tellingly, like so many films and novels before it, *The Men* speaks of rehabilitation using gender-infused battle metaphors. The film opens with a deep-voiced narrator intoning a heroic dedication over the sound of martial drumming:

> In all wars, since the beginning of history, there have been men who fought twice. The first time they battled with club, sword, or machine gun. The second time they had none of these weapons. Yet, this by far, was the greatest battle. It was fought with abiding faith and raw courage and in the end, victory was achieved. This is the story of such a group of men.

It is this "second" and "greatest" battle that structures the majority of the plot. Though the viewer gets a brief glimpse of Bud sustaining his battle wound, the bulk of the film follows a group of paraplegic men recuperating under the watchful and disciplinary eye of Dr. "Bowels and Bladder" Brock (Everett Sloan), a man who barks medical orders like an impatient first sergeant. He is assisted, quite predictably, by

---

[21] Beth Linker, *War's Waste: Rehabilitation in World War I America* (Chicago, 2011).

[22] Frances Bernstein, "Prosthetic Manhood in the Soviet Union at the End of World War II," in this volume.

[23] For more on how amputees became a model of the US rehabilitation effort precisely because they could be more easily normalized and "fixed" through prosthetic wear, see Linker, *War's Waste* (cit. n. 21).

an equally disciplined and trustworthy nurse, Nurse Robbins (Virginia Farmer), who ministers care with great concern, but without overindulgence, a reflection of mid-century mothering ideals.[24]

But for all of Nurse Robbins's motherly coaxing and Dr. Brock's staunch patriarchal demands, Bud remains depressed, withdrawn, and uncooperative. Only when his fiancée, Ellen (Theresa Wright), enters the picture does Bud begin his campaign of self-improvement and remasculinization. Taking his lead from fellow paraplegic Angel Lopez (an actual paraplegic played by a nonactor, Arthur Jurado), Bud engages in upper body training, first with a hospital trapeze bar, then with modified pushups. Once he gains enough strength, Bud graduates to a wheelchair and participates in sports such as water polo, bowling, and wheelchair racing. Eventually, Bud learns how to drive a car, further fostering the masculinization process. As historian Christina Jarvis writes, "The frequent portrayals of [Bud] driving help to rephallicize his body. Just as wartime advertisements and recruitment posters fused soldiers' bodies with weapons in an attempt to portray their strength and 'steel-like qualities,' the speed and technology of the specially designed car help counter some of the weakness associated with [Bud's] disability, rendering him both whole and mobile."[25]

It is important to note that, like other films of the era, *The Men* relies upon its female characters as crucial elements in the rehabilitative process. Though it is Bud who sweats through workouts and checkups, without Ellen—and the symbolic prospect of heteronormative family life she represented—he would have had little motivation to engage in the arduous process of recovery. As already suggested, this tendency has made *The Men* a target for scholars who see it as a participant in the postwar subjugation of women. Norden, for example, tells us that the "Hollywood mentality of the time . . . encouraged women on the domestic front to give up their interests and accommodate the returning veterans"[26] and further notes that "female figures . . . do not represent women, but the needs of the patriarchal psyche."[27]

But while *The Men* replicates many of these postwar tropes, it also differs from other films of the genre in significant ways. For though Bud does recover much of his physical prowess and some of his confidence by the end of the story, a truly full and glorious recuperation never arrives. The catharsis of a happy ending is withheld.

To be sure, much of the film revolves around Bud's relationship with Ellen, particularly her efforts to convince him to resume their engagement despite his paraplegia. There is, however, a sense of unease—and indeed failure—at all the crucial moments when Bud and Ellen attempt to formalize their relationship, first through marriage, and then as a wedded couple. Their wedding ceremony, which takes place not in a church, but in the hospital chapel, is laced with feelings of doom and dread. Ellen marries Bud against her parents' wishes and, thus, in their absence. Just before the wedding vows, Bud, seated in his wheelchair, struggles to stand upright (with the

---

[24] Rebecca Jo Plant, *Mom: The Transformation of Motherhood in Modern America* (Chicago, 2010). Robbins's maternal role is established early in the film as she wheels Bud, immobile on a gurney, into the ward where other paraplegics reside. As she enters the ward, with a diminutive Bud cradled in front of her as if he were in a pram, Robbins chimes to the other bedridden men, "I'm bringing you a playmate."

[25] Christina S. Jarvis, *The Male Body at War: American Masculinity during WWII* (Dekalb, Ill., 2004), 110.

[26] Martin Norden, "Resexualization of the Disabled War Hero in *Thirty Seconds over Tokyo*," *J. Pop. Film & Telev.* 23 (1995): 50–5, on 51.

[27] Ibid.

assistance of leg braces and canes) but quickly falters and has to be held up by Brock. Unlike *The Best Years of Our Lives*, there is no bedroom scene where wedding night vows are physically consummated. Instead, Bud and Ellen's honeymoon night ends in a fight. Rather than Bud carrying Ellen over the threshold, as a nondisabled man would do, Ellen swings open the door and enters the living room with a diminutive Bud trailing behind, his wheelchair's disruptive squeaking announcing his disability. A bottle of celebratory champagne explodes prematurely, Bud's leg begins to tremble, and he erupts with rage. As Ellen bends over to clean up the spill, Bud is shot from below, looking almost inhuman. The whole scene can be easily read as a metaphor for Bud's frustration over his symbolic—and perhaps anticipated—failure to perform sexually and Ellen's fear that she has made a mistake. This catastrophic attempt to enact domestic heteronormativity behind him, Bud returns "home" to the VA hospital, seeking safety and comfort in the fraternity of his fellow paraplegic veterans.

Not until Dr. Brock reveals his own personal commitment to paraplegic reintegration (Brock's wife was paralyzed in a car accident and "medical science couldn't save her") does Bud give his own marriage another try. In the closing scene of the film, Bud appears at his in-laws' house unannounced. Busy with gardening outside, Ellen sees Bud and says, "You've come a long way," clearly referencing both a physical and emotional distance. As Bud wheels himself to the front porch, a set of concrete steps impedes additional movement. "Do you want me to help you up the steps?" Ellen asks. "Please," Bud replies, uttering the final words of the film. Bud and Ellen do reunite, but the peace that they make is an ambiguous and uneasy one, owing much to popular understandings and taboos concerning paraplegia and sexuality in this period.

In fact, the portrayal of paraplegic veterans' sexuality throughout the movie does much to underscore its rather atypical ending. Although the production codes of the era meant that the frank depiction or discussion of sexual subjects was forbidden, sexuality—and in particular male sexuality—appears to be *The Men*'s central anxiety.[28] Take, for example, the first scene that brings the concern of resexualization—and in turn Bud's remasculinization—to the fore. When Ellen first lays eyes on Bud in the hospital, he is prostrate in a bed, with disheveled hair and crumpled sheets. Upon seeing Ellen, Bud reacts with anger and fear, proclaiming, "I'm not a man. I can't make a woman happy." Taken out of context, Bud's assertion of unmanliness could mean many things. Adjusting to his new reality as a paraplegic, Bud might feel incapable of fulfilling his manly duties in an economic sense; though the film takes care to establish that the men were enormously well-provided for by the government, he might doubt that he could hold down a traditional job. He might also doubt his masculinity because of an identification with his dependent, infantile state. He is immobile and, like most paraplegics immediately after injury, has no control over his bladder and bowels. Bud might be defining his manhood in contradistinction to childhood.

But the setting in which Bud utters these lines is key to understanding the symbolic meaning of manliness in *The Men*. As the conversation continues, Bud—still lying in bed—dramatically unveils the lower half of his body, whipping back the hospital bed covers and shouting: "Alright, I'll give you what you want. Want to see what it's like? Alright, look. I said look at me. Now get a good look. . . . Is that what you want?" The

---

[28] Thomas Patrick Doherty, *Hollywood's Censor: Joseph I. Breen and the Production Code Administration* (New York, 2007).

reference is not explicitly sexual, but context makes it clear that Bud's concern is not solely his atrophied legs, but also his potentially incapacitated genitalia. Significantly, the camera never reveals exactly what Ellen sees, allowing viewers' phobic imaginations to fill in the gaps.

Ellen's response, though similarly veiled, also suggests an undercurrent of sexual tension. After Bud reminds her that he will never walk again, Ellen—who has until this point been standing upright next to Bud's bed—lowers herself next to Bud and, leaning forward until they are nearly nose to nose, pleads: "But you could do lots of things. Oh please, please, try. Don't you see? I need you. There will never be anyone else. Oh darling, don't you want us to be happy?" She collapses into his arms, sobbing, while Bud's eyes fill with tears. "Sure I want us to be happy, honey, but I don't know. I don't know." The conversation itself is ambiguous, but Ellen's body language implies that this discussion is a particularly intimate one. In this scene, Ellen establishes herself as a figure both maternal and sexual. This dual role, we are told by film scholars and historians, was seen as necessary for Bud's recuperation. Perhaps even more interesting than the film's negotiation of Ellen's sexuality, however, is the way in which *The Men* deals with Bud's sexuality—and, in turn, the sexuality of disabled men more generally.

To date, the sexuality of disabled veterans has been understudied, as most scholars have assumed that it is identical to that of nondisabled males (itself often erroneously understood as historically stable in its meaning and manifestation).[29] Moreover, male sexuality is frequently assumed to be dichotomous: a man is either potent or impotent, with few intermediary gradations. This rather crude understanding seems to have been fueled by a scholarly fascination with "symbolic castration," a theoretical framework that is applied ubiquitously to novels and films, especially when a disabled male figure is involved. However, much nuance and insight can be gained if we move beyond symbolism and train our eyes on those men who have the power to castrate or cure other men, even if "symbolically," as well as on the men who are being rendered impotent.

THE SCIENCE AND SUBJECTIVE EXPERIENCE OF PARAPLEGIC IMPOTENCE

It is difficult in retrospect to appreciate the depths of uncertainty that accompanied a diagnosis of spinal cord injury in the immediate post–World War II years. Contemporary understandings of paralysis largely came from polio patients, the most prominent of whom was the recently deceased Franklin D. Roosevelt, but spinal cord injury was distinctly different in that it affected both sensory and motor functioning. The working (albeit vague) definition of paraplegia was complete motor and sensory paralysis of the entire lower half of the body.[30] World War II paraplegic Kenneth Wheeler

---

[29] E.g., Thomas Laqueur writes that "it is probably not possible to write a history of man's body and its pleasures, because the historical record was created in a cultural tradition where no such history was necessary." See Laqueur, *Making Sex: Body and Gender from the Greeks to Freud* (Cambridge, Mass., 1992), 22. Erika Lorraine Milam and Robert A. Nye express a similar sentiment, noting that historians have "left largely unexamined the discriminatory hierarchies within all-white male cultures, which advanced the careers of some men while excluding or marginalizing other men (and, by extension, women) on the basis of class, race, religion, or sexual orientation." See Milam and Nye, "An Introduction to *Scientific Masculinities*," in this volume, 3.

[30] If a soldier sustained a high cord lesion and experienced paralysis of the arms and hands, he would sometimes be referred to as a quadriplegic, although many laypeople and spinal cord injured preferred the term "paraplegia."

described the subjective feeling of paralysis as one in which he was "chained to a body that was only half, or less than half a man."[31] Other paraplegic memoirists spoke in terms of being "half dead."[32] Immediately after Bud sustains his injury from a German sniper in the opening scene of *The Men*, he tells his fellow comrades that he is only "half alive."

While paraplegic veterans mourned the possibility of never walking again, the expressions of being only "half alive"—with everything below the waist dead to feeling and purpose—speak also to lost sexual capacity. In this sense, World War II paraplegics joined the ranks of military men who dreaded genital injury above all else. About traumatic wounds to the groin, Wheeler wrote: "Nothing much worse can happen to a man this side of death."[33] The men who fought during the Second World War heard tales of soldiers privileging the safety of the penis over the rest of the body, protecting their genitalia by crossing their legs under shell fire and using helmets to shield sexual organs instead of eyes and heads.[34] This concern for genital integrity can still be seen today. Recounting her work as a nurse in an Army intensive care unit that treats US soldiers injured in Afghanistan, Kathryn Gillespie explains that when a wounded soldier initially regains consciousness, he first asks: "'Is my junk all together?' "They want to check their 'package,' first," she says, "then they check their arms and legs. This all happens probably within 15 minutes of being off sedation."[35]

While not precisely the same kind of injury as genital amputation, paraplegics—and their families and doctors—had similar concerns about sexual potency, in terms of both physical functioning and fertility. The concern for potency is addressed most directly in *The Men* during a scene in which Dr. Brock conducts a question-and-answer session in the hospital chapel with the wives, mothers, and fiancées of paraplegic men in his care. For the film's purposes, this scene was meant to educate the audience on the basic medical facts of spinal cord injury. After several rounds of questions regarding mobility—"the word walk must be forgotten," barks Brock—one woman from a back pew slowly stands and says "but doctor, my husband and I want a large family." With knowing concern in his eye, Brock answers in a low voice, "in some cases it is possible, but we can't discuss it here."

In real life, Dr. Ernst Bors (the actual physician-urologist upon whom *The Men*'s Dr. Brock was based) researched paraplegic sexuality quite extensively, as did other physicians who specialized in the treatment of spinal cord injuries. Bors joined the Army Medical Corps in 1943 and was originally stationed at Hammond Army Hospital in Modesto, California, as assistant chief of urology on the transverse-myelitis ward. Bors earned his medical degree at the University of Prague in 1924 (he was Czechoslovakian by birth) and spent his early career as a surgical assistant at the University of Zurich. He had little experience with paralysis before taking up his post at the Birmingham VA in Van Nuys, just outside of Los Angeles. "In the 20 years that I

---

[31] Keith Wheeler, *We Are the Wounded* (New York, 1945), 13.

[32] Jarvis, *The Male Body at War* (cit. n. 25), 90.

[33] Wheeler, *We Are the Wounded* (cit. n. 31), 175.

[34] Jarvis, *The Male Body at War* (cit. n. 25), 87.

[35] David Brown, "Amputations and Genital Injuries Increase Sharply among Soldiers in Afghanistan," *Washington Post*, 4 March 2011, http://www.washingtonpost.com/wp-dyn/content/article/2011/03/04/AR2011030403258.html (accessed 15 June 2013). Thanks to Nathan Ensmenger for bringing this source to our attention.

had practiced prior to coming to [California]," Bors wrote, "I had seen possibly a total of eight paraplegics—no more."[36]

At the time Zinnemann began filming *The Men*, Bors had just completed a study on fertility in thirty-four paraplegic men at Birmingham.[37] He was not the first physician in charge of a spinal cord unit to do so. Donald Munro, a neurosurgeon at Cushing Veteran Administration Hospital in Boston, had conducted multiple studies on sexuality and paraplegia in the immediate postwar years. Munro made plain that for spinal cord injured patients, the loss of sexual ability was "one of the most difficult psychological [problems]" they had to encounter.[38] Similar to Bors, Munro sought to conduct research on paraplegic sexuality because he had no answers for the questions that his patients posed. "Doctors," Munro lamented, "either admit their ignorance or make up explanations and prognoses out of whole cloth."[39]

The prevailing popular assumption concerning paraplegic sexual capacity was that a severed spinal cord would render a man entirely impotent, flaccid, and unable to sire children.[40] Scientists steeped in animal experimentation and physiology, on the other hand, believed that paraplegics would experience priapism, a reflexive, "persistent abnormal erection of the penis, usually without sexual desire."[41] Indeed, earlier generations of doctors sometimes used symptoms of priapism to diagnose high cord lesions.

Neither of these stereotypes reflected clinical or, for that matter, personal experience on the part of World War II paraplegic veterans. Recounting his recuperation experience in a Memphis VA hospital, paraplegic Terry McAdam wrote in 1955, "The sexual abilities of each patient were eagerly discussed, for strangely enough, some of these boys could still function sexually. . . . In some cases there was an absolute end to any form of sex," McAdam continued, "in some there was a continuing ability to perform the mechanical act itself, but without any physical pleasure on the part of the man. . . . In still others," he described, "the feeling was mysteriously present in advantageous localities even though the level of the injury was far above."[42]

The clinical picture, then, was far from clear. However, systemic ignorance of the actual situation was not helped by the pervasive stigma attached to disability and sex.[43] Alfred Kinsey, often lauded for his progressive views on sexuality, all but ignored

---

[36] "Meet Dr. Bors," *Paraplegia News* 9 (October 1955): 4.

[37] Ernest Bors, "Fertility in Paraplegic Males: Preliminary Report of Endocrine Studies," *J. Clin. Endocrinol.* 10 (1950): 381–98.

[38] Donald Munro, Herbert W. Horne, and David P. Paull, "The Effect of Injury to the Spinal Cord and Cauda Equina on the Sexual Potency of Men," *New Engl. J. Med.* 239 (9 December 1948): 903–11.

[39] Ibid., 903.

[40] See Donald Munro, "The Rehabilitation of Patients Totally Paralyzed below the Waist: With Special Reference to Making Them Ambulatory and Capable of Earning Their Living," *New Engl. J. Med.* 250 (1954): 4–14. In his study of 408 paraplegic patients, urologist Herbert S. Talbot also insisted that the "popular belief that (spinal cord injured) patients are impotent is . . . unjustified." Talbot, "The Sexual Function in Paraplegics," *J. Nervous Mental Dis.* 115 (1952): 360–1. See also S. Leonard Simpson, "Impotence," *Brit. Med. J.* 1 (1950): 692–7.

[41] Munro, Horne, and Paull, "Effect of Injury to the Spinal Cord" (cit. n. 38), 909.

[42] Terry McAdam, *Very Much Alive: The Story of a Paraplegic* (New York, 1955), 65.

[43] Then and now, sex lives of the disabled have conventionally been ignored, stigmatized, and controlled through institutionalization as well as eugenic sterilization. For more on this, see David Serlin, "Touching Histories: Personality, Disability, and Sex in the 1930s," in *Sex and Disability*, ed. Robert McRuer and Anna Mollow (Durham, N.C., 2011), 145–62. See also Irving Zola, *Missing Pieces: A Chronicle of Living with a Disability* (Philadelphia, 1982). Zola was one of the first sociologists to make this point about disability and sexuality. For eugenic sterilization and disability, see Molly Ladd-Taylor, "The 'Sociological Advantages' of Sterilization: Fiscal Politics and Feebleminded Women

the disabled, even though he gathered data for *Sexual Behavior in the Human Male* during the war years. The one mention he makes on the subject of disability in his 800-page book demonstrates that while Kinsey was eager to normalize other forms of able-bodied "deviant sex" (i.e., masturbation, homosexuality, and bestiality), he found the sexual needs of the physically disabled aberrant, beyond the possibility of normalization. He wrote:

> Persons who are deformed physically, deaf, blind, severely crippled, spastic, or otherwise handicapped, often have considerable difficulty in finding heterosexual coitus. The matter may weigh heavily upon their minds and cause considerable psychic disturbance. There are instances where prostitutes have contributed to establishing these individuals in their own self esteem by providing their first sexual contacts. Finally, at lower social levels there are persons who are . . . so repulsive and offensive physically that no girl except a prostitute would have intercourse with them. Without such outlets, these individuals would become even more serious social problems than they already are.[44]

In his quest to find the normal or average American sex life, Kinsey continued to categorize disabled sex as abnormal, stripping away sexual agency from a large swath of the population, while labeling the very existence of disabled persons a "social problem."[45]

Because so few sex researchers devoted their studies to disability in midcentury America, paraplegic sexuality continued to raise more questions than could be answered. To begin, it rekindled an old debate about where male sexuality was located in the body and what anatomical and/or physiological processes were essential to potency. Was male potency located in the endocrine system, where androgens were produced? Was it to be found in the testes, where spermatozoa were stored? Or was potency situated in the psyche, where desire, lust, as well as "perversions" resided? The new population of men with spinal cord injuries inspired yet another direction of inquiry: What role did the neuromuscular system play in potency?[46]

Early findings concerning paraplegics and their potency were actually quite

---

in Interwar Minnesota," in *Mental Retardation in America: A Historical Anthology*, ed. S. Noll and J. Trent (New York, 2004), 281–99.

[44] Alfred C. Kinsey, Wardell B. Pomeroy, and Clyde E. Martin, *Sexual Behavior in the Human Male* (Philadelphia, 1948), 608.

[45] For more about the importance of sexual agency and disability, see Wilkerson, "Disability" (cit. n. 19). Wilkerson, along with other disability scholars, contends that the sex experts and other scholars fail to focus on female disabled sexuality and instead attend almost exclusively to male disabled sexuality, particularly that of the spinal cord injured. We have found, however, little historical analysis of paraplegic male sexuality. For more on this debate, see Anne Finger, "Claiming All of Our Bodies: Reproductive Rights and Disability," in *Test Tube Women: What Future for Motherhood,* ed. Ritta Arditti, Renate Duelli Klein, and Shelley Minden (London, 1984), 281–97. Russell P. Shuttleworth has conducted interesting anthropological work on paraplegic men and sexuality. Shuttleworth, "The Search for Sexual Intimacy for Men with Cerebral Palsy," *Sexual. Disability* 18 (1 December 2000): 263–82. There are also some notable memoirist accounts of paraplegics and sexuality written by men with polio. See, e.g., Zola, *Missing Pieces* (cit. n. 43), 217–9; Lorenzo Milam, *Cripple Liberation Front Marching Band Blues* (San Diego, 1983).

[46] While most impotence studies at the time had moved decidedly away from anatomical explanations, toward the psychological and endocrinal, early clinical scientists who treated paraplegics in VA Hospitals returned to an older model for understanding sexual virility. For more on the history of impotence and how scientists studied it, see McLaren, *Impotence* (cit. n. 13). For work on "sex" hormones, see Nelly Oudshoorn, *Beyond the Natural Body: An Archeology of Sex Hormones* (London, 1995); Chandak Sengoopta, *The Most Secret Quintessence of Life: Sex, Glands, and Hormones, 1850–1950* (Chicago, 2006).

encouraging. As early as 1948, Munro reported that 74 percent of the paraplegic veterans at Cushing Veterans Hospital retained the ability to have erections after injury. In 1950, Bors estimated that 88 percent of his patients at Birmingham had a return of erections within a few months after injury. In the same study, paraplegic veterans reported having sexual dreams, some of them "wet dreams with both orgasm and actual seminal emission."[47] Boston urologist Herbert S. Talbot came to conclusions similar to those of his colleagues. Out of the 408 paraplegics under his care, 66 percent had "erections in response to local stimulation." "About one-third of those having erections," Talbot concluded, "were able to have intercourse."[48]

The experience of injured veterans writing for the magazine *Paraplegia News* confirmed these medical findings. Indeed, for the members of the PVA who had editorial control over *Paraplegia News*, maintenance of sexual capabilities after spinal cord injury was almost a given. Take, for example, a letter that the editor in chief, Robert Moss, wrote in 1951 to the newly paralyzed soldiers coming home from the Korean War: "Most of you will be physically capable of sexually satisfying your mate."[49] The editors of *Paraplegia News* also commissioned wives of paraplegics to offer testimony to their husbands' abilities to maintain a healthy, postinjury sex life. The wife of one former Birmingham hospital paraplegic wrote: "My husband and I have a fairly normal sex life together, and I am satisfied with it."[50] In an attempt to further normalize paraplegic sex, veterans such as Moss insisted that all married couples, "paraplegic or not, have a 'sexual problem.'"[51] Paraplegia was just a different kind of "problem." This pluralistic view of a married sex life led another wife writing for *Paraplegia News* to conclude that the sexual "response of a man and a woman to each other takes many varied and beautiful forms."[52]

The ability to have a postinjury erection, however, seemed cause for little celebration among medical scientists. Even though Munro expressed irritation at the fact that most researchers conflated erectile function and fertility (and insisted that potency should not be defined exclusively by the latter at the expense of the former), he still considered reproduction to be the sine qua non of paraplegic sexuality. The privileging of reproduction over sexual capacity is most apparent in how Munro presented his research findings. Detailed medical case studies of paraplegics who fathered children far outnumber write-ups of men who had regained sexual function but remained sterile.[53] Here is one characteristic example of how Munro recorded the sexual recovery of his patients with paraplegia:

---

[47] Bors, "Fertility" (cit. n. 37), 392.

[48] Talbot, "Sexual Function" (cit. n. 40), 360.

[49] Robert Moss, "Memo: To Paraplegics of the Korean War from Robert Moss, World War 2," *Paraplegia News* 5 (June 1951): 1, 8.

[50] "Have You Been Thinking about Marriage?" *Paraplegia News* 10 (April 1956): 6.

[51] Moss, "Memo" (cit. n. 49), 8.

[52] Judith Hover, "I Married a Paraplegic," *Paraplegia News* 10 (November 1956): 8–9.

[53] The medical profession was not alone in this bias. In the pages of *Paraplegia News*, a magazine written and published by members of the PVA, birth announcements became front-page news. A typical announcement in *Paraplegia News* would provide the name, length, and weight of the baby and also include vital paraplegic statistics on the father, giving the date and place of the injury, and the level at which the spine was wounded. See, e.g., "New Arrival," *Paraplegia News* 7 (April 1953), 1: "Mr. and Mrs. William Marquardt of Worthington OH announce arrival of Denis Marie who will take up her residents [*sic*] in OH. The father, a member of the PVA, is a service-connected quadriplegic."

Patient is a 27 year old injured October 1944 by a bomb fragment which struck him in the region of the sixth to seventh thoracic vertebras. Three months after the injury his urethral catheter, which he had worn until then, was removed. . . . About one month after injury he began to have spontaneous erections lasting from 1–20 minutes. They had no relation to cerebral activity. The patient then discovered that erections could be caused by local stimulation of the penis. About 10 months after his injury, he attempted masturbation for the first time. After this stimulation he obtained a brownish-red ejaculate. A week later he repeated the procedure with essentially the same result. Since then ejaculations have been described as being normal in appearance, substance, and quality. During ejaculation, the patient's spasms increase momentarily, but no other sensation occurs. He was married to a nurse, aged 25 years old in May 1946. Intercourse was timed to correspond with his wife's periods. His wife delivered a normal, full-term child on March 1, 1947.[54]

Although proof of phallic recovery is outlined on a step-by-step basis in this case study, it is clear that siring a child is the point of the clinical lesson.[55] Munro would not have published the details of this case study if the patient had recovered "merely" to the point of having an erection or being able to masturbate.

One reason for the focused attention on the reproductive paraplegic was that fathering a child after such an injury appeared to be a medical oddity. As late as 1952, Talbot concluded that only 7 percent of paraplegic men were capable of siring children.[56] Moreover, medical researchers found no reliable physiological predictors that would indicate whether or not a paraplegic would be fertile. In his attempt to study paraplegic fertility through sperm motility and testicular biopsies, Bors concluded, with disappointment, that there was no evidence of a relationship between biopsy findings and sexual function.[57] Nor did ejaculations or seminal emissions seem to be reliable predictors of paraplegic fertility.

Another reason that nondisabled male physicians heralded reproduction was that siring a child provided incontrovertible proof that medicine could cure paraplegic men of their emasculating injury. Sexual functioning was largely a private matter, with success being agreed upon between a particular husband and wife. Siring a child, on the other hand, offered public, physical proof of a paraplegic's remasculinization and of medicine's role in assisting his sexual recovery.[58]

[54] Munro, Horne, and Paull, "Effect of Injury to the Spinal Cord" (cit. n. 38), 909.

[55] McLaren makes a similar point, noting that "it was taken as a given in Western culture that sex was synonymous with intercourse, a man penetrating his partner. The implication of such a belief is that a man feared impotence, not so much because it might deprive him of pleasure, but because it would prevent him from providing proof that he could perform as a male should. Potency was long linked to maturity. . . . [The association between] sexual virility and youth is a relatively recent phenomenon." McLaren, *Impotence* (cit. n. 13), xiii.

[56] Talbot, "Sexual Function" (cit. n. 40), 360.

[57] Bors, "Fertility" (cit. n. 37), 392.

[58] To the extent that they were studied, the same standard applied to paraplegic women at the time. Even if sexual arousal was briefly mentioned, most investigations of female paraplegic sexuality focused on reproduction and fertility, honing in on menstruation and childbearing. In 1975, E. R. Griffith and R. B. Trieschmann pointed out this problematic orientation, noting that "the literature on women with spinal cord injury deals primarily with the factors of hormonal function, fertility and delivery. Unfortunately, information is limited concerning issues which are relevant to the total sexual functioning of these women" and calling for further research. Griffith and Trieschmann, "Sexual Functioning in Women with Spinal Cord Injury," *Arch. Phys. Med. Rehabilitation* 56 (1975): 18–21. For examples of the focus on fertility, see L. Guttman, "Cardiac Irregularities during Labor in Paraplegic Women," *Paraplegia* 3 (1965): 144–51; H. Goller and V. Paeslack, "Pregnancy Damage and Birth Complications in the Children of Paraplegic Women," *Paraplegia* 10 (1972): 213–7.

While physicians like Bors (and films like *The Men*) insisted that questions of paraplegic fertility were fueled by the "natural" desire of wives to have children, in actuality, the push toward reproduction came from both the medical men and the paraplegic men whom they treated.[59] Evidence from personal letters of women married to paraplegic men as well as studies of college-aged women's views on disability and marriage indicate that most women were more concerned about a man's sexual capabilities than his fertility. In an advice column written for paraplegic veterans considering marriage, one wife insisted that as long as the veteran felt like he could satisfy his wife sexually—even if he could not father children—"she would be content." "While some girls feel that their lives would not be complete without bearing a child," she wrote, "others are just as happy without ever being mothers." "For myself," she admitted, "I would love to have a baby, but . . . if the Lord does not see fit to let us have one, I will love my husband just as much."[60]

Penn State College psychologist Clifford Adams confirmed such testimonials. During the war, Adams developed a questionnaire that asked unmarried female college students whether or not they would marry a disabled veteran. The questionnaire included thirty-three different types of war disabilities to serve as possible scenarios.[61] Adams found that, on the whole, "older girls showed a greater willingness to marry injured men than the younger girls . . . due to the fact," he surmised, "that the older girls are more concerned about their chances of marrying."[62]

Adams's research demonstrated, most tellingly, that college-aged women had clear opinions about which types of disabilities they believed they could adjust to in marriage and which they could not. Of the 500 college-aged women surveyed, a majority found the following disabilities (in rank order) serious enough to end their engagements: (1) impotence, (2) loss of both arms in such a way that they could not be replaced with artificial arms, (3) mental imbalance requiring institutional confinement for several months or longer, and (4) loss of both legs so that they could not be replaceable.

Knowing that impotence could be construed as both sterility and loss of sexual function, Adams drafted a follow-up questionnaire that was more sensitive to definitional variance. When asked to distinguish between fertility and sexual capacity, over 80 percent of the women questioned said that they "would marry, if they could have a normal sexual life, even though there was no possibility of conceiving children."[63] Only 16 percent of the respondents said that they would refuse to marry an ex-soldier who had become sterile. In other words, college-aged women found loss of sexual functioning more disabling than sterility.

In *The Men*, it is Ellen's father who insists on defining potency as reproduction, a mandate that Ellen rejects by the end of the film. Concern about Bud's reproductive capabilities comes to the fore in a scene in which Ellen visits her parents to tell them of her intent to marry Bud. The three dance gently around the topic for some time, until Ellen's father broaches the subject head on. "Love can be very fragile," he tells Ellen,

---

[59] Bors, "Fertility" (cit. n. 37), 392.

[60] "Have You Been Thinking about Marriage?" (cit. n. 50), 6.

[61] Clifford R. Adams, *Preparing for Marriage: A Guide to Marital and Sexual Adjustment* (New York, 1951), 20. Some of the other disabilities listed on the questionnaire included loss of speech, loss of one leg, loss of one arm, general permanent bad health, mental instability requiring long psychotherapy, incurable insanity, and incurable communicable disease.

[62] Ibid., 172.

[63] Ibid.

"Even healthy people can't hold on to it. How long do you think that love is going to last after you've become his nurse. You're a young healthy girl." But the conversation truly breaks down when the subject of procreation emerges. Ellen's father entreats, "Is it so wrong for us to want a grandchild?" As soon as Ellen's father brings up the "problem" of reproduction, she storms out of the house and visits Dr. Brock to seek his counsel. Brock provides few answers to Ellen's queries and fails to disabuse her of the notion that procreation was the primary—if elusive—goal of paraplegic male sexuality. In the end, Ellen decides that procreation "really doesn't matter," and the wedding proceeds as planned. While scholars such as Michel have read this scene as Ellen relinquishing her sexual needs for the purpose of Bud's reintegration, it could just as easily be understood as Ellen choosing to privilege sexual function, just as real-life wives of paraplegic veterans did at the time.

Men living with paraplegia remained divided on how potency should be defined and whether sexual capacity alone legitimated their manhood. Certain members of the PVA insisted that in order for a paraplegic man to readjust and make his way back into society, he had to absorb the fact that "paraplegics are different from the rest of the world in two respects: they are paralyzed and probably sterile."[64] Even if a paraplegic man could not produce children, the argument went, he could still assume a patriarchal role in society by being productive in other realms of life, such as finding a steady job, owning a home, and engaging in adaptive sports like water polo and wheelchair basketball.

Other paraplegics, however, were not as willing to give up sexual agency, seeing it as more essential to defining their masculinity than middle-class notions of productivity in the home and workplace. McAdam's memoir recounts a telling debate between two unmarried patients, Carl Fuller and Jerry Radcliff. When Jerry has a date with a very special "gal" whom he plans to wed, Carl responds: "Damn it, I tell you a paraplegic's got no business getting married."[65] Jerry responds by making a plea for his desire and need for sexual intimacy: "Look, Carl, we know you can't feel nothing, but some of us guys can. Why the hell should we sleep alone from now on?"[66] Exasperated, Carl replies, "Look, for the last time. It hasn't got nothing to do with tail. A damn paraplegic can't support a wife. You guys [will] make some poor woman unhappy and frustrated for a normal life."[67]

Whereas lust, sexual desire, and being able to satisfy a wife were assumed to be essential components of able-bodied, heterosexual manhood in midcentury America, the same rules did not readily apply to disabled men.[68] Take, for instance, the story of Fred B. Woolsey, cartoonist for the *Paraplegia News* from 1946 to 1950. A paraplegic himself, Woolsey edified many readers with his depictions of "delicious looking Florence Nightingales" massaging veterans and pushing them in their wheelchairs (figs. 2, 3).[69]

---

[64] "A P.V.A. Credo," *Paraplegia News* 4 (May 1950): 4.

[65] McAdam, *Very Much Alive* (cit. n. 42), 85.

[66] Ibid.

[67] Ibid., 86.

[68] For more on this subject, see Carolyn Herbst Lewis, *Prescription for Heterosexuality: Sexual Citizenship in the Cold War Era* (Chapel Hill, N.C., 2010).

[69] For one reader's enthusiastic response, see Frank G. MacAloon's letter, "Cartoon Denunciation Outlandish," *Paraplegia News* 4 (August 1950): 2.

What do you mean . . . "you'd rather just watch" . . .?

(*Fred Woolsey of Madison, Wisconsin, retains all reproduction rights of the above cartoon.*

Figure 2. *"Rather Just Watch,"* Paraplegia News *4 (May 1950): 6.*

To be sure, such images contributed to the objectification of women—nurses, in particular—in a pinup style common in postwar America.[70] Yet Woolsey's cartoons also served the function of normalizing the paraplegic male in his sexual desire, insisting that he was no different than a nondisabled man. His cartoons represented the potent, heterosexually functioning male, not the reproductive male.[71]

In his "Head Nurse" cartoon, Woolsey depicted a flustered yet sultry (and scantily clad) ward nurse who had just encountered a paraplegic who had become erect while she was caring for him (fig. 4). The ward nurse, bringing a complaint to the chief nurse, says in the caption, "And then he said 'I must have had a spasm.'" It was common for paraplegic men to experience reflexive muscle spasms in the legs and abdomen during intercourse and masturbation—this was a neurophysiologic response of

[70] The role that nurses play as mothers, wives, and sexual objects is crucial to understanding the history of sexuality for paraplegics. There is a large literature on the history of nursing (see Patricia D'Antonio, *American Nursing: A History of Knowledge, Authority, and the Meaning of Work* [Baltimore, 2010]; Julie Fairman and Patricia D'Antonio, "Reimagining Nursing's Place in the History of Clinical Practice," *J. Hist. Med. Allied Sci.* 63 [2008]: 435–46; Susan Reverby, *Ordered to Care: The Dilemma of American Nursing* [New York, 1987]; etc.) but little about nurses as sex objects. See Linker, *War's Waste* (cit. n. 21), 61–79; D. A. Nicholls and J. Cheek, "Physiotherapy and the Shadow of Prostitution," *Soc. Sci. & Med.* 62 (2006): 2336–48, on 2339; Jane Marcus, "Corpus/Corps/Corpse: Writing the Body at War," in *Arms and the Woman: War, Gender and Literary Representation*, ed. Helen M. Cooper, Adrienne Auslander Munich, and Susan Merrill Squier (Chapel Hill, N.C., 1989), 124–67.

[71] Woolsey described himself as a paraplegic with "complete flaccid paralysis below the waist." "Who's Who? Fred B. Woolsey—Cartoonist," *Paraplegia News* 4 (August 1950), 3.

"Why won't you ever let one of the boys help you up the curb?"

(Fred Woolsey of Madison, Wisconsin, retains all reproduction rights of the above cartoon.)

Figure 3. *"Help Up the Curb,"* Paraplegia News 4 (April 1950): 6.

which both paraplegics and their doctors were aware.[72] The joke, of course, was that a paraplegic could use his disability as an excuse for having an "inappropriate" sexual response to a professional nurse—something unthinkable for a nondisabled veteran according to social mores. In this sense, a paraplegic man could have a sexual advantage over nondisabled men. The "Head Nurse" cartoon also engaged in fantasies of paraplegic sexual prowess, showing how, contrary to assumed paralysis and flaccidity, a paraplegic with priapism could outperform other men.[73]

But the sexual desire of disabled men created unease among the able-bodied public. Sexual desire, except when discussed in the most clinical of ways, was—and still is—anathema to disability. Even in a presumably safe place like *Paraplegia News*, Woolsey received mail condemning his "offensive" cartoons and he eventually quit his post as the magazine's cartoonist because of such complaints. One reader wrote to

[72] See case study in text above, e.g. See also Munro, Horne, and Paull, "Effect of Injury to the Spinal Cord" (cit. n. 38), 909.

[73] John Money, "Phantom Orgasm in the Dreams of Paraplegic Men and Women," *Arch. Gen. Psychiat.* 3 (October 1960): 373–92.

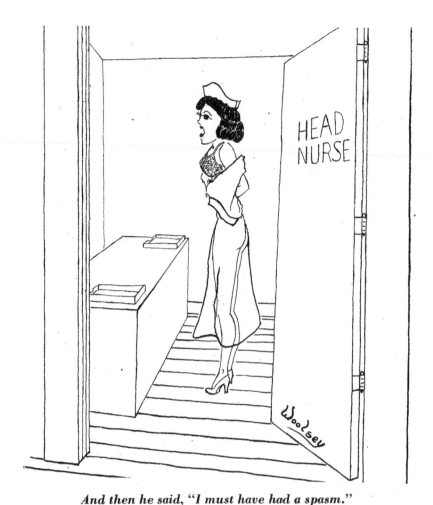

And then he said, "I must have had a spasm."

(*Fred Woolsey of Madison, Wisconsin, retains all reproduction rights of the above cartoon.*)

Figure 4. *"I Must Have Had a Spasm,"* Paraplegia News *4 (February 1950): 7.*

the editorial board saying that although he and his wife "thoroughly enjoyed reading the *News* and ha[d] . . . endorsed the policies set forth therein," they found the "so-called humor of the cartoons offensive and in extremely bad taste."[74] Whereas such pinups would have been seen as "normal" among heterosexual able-bodied men, such lustful images in association with disabled men were taken to be deviant.

This double standard is particularly ironic given the advice of marriage and sex experts at the time. Historian Jessamyn Neuhaus explains that most sex experts of the interwar years geared their manuals toward the education of men, specifically teaching them how to become "good lovers," including the art of pleasuring their wives.

[74] "Cartoons Offend," *Paraplegia News* 4 (August 1950): 2. See also Herbert L. Kleinfield's criticism of the cartoons in the June 1950 issue of *Paraplegia News*.

Husbands were considered to be "bumbling, ignorant, insensitive clods," writes Neuhaus, and thus needed instruction on how to properly stimulate women so that they, too, could reach climax.[75] Sex expert Oliver M. Butterfield believed that the biggest problem for a husband to overcome was his single-minded attention to his penis, and his rather simple animalistic urge for vaginal intercourse. In 1937, Butterfield wrote, the husband should have "more than his own desire to consider," warning against the tendency to be "carried away" by the penis's "automatic and reflex action."[76]

In some ways, the paraplegic man could ostensibly function better than his nondisabled counterparts. Take, for example, the case of "Dave," another ward mate featured in McAdam's memoir. "In Dave's case, he couldn't feel a thing," McAdam writes. "But for some strange reason he could go through the mechanics of the act itself so that it was pleasing to his wife." "At first," McAdam admits, "Dave was unhappy about it, but, because he really loved his wife, he gradually came to feel pleasure in the satisfaction he could give her."[77] At least one paraplegic had, by virtue of his physical incapacity, become very much like the kind of man Butterfield wished for.

But the stigma of a paraplegic being "half-a-man" was prevalent among nondisabled medical professionals, sex experts, and even certain paraplegics themselves. To counsel an able-bodied, virile man who enjoyed full feeling and function to attend to a wife's needs was one thing; to give the same advice to a disabled man who had impaired feeling and function was quite another. In the latter case, the socially accepted binary of male and female blurred. The risk in such sexual relations was that the woman would become a dominant aggressor, while the man would be a passive supplicant whose main goal was to satisfy his partner. This fear of role reversal can be seen in the work of Columbia psychologist Stanley Berger, who in 1951 concluded that paraplegic men had "more difficulty in identifying with their own sex and often gave indications of stronger identification with the female."[78] About one paraplegic man in particular, Berger wrote: "Identifies more with the female figure, with marked confusion over his own social-sexual role in life. Has considerable tension in this area. . . . Confusion over sexual functions; strong oral tendencies. Regards psychosexual role in a very infantile way."[79]

In the eyes of medical experts, paraplegics veered toward the infantile and the feminine when it came to sexual capacity, and thus they remained incomplete men.[80] As a result, paraplegics were judged according to whether they could meet the conventional demands of reproduction in the realm of sexuality, but not the newer, twentieth-century expectations of sexual satisfaction and mate gratification. Added to this was the Cold War belief that child rearing united American citizens. As Elaine Tyler May suggests, the nuclear family increasingly became the locus of both security and hope after World War II. In a country that felt besieged by outside forces beyond its control, parenthood—especially among permanently disabled veterans of that war—became

[75] Neuhaus, "Importance of Being Orgasmic" (cit. n. 7), 460.

[76] Oliver M. Butterfield, *Sex Life in Marriage* (New York, 1937), 101, as quoted in Neuhaus, "Importance of Being Orgasmic" (cit. n. 7), 457.

[77] McAdam, *Very Much Alive* (cit. n. 42), 87.

[78] Stanley Berger, "The Role of Impotence in the Concept of Self in Male Paraplegics" (PhD diss., Columbia Univ., 1951), 154.

[79] Ibid., 113.

[80] Berger's comments about sexuality are particularly evocative in the context of the still-widespread Freudian notion that nonpenetrative sex acts were inherently infantile.

the ultimate signifier of national identity, and the "problem" of infertility assumed a new status in medicine and in the public imagination.[81]

In many ways, the paraplegic man became the symbol of such Cold War fears of infertility and, by extension, homosexual "inversion." Even though medical research indicated that many paraplegics retained sexual function, and wives of these men reported sexual satisfaction, paraplegics were more commonly considered completely impotent, their sexual agency stripped away. The power to define male paraplegic sexuality rested, for the most part, with nondisabled men, from medical professionals who treated paraplegics to future fathers-in-law and Hollywood film crews. In all of these realms, male potency was synonymous with fertility, which left paralyzed men effectively exiled from able-bodied manhood, an exclusion that—while in some ways responsive to the real challenges certain paraplegic men faced—was determined more by cultural fears than physical realities.

PARALYTIC IMPOTENCE AS TOTALIZED DISABILITY AND TROPE

Scholar Angus McLaren writes that, at base, the history of impotence is more about power relations among men than it is about a science of individual sexual performance.[82] This observation certainly holds true for *The Men* and the medical research and institutions upon which the film was based. The nondisabled men who directed the film (Zinnemann), wrote the screenplay (Carl Foreman), and produced it (Stanley Kramer) rendered Bud as a modern-day eunuch, when, in all likelihood, the real Bud (Ted Anderson) and other men like him were potent, at least partially so. Despite what may have been a sincere effort to honor the sacrifices of the war wounded,[83] the filmmakers in effect castrated the "lesser" men they portrayed. In doing so, *The Men* contributed to one of the most powerful—yet misinformed—tropes about paraplegic men: that they were completely sexually impotent, in terms of both function and fertility. This trope continued to appear for the remainder of the twentieth century in films such as Hal Ashby's *Coming Home* (1978) and Stone's *Born on the Fourth of July* (1989).[84]

Then and now, disabled individuals—whatever their particular physical, emotional, or intellectual characteristics—are regularly assumed to be somehow incapacitated in all areas of everyday life. Susan Wendell refers to this as a "totalizing" or "global" view of disability, whereby nondisabled onlookers assume that the disabled are more incapacitated than they really are.[85] Disability scholar Tobin Siebers points out that the totalizing view of disability often extends to able-bodied assumptions about both sexual desire and sexual function in the disabled. Disability, he states, "signifies sexual

[81] Elaine Tyler May, *Homeward Bound: American Families in the Cold War Era* (New York, 1990), 120; May, *Barren in the Promised Land: Childless Americans and the Pursuit of Happiness* (Cambridge, Mass., 1997), 142.

[82] McLaren, *Impotence* (cit. n. 13), xiv.

[83] Foreman, in particular, spoke about how his time in the army engendered a "great sense of responsibility" as well as an attention to "the immense educational potential of films, the impact, both intellectual and emotional, that they had on people when properly presented; the potential that they had for good or evil." Carl Foreman, interview by Joan and Robert Franklin, April 1959, transcript, Popular Arts Project, Columbia University Oral History Collection, New York City, 1674.

[84] *Coming Home,* directed by Hal Ashby (1978; New York, 2002), DVD; *Born on the Fourth of July* (cit. n. 16).

[85] Susan Wendell, *The Rejected Body: Feminist Philosophical Reflections on Disability* (New York, 1996), 25.

limitation, regardless of whether the physical and mental features of a given impairment affect the ability to have sex."[86] As such, the myth of "global" disability does a great deal of work in explaining the continued propensity of popular works to assume full incapacitation of sexual function, despite the fact that a significant number of paralyzed soldiers had partial or full use of their genitals. It also explains why outlets such as *Paraplegia News* were essential—both to paraplegic men and to the public—in reclaiming sexual agency among disabled veterans with a spinal cord injury.

Although the myth of global disability persists in the minds and phobic imaginations of nondisabled filmmakers and onlookers, the reality is much more complex, especially when the historian listens to the voices of those living with the actual impairment. Through outlets such as *Paraplegia News*, veterans expressed their sexual agency in myriad ways. Some boasted a sexual prowess that surpassed nondisabled men, others expressed a quiet satisfaction in their ability to pleasure their wives, and still others wished to sire a child in order to demonstrate their fertility to the rest of the world. With the ever-present threat of having their sexuality stripped away from them—as *The Men* did—paraplegic veterans continually reclaimed their sexual agency through advocacy and print. Rather than having a globally uniform view of masculinity and sexual performance, members of the PVA normalized heteronormative disabled sex by insisting that all heterosexual relations varied, that the sexual response of a man and a woman "takes many varied and beautiful forms."[87]

In the past, the history of postwar sexuality has focused primarily on the experiences of women. As this piece demonstrates, however, an understanding of prevailing ideas about male sexuality—and, in particular, disabled-male sexuality—is equally crucial to comprehending sexuality, gender roles, familial relations, and popular culture during this period. Masculinity and femininity are never defined in isolation from one another, but rather relationally, and both are complicated by questions of ability and disability. As Milam and Nye point out in the introduction to this volume, "social constructions of masculinity function simultaneously as foils for femininity and as methods of differentiating between kinds of men," a process clearly at work in the ubiquitous discussions surrounding paraplegic virility.[88] In an age anxious about shifting gender roles and fluid sexuality, Marlon Brando's Bud and his real-life compatriots were stuck at the intersection of seemingly irreconcilable social roles: the successfully reintegrated soldier, the virile husband, and the asexual paraplegic. These "half men" may have helped win the war, but as *The Men's* title sequence suggested, the aftermath was, indeed, another battle entirely.

---

[86] Siebers, *Disability Theory* (cit. n. 19), 142.
[87] Hover, "I Married a Paraplegic" (cit. n. 52), 9.
[88] Milam and Nye, "Introduction" (cit. n. 29), 3.

Maintaining Masculinity in Mid-Twentieth-Century American Psychology:

Edwin Boring, Scientific Eminence, and the "Woman Problem"

by Alexandra Rutherford*

ABSTRACT

Using mid-twentieth-century American psychology as my focus, I explore how scientific psychology was constructed as a distinctly masculine enterprise and was navigated by those who did not conform easily to this masculine ideal. I show how women emerged as problems for science through the vigorous gatekeeping activities and personal and professional writings of disciplinary figurehead Edwin G. Boring. I trace Boring's intellectual and professional socialization into masculine science and his efforts to understand women's apparent lack of scientific eminence, efforts that were clearly undergirded by preexisting and widely shared assumptions about men's and women's capacities and preferences.

> The scientist . . . is a compulsive, hard-working, methodical person whose work is more important to him than anything else in his world, including his family, from which very often he was isolated in childhood or adolescence; he is a man who is often unable to distinguish work from play and in any case brings his methodical work habits into his play, someone who is bright and open-minded, devoting his life to the discovery of new relationships.
> —Edwin G. Boring, "Eponym as Placebo," 1963[1]

Compulsive, hardworking, methodical. All are descriptors easily applied to Edwin Garrigues Boring (1886–1968), the eminent Harvard psychologist hailed most often by his colleagues not as a gifted scientist but as a meticulous historian, biographer, editor, reviewer, administrator, and critic. Although not a "great man" of science himself, Boring was one of the most influential and respected figures in twentieth-century

*Department of Psychology and Institute for Science and Technology Studies, York University, Toronto, ON, Canada M3J 1P3; alexr@yorku.ca.

I would like to thank Michael Pettit, Wade Pickren, and Andrew Winston for comments on earlier versions of this work.

[1] Edwin G. Boring, "Eponym as Placebo," in *History, Psychology, and Science: Selected Papers*, ed. Robert I. Watson and Donald P. Campbell (New York, 1963), 5–25.

American psychology. As one colleague put it, he was an unprecedented "form-giver and meaning-giver" to the field.[2] Julian Jaynes wrote at the time of Boring's death in 1968 that he was "probably the most famous psychologist in the academic world."[3] Indeed, in his later years, after teaching psychology on educational television, Boring even earned the epithet "Mr. Psychology."[4]

In addition to his role as a "form-giver and meaning-giver" to the field, Boring was also a vigorous gatekeeper.[5] He was frequently asked by colleagues to provide his opinions about prospective job candidates and the names and rankings of those suitable for posts ranging from laboratory assistant to university president. He took on this task with relish and a compulsive conscientiousness, writing literally tens of thousands of letters over the course of his career.[6] Thus, his pronouncements on the field and its individual players—meted out in private and published forms—offer valuable insight into the gendered and racialized contexts in which psychology developed, and to which it contributed, over the course of the twentieth century.

In this essay, I draw on Boring's published work, his prolific correspondence, and the historiography on twentieth-century American psychology to explore how he and others created a culture of scientific masculinity that functioned to regulate the form and extent of women's and men's participation in the field. How was this scientific masculine culture expressed, maintained, and enforced? How did it influence who was seen as appropriately scientific, or even qualified to do science, and who was not? How did it affect the ways in which women navigated and responded to androcentrism, sexism, and discrimination? How was this culture challenged and disrupted?

Boring began his career right around World War I, a period that saw psychology in the United States shift from a fledgling experimental discipline heavily influenced by its European origins to a fully indigenized American science with an increasing emphasis on practical application and social usefulness.[7] By the time Boring's career hit full stride, in the two decades or so after World War II, the field witnessed another major change: the relative parochialism of pre–World War II psychology gave way to, if not big science, at least medium science as psychology (especially the newly professionalizing clinical branch) vied for its share of the ample postwar funding pie.[8]

---

[2] Leonard Carmichael, "On the Science and Art of Science," review of *History, Psychology, and Science: Selected Papers*, by Edwin G. Boring, *PsycCritiques* 9 (1964): 305–6.

[3] Julian Jaynes, "Edwin Garrigues Boring, 1886–1968," *J. Hist. Behav. Sci.* 5 (1969): 99–112.

[4] Stanley S. Stevens, *Edwin Garrigues Boring, 1886–1968: A Biographical Memoir* (Washington, D.C., 1973), 40–76, on 41.

[5] Andrew S. Winston has highlighted Boring's gatekeeping activities, providing a thoughtful analysis of Boring's job references for Jewish psychologists; Winston, "'The Defects of His Race': E. G. Boring and Anti-Semitism in American Psychology, 1923–1953," *Hist. Psychol.* 1 (1998): 27–51.

[6] By his personal estimate, his files for 1919–60 contained about 40,000 letters by him, and at least the same number to him; Boring, *Psychologist at Large* (New York, 1961), 87.

[7] See John O'Donnell, *The Origins of Behaviorism: American Psychology, 1870–1920* (New York, 1985). The development of testing was also a significant factor in this transformation; see Michael M. Sokal, ed., *Psychological Testing and American Society, 1890–1930* (New Brunswick, N.J., 1987); Daniel J. Kevles, "Testing the Army's Intelligence: Psychologists and the Military in WWI," *J. Amer. Hist.* 55 (1968): 565–81; Franz Samelson, "World War I Intelligence Testing and the Development of Psychology," *J. Hist. Behav. Sci.* 13 (1977): 274–82; Robert E. Gibby and Michael J. Zickar, "A History of the Early Days of Personality Testing in American Industry: An Obsession with Adjustment," *Hist. Psychol.* 11 (2008): 164–84.

[8] On American psychology's rise through the post–World War II period, see Mitchell Ash, "Psychology," in *The History of the Social Sciences since 1945*, ed. Roger E. Backhouse and Philippe Fontaine (Cambridge, 2010), 16–37; James Capshew, *Psychologists on the March: Science, Practice, and Professional Identity in America, 1929–1969* (New York: 1999); Ellen Herman, *The Romance of American*

Moreover, as American psychologists emerged from their war experiences and faced a new postwar world, the role that they would play in creating it weighed heavily on their minds.

Intricately and intimately connected to these changes were substantial shifts in the discipline's—and society's—gender dynamics, especially as they pertained to the professional status of women. These dynamics, and the role Boring played in constructing, maintaining, and policing them, provide the focus of my account. My goal is to explore how the gendering of psychological science worked in and through Boring, especially when challenges to traditional gender stereotypes confronted him and the discipline during and around World War II. I argue that Boring's (and others') attempts to police the boundaries of the field and the kind of person who was best suited to succeed in it can help us more richly contextualize contemporary debates about gender and science and understand women's continued underrepresentation in certain scientific fields.[9]

For Boring, an essential aspect of the scientific attitude, and one of the most consistent and recurring themes in his personal and professional writings, was thoroughgoing objectivity. For Boring, objectivity was the valorized bedrock on which science, and perhaps even successful civil society, was based. Additionally, and as importantly, his ideas about scientific eminence and how to achieve it resulted in a veritable recipe for success in science, a recipe that was both clear and uncompromising in its prescriptions as well as transparently gendered.[10] Boring's use of the rhetoric of objectivity is important because in his role as form-giver and meaning-giver in American psychology, what Boring thought of as objectivity and, moreover, who was capable of it, were consequential for individual careers and the character of the discipline itself.

To understand the underpinnings of Boring's views on objectivity, science, and scientific eminence, I first explore how his intellectual and professional outlooks were shaped early in his training in part by what historian Robert A. Nye has described as the "masculine codes" that have regulated professional sociability in science well into the twentieth century.[11] I then analyze Boring's collaboration with fellow psychologist Alice Bryan on the "woman problem." I use the dynamics of this collaboration to show how Boring both acted out and deployed his version of objectivity to prevent what he perceived as "values" from intruding into the scientific process. I show how he reiterated his recipe for scientific eminence to argue that although women might in theory be able to become great scientists, in reality this was highly unlikely. I suggest that responses to this work indicate that Boring was not alone in his preoccupation with and interpretation of the woman problem; many of his colleagues (male and female) reinforced the common perception that women were simply not suited to science. Finally, I briefly consider the contextual factors that may have heightened concerns and worries about psychology as a properly scientific, objective, and de facto masculine enterprise in this period.

Psychology: Political Culture in the Age of Experts, 1940–1970 (Berkeley and Los Angeles, 1996); Alexandra Rutherford, *Beyond the Box: B. F. Skinner's Technology of Behavior from Laboratory to Life, 1950s–1970s* (Toronto, 2009).

    [9] See Stephen J. Ceci and Wendy M. Williams, "Understanding Current Causes of Women's Underrepresentation in Science," *Proc. Natl. Acad. Sci.* 108 (2011): 3157–62; Londa Schiebinger, ed., *Gendered Innovations in Science and Engineering* (Stanford, Calif., 2008).

    [10] For a forthright statement of this recipe, see Boring, "The Woman Problem," *Amer. Psychol.* 6 (1951): 679–82.

    [11] Robert A. Nye, "Medicine and Science as Masculine Fields of Honor," *Osiris* 12 (1997): 60–79.

"AN UNEXPURGATED MALE PSYCHOLOGY": GROWING UP WITH E. B. TITCHENER

By his own description, Edwin Boring was a boy "born unwanted into a matriarchy," the only son in a family dominated by his older sisters, mother, maiden aunts, and grandmothers.[12] His Philadelphia-based family was Orthodox-Hicksite-Quaker-Moravian; accordingly, he was educated at home before attending a Quaker school at age nine, where, as he noted, he turned out to be "somewhat of a sissy."[13] In 1904, he graduated from high school and left Philadelphia to pursue an electrical engineering degree at Cornell University. Although he was one of only three or four boys in his high school graduating class of twenty, at Cornell he was thrust into an almost all-male environment. He later remarked that he knew "almost no girls all through the four years."[14]

Although there were few electives available in his regimented program, for one of them Boring chose elementary psychology. It was in this course, in 1905, that he first met Edward Bradford Titchener, an Oxford-educated Englishman who lectured dramatically in his academic robes, proclaiming that this costume gave him "the right to be dogmatic."[15] Boring scored well in the course and received encouraging feedback from his illustrious teacher. Nonetheless, he continued in engineering, completed his degree, and worked for a year at the Bethlehem Steel Company. He then tried his hand teaching science at the Moravian Parochial School in Bethlehem. There his male students heckled and teased him, apparently shellacking him to his chair at one point.[16] He decided to return to college to earn his formal teaching credentials.

Back at Cornell, however, the allure of the psychology laboratory—and Titchener—asserted itself, and in the fall of 1910 Boring decided to pursue a PhD in the field. In Titchener, Boring found both an intellectual mentor and a strong paternal influence. As he noted, Titchener (like Freud) "needed to play the father role"[17] and Boring was an eager son. It has been remarked that the "imprint of Titchener" never left Boring.[18]

Boring began his graduate studies in psychology at a point in Titchener's career when the "feudal lord," as he was sometimes called, was provoked to actively defend the objectivity of his system of rigorous, controlled, experimental introspection. Titchener had received his training in Leipzig under Wilhelm Wundt. In Wundt's laboratory he took up experimental introspection with zeal and hoped to procure a post back at Oxford to start a laboratory in the Wundtian mode. Oxford, however, was not ready for the upstart new discipline of experimental psychology and refused to provide laboratory facilities. Finding this unacceptable, Titchener took up a proposal proffered by his congenial Leipzig lab mate, Frank Angell, to come to the wilds of upstate New York and replace him as director of the laboratory of psychology at Cornell University.[19]

[12] Boring, *Psychologist at Large* (cit. n. 6), 16.
[13] Ibid., 11.
[14] Ibid., 18.
[15] Ibid.
[16] Stevens, *Edwin Garrigues Boring* (cit. n. 4), 44.
[17] Boring, *Psychologist at Large* (cit. n. 6), 25.
[18] John J. Cerullo, "E. B. Boring: Reflections on a Discipline-Builder," *Amer. J. Psychol.* 101 (1988): 561–75, on 561.
[19] Having a laboratory in which to conduct experiments was essential to Titchener's conception of psychology; see Ryan D. Tweney, "Programmatic Research in Experimental Psychology: E. B. Titchener's Laboratory Investigations, 1891–1927," in *Psychology in Twentieth Century Thought and Society*, ed. Mitchell G. Ash and William R. Woodward (Cambridge, 1987), 35–57. For more on the gendered nature of spaces of scientific knowledge making, see Mary Terrall, "Masculine Knowledge, the Public Good, and the Scientific Household of Réaumur," in this volume.

At Cornell, where he remained for his entire career, Titchener became the unassailable leader of the school of structural psychology. Structuralism was distinguished from the distinctively American functionalist psychology represented by William James, John Dewey, and others by its pure goal of uncovering the basic structures of the normal, conscious, adult, human mind, unsullied by any concern with practicality or application. Titchener held strongly that the realm of science should be clearly differentiated from the realm of application—or technology, as he put it. He devoted his complete attention to experimental psychology undertaken with the method of controlled, experimental introspection. In this method, subjects were presented with stimuli of various kinds (auditory, olfactory, visual) and rigorously trained to provide extensive reports on their mental processes. At its most extreme, subjects might give a fifteen- to twenty-minute report on a stimulus presentation that had lasted a mere two seconds![20]

At the time Boring came most fully under Titchener's influence, Titchener was on the defensive about the objectivity of introspection. He defended his position by reiterating a number of his assumptions about the scientific process, assumptions that deeply affected Boring's later views. According to Titchener, introspectors must be strictly impartial, unbiased, and unprejudiced, letting no preconceived theory influence their reporting of the facts. In his famous four-volume laboratory manual, he cited work that characterized objectively minded versus subjectively minded people as having different personalities. Notably, objectively minded individuals were described as contemplative and critical insofar as they carefully evaluated the range of different opinions before deciding on their own. By contrast, subjectively minded individuals were either quick to react, yielding immediately to suggestion so as not to antagonize, or, conversely, too quick to criticize and take the opposing view.[21]

Although Titchenerian structuralism did not survive well past its founder's death in 1927, several aspects of Titchener's outlook persisted in Boring's orientation even as he (and the field) abandoned the specifics of structuralism itself. A number of these aspects would come to have distinctly gendered implications. First was Titchener's strict distinction between science and technology (i.e., application). Boring also reinforced this distinction throughout his career, although he did note that his appreciation for mental testers increased during World War I because of their resemblance to pure experimentalists in their approach to their task: "Titchener's in-group at Cornell had appreciated mental testers in much the same way that the Crusaders, gathered around Richard Coeur-de-Lion, appreciated Moslems, but this First World War gave me a respect for the testers. I saw clearly that . . . testers closely resemble pure experimentalists in habits of work, in enthusiasm, and in thoroughness."[22] But as mental testing, especially the testing of schoolchildren and psychiatric patients, increasingly became women's work after World War I,[23] Boring came to see it as connected with women's widespread tendency toward application, to prefer "particulars" over grand theory,

[20] Deborah Coon, "Standardizing the Subject: Experimental Psychologists, Introspection, and the Quest for a Technoscientific Ideal," *Tech. & Cult.* 34 (1993): 757–84, on 774.

[21] See Christopher D. Green, "Scientific Objectivity and E. B. Titchener's Experimental Psychology," *Isis* 101 (2010): 697–721.

[22] Boring, *Psychologist at Large* (cit. n. 6), 31.

[23] Laurel Furumoto, "On the Margins: Women and the Professionalization of Psychology in the United States, 1890–1940," in Ash and Woodward, *Psychology* (cit. n. 19), 93–113.

and to be "sociotropic" in orientation—that is, oriented toward helping individuals and solving social problems rather than contributing to basic scientific knowledge.[24]

In 1926, a year before Titchener's death and just a few years after taking up his lifelong appointment at Harvard in 1922, Boring wrote to his sister Alice Boring, who was a professor of biology at Yenching University in China. In reporting on his experience teaching experimental psychology to "the girls" at Radcliffe, he wrote,

> With the girls I began yesterday by telling them what they hoped to get out of the course and that they would not get it. What they hope is that they somehow will be able to improve the world socially by the use of the psychological tool. What they get is a straight elementary science as psychology in these days knows how to be. Poor things, it is pathetic. But when I urged them to withdraw they all turned their cards in, so far as I could see. That's women for you. They always do what they ought to do or what somebody wants them to do. . . . Men really care about law and abstract principles while women care mostly about personal relations.[25]

Boring did not hesitate to address these comments to his sister, a biologist. As he reported in an oral history conducted many decades later, he simply did not regard his sister as a scientist. Referring to all three of his sisters, he said, "None of them was scientific, just none of them was scientific. . . . All three of them, in very different ways, seemed to be utterly stupid scientifically. Alice is a biologist, but even . . . the biologist . . . [is] not at the top of the scientific hierarchy. . . . I don't think of Alice as a scientist."[26]

Thus, very early in his career, Boring's views on the compatibility between science and sex were formulated and expressed: women were more interested in technology than science, women were subjectively oriented (i.e., easily led by suggestion), and women were more interested in concrete particulars than abstract theory. As such, it would be very difficult for women to become successful scientists. The very nature and definition of science, for Boring, was constructed in a way that excluded certain kinds of people. Notably, for Boring this fact was simply inescapable in accounting for the woman problem in American psychology.

BECOMING A MEMBER OF THE CLUB: CODES OF CONDUCT IN THE EXPERIMENTALISTS

Boring was shaped intellectually by his mentor, but he also received through Titchener a good dose of socialization into masculine codes of conduct. As Boring wrote, "Along with Titchener's vivid and indefatigable intellectual activity went his immutable beliefs about decorum, good and bad manners, and loyalty. . . . Those who broke the code of manners . . . he ostracized at least temporarily and often permanently."[27] One venue in which Titchener enforced this etiquette was his scientific club, called,

---

[24] Boring, "Biotropes, Sociotropes, and Teaching," *Amer. Psychol.* 21 (1966): 80–3.

[25] Boring to Alice Boring, 30 September 1926, Edwin G. Boring Papers (hereafter cited as Boring Papers), HUG 4229.25.5, Alice M. Boring Correspondence, 1921–1956, Folder: Alice M. Boring, 1925–26, Harvard University Archives (hereafter cited as HUA).

[26] Oral History Interviews of Edwin G. Boring, 5–9 July 1961, conducted by John Chynoweth Burnham, Cornell University Archives, 3, 5.

[27] Boring, *Psychologist at Large* (cit. n. 6), 23.

during his lifetime, the Experimentalists.[28] The Experimentalists operated according to a masculine code of professional sociability, and Boring's exposure to this code shaped his future attitudes toward and professional dealings with both his male and his female colleagues.[29]

As Titchener originally conceived it, the Experimentalists was a small, invitation-only group of men who would come together once a year to present and discuss experimental work in progress in an informal, collegial—though constructively combative—atmosphere. The group included the heads of some of the major psychological laboratories and their promising graduate students, and the host was responsible for issuing invitations, usually in consultation with Titchener. The Experimentalists began meeting in 1904 and met every year thereafter, save one year during World War I. The two features of Titchener's club that are most salient here are that it was focused exclusively on the kind of experimental psychology of which Titchener approved and that it was exclusively male. Boring began attending the meetings in 1911. He later wrote the following of this inaugural experience: "It was my first meeting, and the occasion when Dodge and Holt attacked Titchener on introspection. My wife-to-be and Mabel Goudge secreted themselves in the next room with the door just ajar to hear what unexpurgated male psychology was like."[30]

Although Titchener never directly articulated why women should be so rigorously excluded and in fact supported the training of female students in general, there are some fairly plain indicators that one of the primary functions of the meetings was to provide a gentlemen's club atmosphere in which a particular version of scientific psychology could be cultivated.[31] In one of his first letters announcing his intentions and soliciting interest, he wrote, "For many years I wanted an experimental club—no officers, the men moving about and handling [apparatus], the visited lab to do the work, no women, smoking allowed, plenty of perfectly frank criticism and discussions, the whole atmosphere experimental, the youngsters taken in on an equality with the men who have arrived."[32] Although Titchener was clear that creating this kind of

---

[28] After Titchener's death in 1927, the Experimentalists became the Society of Experimental Psychologists. Boring wrote two histories of the group; Boring, "The Society of Experimental Psychologists, 1904–1938," *Amer. J. Psychol.* 51 (1938): 410–23; Boring, "Titchener's Experimentalists," *J. Hist. Behav. Sci.* 3 (1967): 315–25. See also C. James Goodwin, "On the Origins of Titchener's Experimentalists," *J. Hist. Behav. Sci.* 21 (1985): 383–89; and Goodwin, "Reorganizing the Experimentalists: The Origins of the Society of Experimental Psychologists," *Hist. Psychol.* 8 (2005): 347–61.

[29] Boring continued the practice of a males-only club when he ran the psychological colloquium at Harvard in the late 1920s. A student later recalled that women were banned because Boring "did not believe that men could discuss things frankly in the presence of women." See Mildred Mitchell's autobiographical account in *Models of Achievement: Reflections of Eminent Women in Psychology*, ed. Agnes N. O'Connell and Nancy Felipe Russo (New York, 1983), 121–39, on 127.

[30] Boring, "Titchener's Experimentalists" (cit. n. 28), 322. Boring's "wife-to-be" was Lucy May Day, also a graduate student at Cornell, who received her PhD in experimental psychology a year before her husband-to-be; see Laurel Furumoto, "Lucy May Boring (1886–1996)," *Amer. Psychol.* 53 (1998): 59.

[31] As an Englishman, Titchener would have been exposed to the gentlemen's club culture of late Victorian Britain. See Amy Milne-Smith, *London Clubland: A Cultural History of Gender and Class in Late-Victorian Britain* (New York, 2011). He was also educated at Oxford, which enforced its own forms of masculinity; see Paul R. Deslandes, *Oxbridge Men: British Masculinity and the Undergraduate Experience, 1850–1920* (Bloomington, Ind., 2005). For more on the importance of masculine fraternities and networks, see Michael S. Reidy, "Mountaineering, Masculinity, and the Male Body in Mid-Victorian Britain," and Erika Lorraine Milam, "Men in Groups: Anthropology and Aggression, 1965–84," both in this volume.

[32] E. B. Titchener, as cited in Boring, "Titchener's Experimentalists" (cit. n. 28), 317.

atmosphere required the strict exclusion of women, what was also implicit was that it would require the exclusion of certain kinds of men—men who could not handle apparatus, who were not based in the laboratory, and who could not withstand frank criticism. In a 1961 oral history interview, Boring noted that "the historical antiquity of the experimentalists in the development of psychology tells against women and against Jews because Jews are not good manipulators of apparatus. I swear there's something there. Again and again. It may be that they just love human beings so that they haven't time for brass instruments, but this we just take for granted in the laboratory if a Jew comes in."[33] Thus, even male psychologists—in this case male Jewish psychologists—who could not live up to the masculine ideals of apparatus-driven laboratory science were outcasts.

As historian of psychology Laurel Furumoto has shown, correspondence at the founding of the group revealed some of the other members' opinions on the matter of women, and these reinforced Titchener's masculine ideal. E. C. Sanford, of Clark University, abjured that on the one hand, on purely scientific grounds, several women had a right to be included. But, he wrote "on the other hand they would undoubtedly interfere with the smoking and to a certain extent with the general freedom of a purely masculine assembly."[34]

In 1927, Titchener died, and the task of deciding the fate of the Experimentalists fell to a committee of five, including Edwin Boring. In that year, no decision was reached on the issue of opening the membership to women, although two women, Margaret Floy Washburn and June Etta Downey, were both discussed as suitable candidates. The next year, a larger committee was convened to decide the issue, and the decision was made to elect Washburn and Downey to membership. Did this indicate that the masculinity of the assembly was no longer seen as imperative to uphold? Two observations suggest otherwise. Downey never attended a meeting and died three years after she was elected. As Boring later described her, "She accepted the inferiority of the female sex pretty well. They don't make experimentalists."[35] Washburn did attend several meetings before her death in 1937, hosting the meeting in 1931 at Vassar College (fig. 1). However, another twenty-one years would pass before another woman, Eleanor Gibson, was elected.

There are also indications that strict codes governed proper decorum for the women who did infiltrate this masculine enclave. In the spring of 1934, Washburn wrote to Boring upon her return from a Harvard meeting of the Society of Experimental Psychologists where Boring had played host. Evidently, during this visit, Washburn had committed the egregious blunder of passing through the men's-only entrance to the Harvard Faculty Club, where Boring had arranged for the group to dine (that the dinner was held at a male-only venue is some indication that the masculinity of the assembly was still firmly in place). As she retrospectively wrote to Boring, "Please believe that I hadn't the smallest intention of claiming a right in the Faculty Club's penetralia. I have never done the slightest thing to advance the cause of feminism. The fact was that having got into that room in the search for the rest of you, I had not

[33] Oral History Interviews of Edwin G. Boring (cit. n. 26), 234.

[34] As cited in Laurel Furumoto, "Shared Knowledge: The Experimentalists, 1904–1929," in *The Rise of Experimentation in American Psychology*, ed. Jill G. Morawski (New Haven, Conn., 1988), 94–113, on 104.

[35] Oral History Interviews of Edwin G. Boring (cit. n. 26), 234.

Figure 1. *The members of the Society of Experimental Psychologists at their 1935 meeting at Yale University. Margaret Floy Washburn is seated in the second row, fourth from the left. Edwin Boring is also seated in the second row, in the last seat on the right. Reprinted courtesy of the Archives of the History of American Psychology, Cummings Center for the History of Psychology, University of Akron.*

courage enough to leave it, before the eyes of the strangers in the other rooms, in the custody of one of the attendants."[36]

In his reply, rather than apologizing outright for a situation that had clearly caused his distinguished guest considerable embarrassment and humiliation, Boring wrote, "I am so sorry that you have the issue of the Faculty Club on your mind. I am afraid I was very awkward about it, but I do absolve you now completely of the aggressive feminism which I did momentarily attribute to you."[37] Despite Washburn's reassurance that she had "never done the slightest thing to advance the cause of feminism," the impact of this event reverberated fifteen years later in a letter Boring wrote to another female colleague, Helen Peak.[38]

In 1949, Peak and Boring served on the Policy and Planning Committee of the American Psychological Association (APA). In a letter to Boring after a committee meeting, Peak commented candidly on a small outburst provoked by her frustration with the direction of the meeting: "My slightly silly remark at dinner inquiring whether we should talk shop or be frivolous was an ineffectual effort to steer us back to our job. You see I just can't get over my upbringing where women are supposed not to take charge openly. I am full of all sorts of aggressive feelings on such occasions

[36] M. F. Washburn to E. G. Boring, 10 April 1934, Boring Papers, HUG 4229.5, Correspondence, 1919–1969, First File, 1919–1956, Box 61, Folder: Washburn, M. F., 1933–34, #1400, HUA.

[37] Boring to Washburn, 12 April 1934, ibid.

[38] Helen Peak (1900–85) received her MA from Radcliffe College in 1924 and her PhD from Yale University in 1931. After a position at Randolph-Macon Women's College and jobs at the Office of War Information and Connecticut College, she was chosen to be the first Kellogg Chair at the University of Michigan. See Daniel Katz, "Helen Peak, 1900–1985," *Amer. Psychol.* 42 (1987): 510.

which I often don't conceal but which rarely become effective."[39] In his reply Boring wrote complicitly, "I do not know what to do about women any more than you do. On committees they ought not to have any sex, but they do. Just setting your teeth and barging ahead does not help. Miss Washburn did that. I was awfully fond of her, but at meetings of the SEP [Society of Experimental Psychologists] I was always aware of her because she would insist on going in at some man's entrance to a men's club or something of that sort, and just made a nuisance of things."[40] Boring then went on to note that Ada Comstock, the president of Radcliffe College who had initiated classroom coeducation at Harvard in 1943, managed to conceal her sex "better than most women." As he wrote approvingly of Comstock's professional decorum, "I think on committees I just wasn't aware that we had a woman in the room." He continued, "I don't know about you yet. Ruth Tolman does beautifully, mostly playing the asexual role, sometimes being humorously feminine, occasionally taking command of a situation."[41]

In sum, Boring's discomfort with Washburn, his uncertainty about Peak, and his acceptance of Comstock and Tolman convey the complex terrain that women navigated in order to be accepted into the professional and scientific culture. To accomplish the feat of having "no sex," women actually drew on a variety of self-presentational strategies to be accepted into a community that was distinctly, if invisibly, gendered masculine. Because this masculinity—imbuing both the way business was conducted and the nature of the business itself—was strictly enforced, women had to be extremely vigilant about all of their professional decorum. Even small transgressions of the masculine code threatened to mark them as "sexed" outsiders.

As historian Margaret Rossiter has noted, women's progress in science faltered considerably in most fields in the 1930s.[42] Robert Nye has suggested that one underappreciated factor to help explain women's uneven progress in science has been "the exclusionary role played by the male honor culture in the multiple formal and informal settings where professional sociability controlled behavior, expectations, and opportunities."[43] Through his socialization with Titchener and his participation in the Experimentalists, Boring was exposed to some very powerful norms governing masculinity and scientific conduct that, in combination with personal values and reigning gender ideologies, heavily determined his subsequent attitudes toward both male and female colleagues but clearly had important implications for women. These implications did not go unnoticed. By the early 1940s, some female psychologists began to call attention to the absence of women on committees formed to organize psychologists for the war effort. This catalyzed more general concern about what came to be called the "woman problem," a topic to which Boring would quickly turn his attention.

[39] Helen Peak to Edwin Boring, 29 January 1949, Boring Papers, HUG 4229.5, Box 46, Folder: Peak, Helen, 1949–50, #1018, HUA.

[40] Boring to Helen Peak, 31 January 1949, ibid.

[41] Ibid. Helen Peak, Ruth Tolman, and Florence Goodenough actually formed an informal alliance to combat what they perceived as the feminist militancy of some of their peers in the National Council of Women Psychologists; see Ann Johnson and Elizabeth Johnston, "Unfamiliar Feminisms: Revisiting the National Council of Women Psychologists," *Psychol. Women Quart.* 34 (2010): 311–27.

[42] Margaret W. Rossiter, *Women Scientists in America: Struggles and Strategies to 1940* (Baltimore, 1982).

[43] Nye, "Medicine and Science" (cit. n. 11), 75.

"WOMEN SIMPLY DO NOT GO IN FOR THIS SORT OF THING": SCIENCE AND THE "WOMAN PROBLEM"

Between 1944 and 1947, three articles coauthored by Boring and Columbia University psychologist Alice Bryan appeared in the *Psychological Bulletin* and the *American Psychologist*. All three addressed what Boring and many others referred to as the woman problem, namely, the tendency for women to be accorded less recognition than men in the professions and in public life. In psychology, the woman problem was specified as the underrepresentation of women in the highest offices of psychology's professional organizations and women's general lack of eminence compared to men. In turning his attention to the problem, Boring and his colleagues—both male and female—reinforced a vision of science and a recipe for scientific eminence that was set in distinctly masculine terms.

How did Boring's concern with the woman problem, and his collaboration with Alice Bryan, come about? As the United States' involvement in World War II began to heat up in the early 1940s, the Emergency Committee in Psychology (ECP) was formed to help plan how psychologists could best contribute to the war effort. As the work of the ECP unfolded, female psychologists were noticeably absent from any of the committees and plans. A group of female members of the American Association for Applied Psychology (AAAP) held an informal discussion about the issue at the meeting of the association in Evanston, Illinois, in September 1941.[44] They decided to write a letter to Robert Brotemarkle, the AAAP representative to the ECP. In composing their letter, they were careful to avoid any suggestion that they were expecting special treatment as women. As Harriet O'Shea, one of the leaders of the group, wrote to Brotemarkle:

> The women psychologists of the country have no wish to be considered as a separate group but it is, on the contrary, their particular hope that they will be neither favored nor excluded from service because they are women. . . . It is our hope that the Emergency Committee in Psychology will succeed in keeping psychologists active throughout the entire defense structure, regardless of whether the psychologist be a man or a woman but only in terms of the ability of the psychologist.[45]

Despite the moderation with which they expressed their concern, they did not receive a favorable response. They were admonished to sit tight, "be good girls," and wait until plans could be made that would include them.[46] When these plans did not materialize, the women decided to form their own organization, the National Council of Women Psychologists (NCWP), to mobilize women psychologists for the war effort and to lobby for women's inclusion in established plans.[47] Alice Bryan emerged as a leader of this group and eventually became their elected representative to the ECP.

---

[44] Discussants were Alice Bryan, Edwina Cowen, Elaine Kinder, Harriet O'Shea, and Millicent Pond. See Harriet O'Shea to Robert Brotemarkle, 17 September 1941, Alice I. Bryan Papers, 1921–1992, Box 3, Folder III-B-8 Prof. Activities, Columbia University Archives (hereafter cited as CUA).

[45] Ibid.

[46] Gladys C. Schwesinger, "Wartime Organizational Activities of Women Psychologists. II. The National Council of Women Psychologists," *J. Consulting Psychol.* 7 (1943): 298–301, on 299.

[47] On the establishment of the NCWP, see James Capshew and Alejandra Laszlo, "'We Would Not Take No for an Answer': Women Psychologists and Gender Politics during WWII," *J. Soc. Issues* 42 (1986): 157–80; Mary Roth Walsh, "Academic Professional Woman Organizing for Change: The Struggle in Psychology," *J. Soc. Issues* 41 (1985): 17–28.

In 1942, Bryan was asked to serve as the only woman on the ECP Subcommittee on Survey and Planning, charged with reorganizing the American Psychological Association. Here, she met Edwin Boring, and the seeds for their collaboration were sown.

Along with other male psychologists, Boring was becoming increasingly provoked by Bryan's repeated assertions that women did not hold proportionate representation in APA offices. Impressed with Bryan's levelheaded approach to committee work—although he complained that she got too mired in particulars—Boring suggested that they collaborate on an empirical study of the problem. In pitching the project to Bryan, he wrote, "I should like to do the job, because it interests me, because I always want to bring *objective* analysis to bear on issues that are emotionally tinged."[48]

Boring and Bryan's joint study of what Boring referred to as an "emotionally tinged" issue resulted in the aforementioned three articles published between 1944 and 1947.[49] Their 1944 report, "Women in American Psychology: A Prolegomenon," presented data that clearly documented women's underrepresentation in the highest offices of the APA. They contextualized this finding by comparing women's underrepresentation in the APA with their proportionately smaller underrepresentation in the AAAP. As they wrote:

> One of the reasons why women play a less important role in the American Psychological Association than in the American Association for Applied Psychology is that the older association is devoted to "scientific" psychology, and fewer women become distinguished "scientists" than do men. . . . As to whether the exclusion of women from wide participation in the established sciences is due to nature or nurture, we venture no opinion, but certain it is that modern civilization tends not to place them in "science," not in any large numbers. Here are some of the facts.[50]

They then proceeded to review data about women's membership in the National Academy of Science and their inclusion in *American Men of Science*. Because very few women received either mark of distinction, they concluded that the paucity of women must have greater cultural significance than the fact that men may be more likely to vote for men. They proceeded to explore the nature of various committees within the APA, suggesting that some were more "male" and some were more "female." They wrote:

> It might be said that many of the committees of the American Psychological Association seem to have functions that would ordinarily be thought of as male. A Committee on Publicity and Public Relations might have women on it, might have a woman chairman, but is not the articulate public mostly male and could not a man thus make the best contact with it? . . . Of course, a Committee on War Services to Children might be, and is, female. Women know about children, or are supposed to. All in all, it is our impression that the Association has more committees concerned with activities that are traditionally considered male than it has for the traditionally female interests.[51]

---

[48] Boring to Alice I. Bryan, undated, Alice I. Bryan Papers, Box 3, Folder: Women in American Psychology, CUA; emphasis in the original.

[49] Alice I. Bryan and Edwin G. Boring, "Women in American Psychology: Prolegomenon," *Psychol. Bull.* 41 (1944): 447–54; Bryan and Boring, "Women in American Psychology: Statistics from the OPP Questionnaire," *Amer. Psychol.* 1 (1946): 71–9; Bryan and Boring, "Women in American Psychology: Factors Affecting Their Professional Careers," *Amer. Psychol.* 2 (1947): 3–20.

[50] Bryan and Boring, "Prolegomenon" (cit. n. 49), 453.

[51] Ibid., 453–4.

Although ostensibly skirting the issue of nature versus nurture, "male" was nonetheless equated with the articulate public and "female" with knowledge about children, indicating that Victorian attitudes about separate spheres had persisted despite decades of feminist activism and the accomplishments of the "new women" of the early twentieth century.[52]

What role did Bryan play in these pronouncements? Although her archived papers do not include her end of the correspondence concerning the manuscript, it is clear that the reasons offered for women's underrepresentation were proposed by Boring and reflected his long-standing belief that women were uninterested in science and more suited to technology. In a letter to Bryan about the emerging manuscript, he wrote "what we need to work out is the matter that women are more in the professional positions and less in the scientific positions, and that that makes for their distinction in responsibility in the APA and also in starring in American Men of Science. . . . I think it is clear that the National Academy of Sciences defines science as the natural sciences define it, and that women simply do not go in for this sort of thing."[53]

It is unlikely that Bryan would have agreed that women "simply do not go in for" science, so the question remains as to whether her acquiescence to Boring's views was a strategy to minimize conflict and avoid the label of "belligerency" that could easily have been used to discredit her work. In a talk called "The Dilemma of the Professional Woman" delivered in 1946 at McGill University, Bryan was clear that she saw the woman problem as resulting from straightforward sex discrimination. In describing the plight of women working across a variety of professions, she wrote:

> Among these female professional workers may be found some of the most gifted and energetic women that our society can produce. . . . Society is permissive with respect to acknowledging women's right to practice these professions on an equal basis with men. At the same time, a good many discriminations exist that make it considerably more difficult for women than for men to find a place in professional work. . . . Discrimination against the professional woman is increasing rather than decreasing in the post-war world.[54]

Bryan's silence in published reports about her views on sex discrimination might be understood as the enactment of a form of double-consciousness.[55] Acutely aware of the precarious nature of her position as a woman and as a feminist in a "man's world," as she later described academic psychology,[56] she likely developed a distinct set of strategies for navigating her relationship with Boring. There are multiple indications that she would have been very aware of the need for caution. In his autobiography,

[52] On female social scientists' attempts to challenge the separate spheres argument, see Rosalind Rosenberg, *Beyond Separate Spheres: The Intellectual Roots of Modern Feminism* (New Haven, Conn., 1982). On the characteristics of the "new woman," see Estelle Freedman, "The New Woman: Changing Views of Women in the 1920s," *J. Amer. Hist.* 61 (1974): 372–93.

[53] Boring to Alice I. Bryan, 10 January 1944, Alice I. Bryan Papers, Box 5, Folder 10, CUA.

[54] Alice I. Bryan, "The Dilemma of the Professional Woman," unpublished talk, probably delivered on 23 January 1946 at McGill University, Alice I. Bryan Papers, Box 9, Folder: Articles on Women, Professional Careers, CUA.

[55] W. E. B. Dubois, *The Souls of Black Folk* (New York, 1964), 16.

[56] As Bryan wrote much later in her life in an unpublished reflection, "My second deficiency was the fact that I was a woman in a university 'that was still very much a man's world.' To have attempted to challenge and change that 'world' would have required the faith and self-delusion of a Don Quixote with an impossible dream." Bryan, "Public Library Inquiry: Purpose, Procedures and Participants," unpublished talk delivered at the American Library Association, June 1992, Alice I. Bryan Papers, Box 11, Folder 22, CUA.

Boring characterized their collaboration as one in which Bryan, with her feminist convictions, and he, with his position that women, for both biological and cultural reasons, "determined most of the conditions about which she complained," could potentially moderate each other's positions and reveal "the truth."[57] When they began working together, Boring offered to help review drafts of a research methods book Bryan was also working on. He wrote to her, "My belief is that you are not, temperamentally, a scientist. You are basically a humanist and a person-helper. . . . So your knowledge, your writing, your drive, and my scientific orientation are what is going to make this book correct."[58] Faced with the explicit knowledge that an esteemed colleague considered her temperamentally a nonscientist (despite her obvious interest and expertise in research methods), it is no wonder that Bryan carefully curated her professional presentation.

Even before the first of their three papers came out, Boring was clearly worried that his colleagues might think that his objectivity had been compromised by engaging with the woman problem. As he wrote to his colleague and confidant Richard Elliott:

> Jane Morgan told me in Washington Monday that I was being gossiped about unfavorably because of this paper with Alice. . . . The gossip seems to be that Alice has seduced me to belligerancy [sic] on the part of women, that my objectivity has been done away with by her aggression, or maybe it is her charm. . . . Jane Morgan, by the way, says she is disgusted with women psychologists. They are, she says, a) badly trained, b) lazy not trying to get better trained, and c) so filled up with the matter of their rights and the question of discrimination against them that they do not have time to get to being better psychologists.[59]

In his reply, Elliott reported the opinions of two other distinguished colleagues on the matter of the woman problem: "I asked John Anderson what he thought and he replied that he does not believe the women are owed anything. Miss Goodenough, too, is assured that women are poor scientists and make up for their felt deficiency by aggressive political participation."[60]

By the time they were preparing their third publication, "Women in Psychology: Factors Affecting Their Professional Careers," responses to the series from some of Alice Bryan's NCWP colleagues were clearly disapproving. As Boring put it in another letter to Elliott:

> Some people think I am biased. Some people think I am a woman-hater who is leading Alice astray. . . . Alice is probably a little shaken in her trust in me because her NCWP friends have come down on her. In general she sticks by me but she is consulting some of her friends. I am giving this to Ruth Tolman, Helen Peak, and Thelma Alper and asking

---

[57] Boring, *Psychologist at Large* (cit. n. 6), 72.

[58] Boring to Alice I. Bryan, 13 August 1943, Alice I. Bryan Papers, Box 11, Folder: Scientific Research Methods for Librarians, CUA.

[59] Boring to Richard Elliott, 25 April 1944, Boring Papers, HUG 4229.5, 1944, Box 18, Folder: Elliott, R. M., 1944–45, #379, HUA.

[60] Richard Elliott to Edwin G. Boring, 1 June 1944, ibid. John Anderson was the president of the American Psychological Association in 1942–3 and the editor of *Psychological Bulletin*. Florence Goodenough, an eminent developmental psychologist at the University of Minnesota's Institute of Child Welfare, agreed to be the first president of the NCWP even though she was staunchly opposed to special treatment for women or identifying them as "women psychologists."

them to tell me what they think. But I ought also to have a male mind which has thought along these lines, and I turn to you."[61]

Elliott replied, unsurprisingly, that he thought the draft was "courteous, temperate, and valuable."[62]

OBJECTIVITY COMPROMISED?

A close comparison of the data collected by Bryan and Boring for their third article with Boring's interpretation of these data actually reveals some interesting discrepancies. These discrepancies support the thesis that Boring was heavily invested in presenting women as nonscientists, perhaps even to the point of compromising objectivity. Based on their data, he reiterated his conviction that there was "men's work" (scientific research) and "women's work" (application) in psychology, and that women were on the whole more suited to and interested in applied work from the outset. As he wrote to Elliott, "It is clear from the questionnaire, from our previous study, and from everything else, that there is woman's work in psychology that is somewhat different from men's work. The women do not do so much research and writing. They are interested in the particular cases and in work that has personal relations with individuals, especially those who need help."[63]

In their lengthy questionnaire, sent to 440 female psychologists and 440 matched male psychologists, Bryan and Boring gathered a large amount of data about the attitudes and interests of psychologists. Notably, and in contrast to Boring's assertion to Elliott above, when asked what factors influenced their choice of psychology as a career, men and women evinced a very similar pattern of responses: "interest in the subject matter" was selected by 80 percent of the women and 82 percent of the men, "interest in science" by 54 percent of the women and 60 percent of the men, and "interest in research" by 55 percent of the women and 57 percent of the men. But what of the areas to which Boring felt women were so overwhelmingly predisposed? "Interest in people" was chosen by 63 percent of the women and 52 percent of the men, and "desire to serve humanity" was chosen by 28 percent of the women and 27 percent of the men. Although there were some small differences in percentages, Boring's long-standing conviction that women are more interested in people and particulars than real science does not appear to have been grounded very well in the actual data he himself collected and cited.

Yet in his collaboration with Bryan and in his correspondence with other colleagues, Boring repeatedly asserted his own objectivity and the importance of keeping facts separate from convictions, values, and emotions (despite occasionally muddling them himself). For Boring, intense feeling was often a problem for objectivity. As he wrote Helen Peak in the aftermath of the 1940s "woman problem" publications, "You see, I have always meant to be impersonally objective about these things myself, and the present situation interests me especially because it raises a question as to whether I am unbiased. Well, if I was not I soon shall be! The bias comes in (if it does at all) with

---

[61] Boring to Richard Elliott, 15 November 1946, Boring Papers, HUG 4229.5, Box 18, Folder: Elliott, R. M., 1946, #380, HUA.
[62] Elliott to Edwin G. Boring, 26 November 1946, ibid.
[63] Boring to Richard Elliott, 15 November 1946, ibid.

the emotional overdetermination of some of the people in the NCWP and it shows by my dismissing the problem too casually."[64]

In 1951, provoked by psychologist Mildred Mitchell's resurrection of the woman problem,[65] Boring decided once again to engage with the issue and published his own views in an article that appeared in the *American Psychologist*. It offers another revealing window on the construction of scientific masculinity and its perils for women.

In this paper, Boring first spelled out his recipe for attaining success and prestige in science. In brief, this recipe was as follows: (1) get a PhD; (2) do good research and publish it (thus gaining recognition); (3) add some administrative work to get some larger perspective; (4) write a synthetic book about big ideas/theories; and (5) work up the administrative ladder to become a dean or college president. Boring pointed out that this recipe applied to every past president of the APA. He advised women to follow this formula but noted that top administrative jobs would be very hard for women to get because they would face sex discrimination. The solution was to write a big book, but here again, women would be foiled. As he wrote: "With top-level administrative jobs so hard for her to get, why then does she not write books? Sometimes she does, but the book that brings prestige should deal with broad generalities, and there is some indication that the women of our culture are more interested in the particular, and especially in the young, helpless and distressed."[66]

Boring then ventured that another obstacle for women in science was that women suffered from "job concentration" difficulties. As he put it:

> Beardsley Ruml has spoken humorously of the 168-hour week for the fanatic—who lives primarily for his job—he who eats, sleeps, and finds recreation only because he wishes to work better. . . . Now it has been remarked that these people make poor parents, and presumably they usually do. Thus it comes about that the Woman Problem is found to be affected by philosophy of living. Inevitably there is conflict between professional success and success as a family man or a family woman.[67]

Boring then went on to frame the female scientists' dilemma in terms of two sets of choices: (1) she can choose to devote herself to generalities and basic science or to particularities and applied work, and (2) she can choose job concentration or be a wife and mother. As he put it, "If she chooses less job-concentration in order to be a better wife or a better mother, then she is perhaps choosing wisely, but she is not choosing the maximal professional success of which she would be capable. She is in competition with fanatics—the 168-hour people—and she had better accept that bit of realism about job-concentration."[68] Boring's assumption that scientific success required the kind of job concentration that he well knew would be very difficult for many women, given the traditional gender roles of the times, effectively made job concentration a masculine virtue.

[64] Boring to Helen Peak, 25 November 1946, Boring Papers, HUG 4229.5, Box 46, Folder: Peak, Helen, 1946, #1017, HUA.

[65] Mildred B. Mitchell, "Status of Women in the American Psychological Association," *Amer. Psychol.* 6 (1951): 193–201.

[66] Boring "The Woman Problem" (cit. n. 10), 680.

[67] Ibid. For a discussion of fanaticism in the masculine culture of computer programming, see Nathan Ensmenger, "'Beards, Sandals, and Other Signs of Rugged Individualism': Masculine Culture within the Computing Professions," in this volume.

[68] Boring, "The Woman Problem" (cit. n. 10), 681.

Further insight into Boring's construction of science as masculine is provided by his correspondence with psychologist Jane Loevinger, exchanged in the summer of 1951. Loevinger had read a letter to the editor, authored by Boring, about graduate training in psychology and objected to his emphasis on job concentration. She noted somewhat offhandedly that "surely as psychologists we should not be strangers to the fact that even a father who devoted 60 hours a week to his work would raise a pretty sorry lot of kids."[69] In his unsolicited response to Loevinger's letter, it is clear that the "time factor in fatherhood" to which she had alluded struck a sensitive nerve. Boring reiterated his point about job concentration being required for scientific prestige and added a rather chilling elaboration of this assessment given his personal domestic situation.[70] Boring wrote:

> Have a happy marriage because then you will have well-adjusted children [who] will themselves make happy marriages (cf. Terman & Miles) and so on recurrently ad infinitum. The people with the best minds I know, do not believe that they want to participate in any such temporary success as having good minds sacrificed to get good minds in children. . . . Perhaps we need division of responsibility with a scientific unmarried priesthood.[71]

In suggesting that devotion to marriage and family were incompatible with a successful scientific life, and glibly referring to an unmarried scientific "priesthood," Boring was clearly conveying a set of masculine scientific virtues. He then forwarded Loevinger a draft of his woman problem article for her reaction. Her response, predictably, was another objection to his fanaticism argument. She took issue with his characterization of the relationship between fanaticism and prestige, arguing that creativity and wisdom were more important than a 168-hour workweek to acquiring prestige—for both men and women. Further, she expressed her view that since so many psychologists studied such trivial problems it would not matter whether they studied them for 1 hour or 168 hours a week. As she put it, "I cannot view with equanimity the prospect that another generation of psychologists would devote themselves to such propositions as that there are no differences between the sexes and that double alternation problems are difficult for white rats to learn."[72] In another letter she also noted that job concentration for married women was not only a matter of pulling the same number of hours as men, but that even with 168 hours at their disposal women were still plagued by fewer resources, less geographic mobility, less institutional research support, and lower salaries than their male colleagues.[73]

In response to Loevinger's letter, Boring wrote pointedly, "If this woman-topic has importance enough to justify letters and publication, it has importance enough to merit clear and unmuddled thinking." Boring then criticized her suggestion that fanaticism

[69] Jane Loevinger to Edwin Boring, 6 July 1951, Boring Papers, HUG 4229.5, Box 37, Folder: Li–Lz, 1951–52, #792, HUA.

[70] One of Boring's daughters, Barbara, committed suicide by an overdose of sleeping pills on 19 February 1950, evidently her seventh attempt since the previous August. It is clear from his oral history that Boring later worried that he may have been a neglectful father.

[71] Boring to Jane Loevinger, 15 July 1951, Boring Papers, HUG 4229.5, Box 37, Folder: Li–Lz, 1951–52, #792, HUA.

[72] Jane Loevinger to Edwin G. Boring, 10 August 1951, ibid.

[73] Jane Loevinger to Edwin G. Boring, undated, Jane Loevinger Papers, Box M1220, Folder: Loevinger, Archives of the History of American Psychology.

was not a requirement for prestige, essentially accusing her of confusing values with the facts, and concluding, "I think you are unable to consider this problem dispassionately and objectively."[74]

Regardless of the exact nature of Loevinger and Boring's disagreement about what might be required for prestige, Boring's insistence that Loevinger could not be dispassionate and objective about the issue arose in part because of the intrusion of what he saw as "values" into her argument. Boring was not blind to the cultural restrictions placed on women, but he was unwilling or unable to allow that values could coexist with an "objective study" of the woman problem, even though his own values clearly imbued almost every stage of his inquiry. Was objectivity a masculine virtue for Boring? While women could, theoretically, be as objective as men, it appears that, for Boring, women's objectivity was automatically compromised if they expressed a position on an issue, no matter how logically reasoned.

Interestingly, a couple of decades later Alice Bryan articulated a very different version of objectivity. In a letter to a colleague in which she commented on an article they had both read in *Daedalus*, she wrote of the author of the article: "He assumes that the objectivity required for testing the validity of alternative hypotheses in a given field of research precludes the possibility of a scientist's taking a stand on crucial issues involving value judgments. In my view, this distinction, and the assumption on which it is based, is untenable."[75] But for Boring, one year in the grave when this was written, objectivity and values required strict differentiation, and science was about objectivity. As Julian Jaynes wrote of Boring in a lengthy obituary, "He dreamed of a world of pure science and pure objectivity."[76] As I have shown, this "world of pure science" was also a decidedly masculine world. The women and men who succeeded in it quickly adopted the codes of conduct required to meet its myriad gendered expectations. For women, this was an especially treacherous undertaking.

"FEMINISM IS AS GREAT A HAZARD AS PREJUDICE": MAINTAINING MASCULINITY IN POSTWAR AMERICAN PSYCHOLOGY

In 1949, after her collaboration with Boring had ended, Alice Bryan received a request from colleague May Seagoe. Seagoe was soliciting contributions to a special issue of the *Journal of Social Issues* on the professional problems of women and naturally thought of Bryan's work on the status of women in psychology. In pitching her invitation to Bryan, Seagoe was careful to write, "The point of view throughout should be completely objective; feminism is as great a hazard as prejudice."[77]

In equating feminism with prejudice in terms of the magnitude of its threat to objectivity, Seagoe reiterated the oft-invoked incompatibility of values with science expounded by Boring and others when dealing with "women's issues" and highlighted the dangers to credibility faced by female scientists harboring even the mildest of feminist commitments. Even in the context of soliciting articles on challenges affecting

---

[74] Boring to Jane Loevinger, 13 August 1951, Boring Papers, HUG 4229.5, Folder: Li–Lz, 1951–52, #792, HUA.

[75] Alice Bryan to Jack Dalton, 14 September 1969, Alice I. Bryan Papers, Box 9, Folder: Bryan, CUA.

[76] Jaynes, "Edwin Garrigues Boring" (cit. n. 3), 108.

[77] May Seagoe to Alice Bryan, 19 October 1949, Alice I. Bryan Papers, Box 7, Folder III-E-4 Professional Activities, Publications, Correspondence, General, CUA.

women, one again sees the complex minefield traversed by female psychologists. Acutely aware of sex discrimination and its consequences, and eager to write about women's professional problems, they had to do so in a way that was as explicitly fact oriented and as devoid of overt feminism as possible. The multifaceted (and at times self-abnegating) strategies used by these women to assert their views while resisting marginalization within a heavily masculine scientific culture belie their acute awareness of the "man's world" of scientific psychology that was so vehemently enforced by gatekeepers like Edwin Boring. However, unlike in previous decades when their numbers and their disciplinary status were more precarious, some of these women were temporarily successful in pushing the issue of the professional problems of women onto the agenda.[78] In doing so, they exposed the heavily value-laden culture of masculine science.

I have shown that Boring's interest in and opinions about the woman problem were intertwined with his lifelong concern about maintaining the purity of psychology as an experimental science. This concern reached its first peak following World War I. Boring made vigorous efforts at that time to quell the influence of the rising tide of testing and other applied pursuits and buttress the status of experimental psychology. In 1921, a set of recommendations for new membership criteria for the APA was put forward: members would have to hold PhDs, occupy academic positions, and publish research. In 1924, a separate class of membership was established for those who could not meet these criteria, effectively creating an elite of academic psychologists with full membership and an underclass of applied psychologists with only associate status. Given that research positions were rarely open to women at this time, and many were explicitly funneled into applied work, this division was also a gendered one. As historian John O'Donnell has pointed out, "All of the members of the 1924 committee and all but one of the 1921 deputation were affiliates of Titchener's society. Both committees were chaired by E. G. Boring."[79] I have documented that Boring's views on women's suitability for science were already formed by this time, likely through a combination of his socialization in the Experimentalists and his own personal prejudices, which appear to have been widely shared.

Thus, Boring was instrumental, along with other members of Titchener's elite group, in creating a structure that kept the kind of psychology to which he believed women and some men were naturally predisposed (the psychology of people and particulars) subservient to experimental psychology. He then used his later writings on the woman problem to repeatedly assert that because women were more likely to find themselves in professional psychology, and were unlikely to be able—for both cultural and temperamental reasons—to engage in the kind of science required for achieving recognition and prestige, they would be better off accepting reality rather than trying to change it.

But in the post–World War II world, the fate of pure psychology once again seemed

---

[78] I say "temporarily" because the feminist solidarity evident among some women in psychology and the attention they brought to women's issues appear to have declined in the 1950s and 1960s, to reemerge in the late 1960s with the birth of an explicitly feminist psychology. Winifred Breines has argued that in the 1950s the work of social scientists expressed unwarranted optimism about gender equality, neglecting the actual constraints on women to which Bryan and others referred; Breines, "The 1950s: Gender and Some Social Science," *Sociol. Inq.* 56 (1986): 69–92.

[79] John M. O'Donnell, "The Crisis of Experimentalism in the 1920s: E. G. Boring and His Uses of History," *Amer. Psychol.* 34 (1979): 289–95, on 293.

uncertain. By the end of the 1940s, the rise of clinical psychology, like testing in the post–World War I context, threatened to sideline pure experimentalism. Boring's strategies for maintaining the separate spheres of gendered work where "scientific" equaled "masculine" and "applied" equaled "feminine" continued. As he wrote in 1946 in response to a colleague's query about the appropriate role of experimental psychology in the undergraduate curriculum and its relationship to clinical psychology, "I think we are getting to think of clinical psychology as a sort of feminine activity, something designed for persons of both sexes who are interested in personalities and often not very adept with mechanical skills. You are not likely to change those people after they get to college, so why not train them in what they are temperamentally fitted for?"[80] By designating clinical psychology a "feminine activity" (albeit designed for people of both sexes), Boring appeared to be leveraging the undesirability of the feminine to keep clinical psychology in its place, just as he had intentionally worked to keep the status of applied psychology subservient to experimental psychology decades before. He was also suggesting that those who were temperamentally suited for such work be funneled into it even earlier—at the undergraduate level.

But the efforts of Boring and his cronies to keep psychology "pure" were anachronistic. Whereas 1920s American psychology could be controlled by a small group of male elites, post–World War II psychology was not so easily managed. The American Psychological Association was no longer one scientific organization but had absorbed its rival, the American Association for Applied Psychology, and adopted a divisional structure. Boring, through his work on the Subcommittee on Survey and Planning in Psychology of the National Research Council, was instrumental in these changes but remained concerned about the status of psychology as a natural science. He was not alone in his concern.

In 1945, Robert Yerkes wrote an apparently unsolicited letter to the president of Cornell University advising on the nature of a proposed new hire for their psychology department:

> Word has come to me from various sources that you are about to select a leader for psychology in your great institution. . . . There is today in many quarters [a?] tendency and effort to constitute psychology a social science and to provide for its development among the social sciences and humanities. With all of the resources at my command, . . . I wish to argue for the complete desirability, logically and practically, of now constituting basic instruction and research training in psychology [as] a natural science, and of developing it as an extension and supplement of the life sciences and especially of physiology.

Only then, according to Yerkes, could we hope to achieve "a psychology worthy of the name science."[81]

As historian Margaret Rossiter has discussed, because feminization is regarded as inversely proportional to prestige, "malcontent" scientists in soft fields have often developed strategies to make their disciplines "harder." These strategies include making the field more experimental or apparatus/instrument heavy, or, conversely, more abstract (say, through mathematization).[82] Yerkes's insistence that psychology be a

[80] Boring to Anne Anastasi, 19 March 1946, Boring Papers, HUG 4229.5, Box 1, Folder: A, 1946 #15, HUA.

[81] Robert Yerkes to Edmund Day, 20 February 1945. Boring Papers, HUG 4229.5, Box 66, Folder: Yerkes, R. M., 1945, #1513, HUA.

[82] Margaret Rossiter, "Which Science, Which Women?" *Osiris* 12 (1997): 169–85, on 179.

natural science conducted in apparatus-filled laboratories, closer to physiology than the social sciences, clearly demonstrates a "hardness" strategy at work.

In sum, Boring and his colleagues worked hard to uphold psychology's reputation as a natural science, a reputation that was grounded in a set of thoroughly masculine ideals that strictly separated science from technology and accorded the former higher prestige. Boring's assessment of the woman problem and his concern about psychology's natural science status emerge as tightly connected and mutually reinforcing. As historian of psychology Ian Nicholson has noted, "It is important to acknowledge the extent to which questions of gender are subtly and in some cases quite explicitly woven into scholarly debates about the meaning of science and the boundaries of psychology."[83]

The reactions of Helen Peak, Alice Bryan, Jane Loevinger, and a host of other women who were faced directly with pronouncements on science and women's suitability for it demonstrate again the multiple strategies women used to deal with the masculine gendering of their field. As Sally Kohlstedt and Helen Longino have noted, "Many women scientists ignore or do not resist the gendering practices that surround them, while others are acutely sensitive and resistant, and still others manage to be conscious of but successfully negotiate the treacherous gender shoals in which they work."[84] Helen Peak expressed a poignant consciousness of the ways her assertive professional conduct might come into conflict with gender stereotypes and cause trouble with her male colleagues. Working this out privately, she took great care to insist on no special treatment and adhered scrupulously to the rules of the masculine game. Alice Bryan lived a double life, one among her female colleagues in the NCWP, and one in her collaboration with Boring. While there are occasional glimpses of the conflict between them, this conflict was artfully negotiated, and Bryan appears to have maintained her credibility with both camps. Jane Loevinger, although ambivalent about feminism, confronted Boring's pronouncements on the woman problem directly, perhaps perceiving that she had little to lose by challenging his views.

By focusing on this pivotal period in American psychology, I have illuminated some of the microhistorical processes that have contributed to constructing the image of the model male scientist, the reverberations of which we continue to contend with today. In a recent analysis of its own reviewing and publishing practices, the prestigious journal *Nature* revealed that of the thirty-four researchers profiled in the journal in 2011 and 2012, only 18 percent were women.[85] They attribute this not to conscious discrimination against women, but rather to unconscious factors that bring men most readily to the editors' minds when they think about who is doing interesting or relevant scientific work. In addition to these unconscious factors, it appears that women's "preferences" and "choices" are still framed as problems—albeit ones that today we are committed to solving. By being apparently unwilling or unable to spend 168 hours at work because of the pull of domestic responsibilities, and obstinately preferring "people science" over abstract theory, today's female scientist continues to be marked in a way her male colleagues are not. Our contemporary understanding of the

[83] Ian A. M. Nicholson, "'Giving up Maleness': Abraham Maslow, Masculinity, and the Boundaries of Psychology," *Hist. Psychol.* 4 (2001): 79–91, on 89.

[84] Sally Kohlstedt and Helen Longino, "The Women, Gender, and Science Question: What Do Research on Women in Science and Research on Gender and Science Have to Do with Each Other?" *Osiris* 12 (1997): 3–15, on 12.

[85] "*Nature*'s Sexism," *Nature* 491 (22 November 2012): 495.

"preferences" and "choices" demonstrated by female scientists must be informed by an ample historical record that offers up repeated examples of how gender stereotypes and structural barriers have operated to constrain and limit these choices and preferences so that, over time, they come to appear natural or essential.

Finally, despite contemporary appearances, history also shows us that there have been many different ways of being a scientist—for both men and women. If we can expose the model scientist for what he is—a deeply gendered construction—we open up possibilities for constructing science and scientists in new ways, ways that may make women not a problem for science, but part of the solution.

Sexual Violence, Predatory Masculinity, and Medical Testimony in New Spain

*by Zeb Tortorici**

ABSTRACT

This essay examines the medical and legal construction of predatory masculinity in New Spain by contrasting criminal cases of rape [*estupro*] with those of violent or coercive sodomy [*sodomía*]. In the context of male-female rape, the rulings of most criminal and ecclesiastical courts imply that predatory masculinity was a "natural" manifestation of male sexual desire, whereas in cases of sodomy and nonconsensual sexual acts between men, courts viewed such desire as "against nature." The processes by which the colonial state prosecuted certain sexual crimes simultaneously criminalized and validated predatory masculinity. By analyzing the roles of the medics, surgeons, and midwives who examined the bodies of the male and female victims in these cases, this essay argues for a commonality in the authoritative judgments based on medical evidence, whether conclusive or inconclusive.

In 1684, in the largely indigenous town of Yanhuitlan in the region known as the Mixteca Alta (in what is today the southern Mexican state of Oaxaca), Diego García, a Mixtec Indian, filed criminal charges against Domingo de Silva for the *estupro* or "rape" of his eight-year-old daughter, Angelina García.[1] According to the 1732 *Diccionario de la lengua castellana*, the term *estupro* had a very specific meaning: "concubinage and illicit, forced intercourse with a virgin or *doncella* [unmarried maiden]."[2] While the term *estupro* might be seen as roughly synonymous with the

*Department of Spanish and Portuguese Languages and Literatures, New York University, New York, NY 10003; zt3@nyu.edu.

I am especially grateful to Erika Lorraine Milam, Robert A. Nye, Lee M. Penyak, and two anonymous readers for their incisive questions, comments, and suggestions, which improved this essay. This essay benefited greatly from conversations with the participants of the "Masculinities in Science/ Sciences of Masculinity" conference at the Philadelphia Area Center for History of Science in May 2012, especially Leah DeVun and Nathan Ha. I want to express my gratitude to Santa Arias, Greg Cushman, Ivonne del Valle, Laura Gutiérrez, María Elena Martínez, Anna More, Ana Paulina Lee, Rachel O'Toole, Adam Warren, Marta V. Vicente, and Pamela Voekel for their feedback at the University of Southern California–Huntington Library conference on "Race and Sex in the Eighteenth-Century Spanish Atlantic World." Close readings by Marcela Echeverri, Anne Eller, Yuko Miki, and Tamara Walker have also been particularly helpful. Generous funding from the Department of History at the University of California, Los Angeles, the Department of History at Stanford University, and the American Council of Learned Societies facilitated research for this article.

[1] Archivo Histórico Judicial de Oaxaca (hereafter cited as AHJO), Sección Teposcolula, Serie Penal, leg. 18, exp. 34, fols. 21. Girls in the early modern Iberian world were considered to be of marriageable age when they were twelve years old.

[2] *Diccionario de lengua castellana* (Madrid, 1732), s.v. "estupro": "Concúbito y ayuntamiento ilícito y forzado con virgen o doncella," http://www.rae.es/recursos/diccionarios/diccionarios-anteriores

English-language "rape," there are subtle yet important differences and ambiguities embedded in the legal category of *estupro*. As seventeenth-, eighteenth-, and early nineteenth-century Mexican criminal cases of *estupro* show, popular usages of the term do not always reflect the specificity of legal definitions or the medieval Spanish juridical notion that only virgins could be "raped."[3] This unique 1684 criminal case of *estupro* serves as an entrée to the legal regulation of criminalized sexuality—specifically, sexual violence that manifested itself in acts legally defined as either *estupro* or sodomy, depending partly on the gender of the victim—and affords us insight into the interactions between morality, masculinity, and increasingly medicalized conceptions of sexual desire under Spanish colonial rule.

Although the crimes of both *estupro* and sodomy were legally distinct categories that were known by different terms, required different expert judgment, and resulted in different punishment, their juxtaposition here allows us to see how criminal courts and medics interpreted male sexual violence in dichotomous terms that revolved around "natural" and "unnatural" manifestations of desire. In the language of the criminal courts, sodomy was deemed the "sin against nature" [*pecado contra natura*], whereas cases of male-female *estupro* putatively represented more "natural," procreative modes of male desire. The comparison of *estupro* and nonconsensual sodomy thus enables us to reconstruct conceptions of masculinity—particularly violent and predatory masculinities—that were at the center of both of these acts.[4] Violent acts of sodomy and *estupro* were mediated through medical expertise and medical hierarchies, which determined to a large extent who had the ability to judge and interpret the signs of the violated body in the context of court-mandated medical examinations, be they licensed medics, midwives, surgeons, or informal healers.

In New Spain, the crimes of *estupro* and sodomy fell primarily under the jurisdiction of the secular criminal courts as opposed to local ecclesiastical courts or those of the Inquisition, for reasons that are outlined below. This fact allows for an interesting point of comparison, though, since as numerous scholars of colonial Latin America have shown, both episcopal courts and the Holy Office of the Inquisition frequently

---

-1726-1996/diccionario-de-autoridades (accessed 4 June 2015). Colonial concepts of sexual coercion and *estupro* cannot be easily translated as "rape." According to certain Spanish laws, prostitutes could not technically be "raped," nor could married women be "raped" by their husbands. This, however, does not mean that there was no legal recourse for women whose husbands treated them violently or sexually abused them, yet such topics fall outside of the scope of this article. Additionally, Sonya Lipsett-Rivera notes that in the Spanish legal lexicon, "forcible sex" was divided into three categories: *rapto* (the extraction of daughters from their parents' homes, with the intent of marriage and usually with the consent of the woman involved), *estupro* (technically forcible sex with a virgin, though later legal codes including the 1805 legal code of the Spanish colonies known as the *Novísima recopilación* ambiguously equated *estupro* with the seduction of "honest women"), and *violación* (a nineteenth-century legal category that referred specifically to forced sex, or to what we would today term "rape," but was rarely used in colonial archival records); Lipsett-Rivera, "The Intersection of Rape and Marriage in Late-Colonial and Early-National Mexico," *Col. Latin Amer. Hist. Rev.* 6 (1997): 559–90, on 567–8.

[3] Such findings regarding the ambiguities of *estupro* have been examined by a number of scholars. See, e.g., Carmen Castañeda, *Violación, estupro e sexualidad: Nueva Galicia, 1790–1821* (Guadalajara, 1989); Eugenia Rodríguez, "'Tiyita bea lo que me han hecho': Estupro e incesto en Costa Rica (1800–1850)," *Anuario Estud. Centroamer.* 19 (1993): 71–88; and Eugenia Rodríguez Saenz, "Pecado, deshonor, y crimen. El abuso sexual a las niñas: estupro, incesto y violación en Costa Rica (1800–1850, 1900–1950)," *Nueva Época* 2 (2002): 77–98.

[4] For an analysis of consensual sodomy, which falls outside the focus of this article, see Zeb Tortorici, "'*Heran Todos Putos*': Sodomitical Subcultures and Disordered Desire in Early Colonial Mexico," *Ethnohist.* 54 (2007): 36–67; and Tortorici, "Against Nature: Sodomy and Homosexuality in Colonial Latin America," *Hist. Comp.* 10 (2012): 161–78.

tried cases of sodomy and *estupro* throughout New Spain, whether or not the cases fell within their proper jurisdiction.[5] Local criminal courts relied heavily on medical testimony from formally trained medics and informal healers to ascertain the facts of *estupro* and forced sodomy, which resulted in the construction of particular narratives of gender, masculine violence, and the body. Although the court system was ostensibly punishing male sex offenders, the archival evidence drawn from criminal courts shows how medical testimony coalesced with legal pronouncements to paradoxically reinforce male sexual prerogatives, including instances of sexual violence and predatory masculinity. Analyzing the links between the criminal court system, sexual violence, medical testimony, and gendered behavior under colonialism permits us to understand the subtle contradictions inherent in Spanish colonial rule concerning the simultaneous criminalization and validation of predatory masculinity.

Medical experts both responded to the discourses of deviancy (and colonialism) and reconstructed them, interpreting and classifying the colonial male body through the categories of normalcy, criminality, and eventually pathology. Scientific observations of bodies by physicians, surgeons, and midwives were indispensable to colonial governance, in conjunction with the religious and judicial discourse with which they converged. Together, these cases produced narratives of violent and predatory masculinity that reinforced the prevailing models of ideal masculine behavior. If we read criminal cases of rape and sodomy "along the archival grain," as Ann Stoler suggests, we can better expose how these forms of legal and medical expertise on the body functioned at multiple sites to construct narratives of pathology and normativity. Stoler, in exploring the archive ethnographically, explains, "the ethno-graphic is about the graphic, detailed production of social kinds, the archival power that allowed its political deployment, and the grafting of affective states to those inventions."[6]

In recent years, historians have used ethnography and archival evidence to paint a more detailed picture of the medical profession and healing practices—both formal and informal—in colonial Latin America. Surprisingly little of the current historical research on the Spanish colonial world, however, has focused on the expert testimony of medics, surgeons, barbers, and midwives. While historians of crime and criminality in the early modern world and the nineteenth century regularly employ medical testimony to re-create the details and events surrounding sexual crimes, they often do so without offering a sustained analysis of how such medico-scientific narratives functioned within the broader discourses of gender, criminality, and deviancy.[7] The cases examined here reveal who had the authority to judge bodies, on what basis, and according to what criteria. They elucidate the gendered landscape of ideal and deviant

[5] Asunción Lavrin, e.g., discusses records of *estupro* and "rape" cases located in the Archivo de la Sagrada Mitra de Guadalajara and the Archivo del Antiguo Obispado de Michoacán. See Lavrin, "Sexuality in Colonial Mexico: A Church Dilemma," in *Sexuality and Marriage in Colonial Latin America* (Lincoln, Neb., 1989), 47–95, on 71–2. I wish to thank an anonymous reviewer for *Osiris* for directing me to these cases. Ecclesiastical courts also dealt with cases of extreme abuse (including sexual violence) against women by their husbands, which could potentially lead to divorce.

[6] Ann Laura Stoler, *Along the Archival Grain: Epistemic Anxieties and Colonial Common Sense* (Princeton, N.J., 2009), 53.

[7] For an important discussion of midwives as medical examiners in late colonial and early national Mexican courts, see Lee M. Penyak, "Midwives and Legal Medicine in México, 1740–1846," *J. Hispanic High. Educ.* 1 (2002): 251–66.

sexual behavior and the hegemonic and nonhegemonic masculinities that existed and overlapped in New Spain.[8]

This essay examines the ways that licensed physicians, surgeons, barber surgeons, and midwives—occasionally referred to as *peritos*, or "experts," in the trial transcripts—read the signs of violated female and male bodies. I first explore the crucial issues regarding the male/female and formal/informal hierarchies of medical knowledge in the early modern and nineteenth-century Iberian world to provide a context for understanding the interactions, tensions, and complicities between licensed medical practitioners, midwives, surgeons, criminal courts, victims, and offenders. I then provide a microhistorical analysis of predatory masculinity in criminal cases of *estupro* and nonconsensual sodomy. I posit that we cannot make sense of isolated cases of sexual violence without looking at the larger apparatus of legal machinery in Spain's overseas colonies and at the natural/unnatural dichotomy, which differentiated reproductive sex and nonprocreative sex in the logic of criminal authorities. In addition to the original crimes of sexual violence, the conjunction of law and medical expertise in such cases often required invasive (yet necessary and inevitable) medical examinations, which may also have had the effect of reinforcing predatory masculinity in both theory and practice.[9] Thus, multiple violences were enacted on the bodies of colonized subjects. First, however, we turn once again to the 1684 *estupro* of Angelina García and to the criminal court system and legal environment in New Spain.

A 1684 CRIMINAL CASE OF *ESTUPRO*

In the wake of the Spanish conquest of the Aztec capital of Mexico-Tenochtitlan (1519–21), the region of the Mixteca Alta, where Angelina García and her family resided in the town of Yanhuitlan, and the nearby Valley of Oaxaca were incorporated into the extensive viceroyalty of New Spain, which included much of what are today Central America, Mexico, the Caribbean, parts of the United States, and the Philippines. The Mixteca was one of the most densely populated areas of Mesoamerica, and the town of Yanhuitlan, where Diego García denounced Domingo de Silva, eventually became an important center of Spanish activity and colonial administration.[10] Over time, the local networks of power in this multiethnic and multilingual region grew increasingly complex. This was due, in part, to the growing colonial presence of several groups: Spaniards, enslaved Africans and mulattoes, culturally and racially mixed mestizos, Nahuatl-speaking native peoples from central Mexico (who accompanied the Spanish throughout New Spain as allies, helping to "pacify" the native

---

[8] The contested notion of "hegemonic masculinity" as "a pattern of practice" is useful in thinking through the importance of context in defining male predatory masculinity. R. W. Connell and James W. Messerschmidt note that "hegemonic masculinity was not assumed to be normal in the statistical sense; only a minority of men might enact it. But it was certainly normative. It embodied the currently most honored way of being a man, it required all other men to position themselves in relation to it, and it ideologically legitimated the global subordination of women to men." Connell and Messerschmidt, "Hegemonic Masculinity: Rethinking the Concept," *Gend. & Soc.* 19 (2005): 829–59, on 832.

[9] Even when midwives examined the bodies of rape victims, medical examinations could reinforce predatory masculinity given that midwives were looking both for signs of sexual violation and for evidence that the girl who had been violated had her hymen intact prior to the *estupro* and therefore could be deemed a virgin.

[10] On criminal justice and local governance in the Mixteca Alta, see Kevin Terraciano, *The Mixtecs of Colonial Oaxaca: Ñudzahui History, Sixteenth through Eighteenth Centuries* (Stanford, Calif., 2001).

inhabitants of regions like the Mixteca Alta), and secular criminal and ecclesiastical colonial authorities.

This 1684 case is representative in that criminal investigations in indigenous communities throughout New Spain were initiated at the local level. Diego García lodged his initial complaint with the indigenous officials of the *cabildo*—the Spanish-style town council and local municipal government—who, due to the severity of the crime, forwarded the written record to the Spanish colonial administrative official, the *alcalde mayor*, in the nearby city of Oaxaca. Throughout New Spain, native peoples had access to legal defense, interpreters, and notaries, which were appointed by the courts to native peoples free of charge, and they became familiar with Spanish legal procedures and the colonial judicial landscape through the *cabildo*.

Because this essay deals at length with case studies of criminal investigations and trials from New Spain, it is necessary to review procedural norms and ideals. Following the initial denunciation, the first phase of judicial proceedings, the *sumaria*, was the fact-gathering stage in which the *alcalde* sought to determine the particulars of the case. Judicial officers elicited testimony from the witnesses and the victims. After gathering information (which may have been inconsistent as one testimony was compared with the other), the *alcalde* would request a statement from the suspect, who typically maintained his innocence and offered his version of the events. During this phase of the investigation, questioning could also be accompanied by force. Torture, while used sparingly in colonial Mexico, was seen as an important means of securing a confession. During the second phase of the criminal investigation, the *juicio plenario*, both the prosecution and the defense produced witnesses, who were often cross-examined. In this phase, the court appointed a *defensor* to give the defendant legal counsel and defense. During the final stage of judicial proceedings, the *sentencia*, the criminal court either absolved the defendant of the charges or pronounced sentence. It is crucial to note, however, that in many cases, the investigation and trial did not proceed according to this ideal plan. Inconsistencies abounded, and the procedural steps are sometimes difficult to reconstruct in the transcripts of the cases themselves.[11]

After Diego García's initial denunciation, witnesses and the victim were called forth to record their statements and have them ratified by the notary. According to Angelina García's testimony, given in her native Mixtec and recorded in Spanish by a court-appointed interpreter, she was tending turkeys on the outskirts of town during a local festival when de Silva "grabbed her by the arm and led her into a ravine . . . [where he forcibly] took off her underwear and took the handkerchief she had tied in her hair, he covered her mouth with it because she was going to scream; pushing her to the ground, he opened her legs and put his member in her womanly vessel [*vasso mujeril*]."[12] As it is doubtful that the young girl would have referred to her genitalia with the term *vasso mujeril*, this is likely an instance of the scribe's insertion of official terminology into the testimony of the victim, thus partly shaping the narratives he mediated and transcribed. As Kathryn Burns asserts, "notaries produced a shaped, collaborative truth—one that might shave, bevel, and polish witnesses' words a bit

---

[11] Charles R. Cutter, *The Legal Culture of Northern New Spain, 1700–1810* (Albuquerque, N.Mex., 2001), 125.

[12] AHJO, Sección Teposcolula, Serie Penal, leg. 18, exp. 34, fol. 3: "el entonces la coxio de un brasso y vajo a una barranquilla y . . . se quito los calzones y un paño que tenia amarrado en la caveza y porque queria gritar le tapo la voca con el y tendiendola en el suelo le abrio las piernas y metio por el vasso mujeril el miembro genital."

here, a bit there, as they were 'translated' into writing."[13] In the girl's own words, de Silva left her "covered in blood" and "very hurt," after which she painfully made her way home and told her mother what had happened and who had perpetrated this act. The court subsequently called upon a *partera*, or midwife, to examine her body for evidence of *estupro* and to testify to the court about what she discovered. According to the midwife, "she found the said [girl] open, with the [internal] membranes torn and her hips severely damaged, and from the mouth of the mother [vagina], which was in need of treatment, certain matter was being discharged."[14] The court stated its preference for a midwife to make the examination, rather than a male surgeon or barber surgeon, because of the victim's young age.

While de Silva, an indigenous man in his early thirties, initially denied his involvement in the crime, rumors quickly circulated throughout the community, in part because his own aunt testified that he had admitted his crime to members of his family. In light of the strong evidence against him, the court opted to employ torture [*tortura de cordeles*] as a means to extract the suspect's confession. On 17 October 1684, officials had the defendant stripped, put a collar around his neck, and gradually tightened ropes around each of his arms and legs. The torture had the desired effect: de Silva confessed that he'd been drunk on *pulque* when he approached the girl, who, according to him, "without any resistance" permitted his advances.[15] Taking into consideration the brutality of the sexual assault as well as the girl's young age and her loss of "honor," the judge ruled that Domingo de Silva was to be taken by a beast of burden from prison to the central plaza of the city, hands and feet tied, and a noose around his neck, where he was to be hanged for his crime "until he died naturally."

Domingo de Silva's fate merits particular attention since the death penalty was pronounced and, presumably, carried out for the crime of *estupro*. Usually, criminal sentences handed out to perpetrators of *estupro* in New Spain between the seventeenth and early nineteenth centuries tended to be comparatively lenient, consisting of public shunning, temporary imprisonment and forced labor on public works projects, financial reparations to the victim or to her parents (were she still a child), and occasionally corporal punishment. In cases where the perpetrator was married, as Lavrin shows, "The man was then ordered to return to his legal wife and to carry out marital life without any further cause of public scandal."[16] In cases where the perpetrator was unmarried, contracting marriage with the victim (or a marriage offer) was also a potential and relatively common outcome.[17] Capital punishment, however, was exceedingly rare.

While this rape case from 1684 is unique in its outcome, certain elements are typical of the judicial process in colonized rural areas. Scholars have recently demonstrated

[13] Kathryn Burns, *Into the Archive: Writing and Power in Colonial Peru* (Durham, N.C., 2010), 34.

[14] AHJO, Sección Teposcolula, Serie Penal, leg. 18, exp. 34, fol. 4: "allava estar la dha abierta y rotas todas las telas y descompuestas las caderas y que por la boca de la madre echava materias de que nesesitava de curacion."

[15] *Pulque* is a thick fermented alcoholic beverage made from various species of the maguey or agave plant. It is defined in the Real Academia's *Diccionario de la lengua Castellana* (Madrid, 1737) as "the juice or liquor of the maguey made by cutting its trunk when it is ready to be opened and then leaving a large cavity where it is then distilled. This drink is highly esteemed in New Spain where they are used to adding certain ingredients in order to give it a greater punch."

[16] Lavrin, "Sexuality in Colonial Mexico" (cit. n. 5), 71.

[17] See Castañeda, *Violación, estupro e sexualidad* (cit. n. 3) and Lipsett-Rivera, "Intersection" (cit. n. 2) for a discussion of several rape cases that resulted in the contraction of marriage or an offer of marriage from the perpetrator.

the fundamental role of native intermediaries in sustaining the Spanish project of colonialism, while simultaneously challenging its many abuses. Yanna Yannakakis, for example, frames the native leaders of the *cabildo*, who mediated the competing demands of Spaniards and indigenous people, as the "cultural intermediaries and political brokers" who, through their engagement with politics and the economy, "held the colonial order in balance."[18] *Cabildo* leaders in towns like Yanhuitlan made the colonial economy function by collecting tribute and overseeing native labor, and they also enabled individuals like Diego García, who initiated the 1684 criminal proceedings against Domingo de Silva, to gain access to the colonial judicial system. That this case of *estupro* ended up in the courts shows how indigenous villagers recognized that they had a right to use the colonial criminal court to demand justice. The institutional features of this case also show that a variety of intermediaries were necessary in making the Spanish colonial system work. First, the *cabildo* leaders pushed the case from the native town council to the higher criminal courts in Teposcolula and Oaxaca. Then, the intermediary translated the testimonies from Mixtec to Spanish, and the notary transcribed the narratives. Finally, the midwife provided expert testimony.

In this criminal case, as in so many others, the role of the medical practitioner as an intermediary was central to the judicial corroboration of the physical evidence. Medical practitioners and healers, along with the intermediaries cited above, actively constructed narratives of deviant and hegemonic masculinity (and femininity) from their readings of the body. In this case, the midwife's medical testimony contributed to the court's decision to pronounce the death penalty and demonstrates the influence of medical intervention in the colonial judicial process. This case, however, must be understood within the broader historical context of medical expertise and its intersections with masculinity, sexual violence, and the natural/unnatural dichotomy. The first step in assessing how medical discourse intersected with legal discourse through local networks of power is to delineate the gendered categories and social functions of those whose expert testimonies were sought by courts in criminal trials.

MALE MEDICAL AUTHORITY AND LOCAL NETWORKS OF POWER

Within the larger world of male medical authority, the Royal Protomedicato (the Royal Medical Court) officially licensed surgeons in the Iberian world. First conceived by the monarch Alfonso X in the thirteenth century and fully implemented in the Iberian Peninsula and its colonies by the mid-sixteenth century, this institution examined "all who aspired to become physicians, surgeons, bonesetters, apothecaries, dealers in aromatic drugs, herbalists, and any other persons who 'in whole or in part' practice these professions—women as well as men."[19] The supply of licensed surgeons, however, was far below the demand for medical services. Thus, in practice, the majority of the population received medical care from unlicensed surgeons, phlebotomists (popularly known as "barbers"), folk healers [*curanderos*], and midwives rather than from formally trained and licensed surgeons or physicians.[20] In contrast

---

[18] Yanna Yannakakis, *The Art of Being In-Between: Intermediaries, Indian Identity, and Local Rule in Colonial Oaxaca* (Durham, N.C., 2008), 3.

[19] John Tate Lanning, *The Royal Protomedicato: The Regulation of the Medical Professions in the Spanish Empire* (Durham, N.C., 1985), 17.

[20] Sherry Fields, *Pestilence and Headcolds: Encountering Illness in Colonial Mexico* (New York, 2008), 43.

to university-trained physicians who, in theory, dealt only with internal medicine, such as illness arising from epidemic disease and fevers, surgeons primarily treated external wounds—upholding an established professional difference between internal and external medicine. Barber surgeons performed amputations and other procedures such as letting blood and pulling teeth, and they also treated venereal disease, broken bones, skin conditions, kidney stones, and the like.[21] While *protomédicos*—royal medical officers—often complained that their authority outside of Mexico City was undermined by local authorities who protected informal practitioners, licensed medical professionals were scarce and could not provide the necessary health care for the population. Out of necessity, municipal authorities handled this problem by implicitly allowing unlicensed healers to practice medicine.[22]

The informal and unlicensed healers of the sixteenth and seventeenth centuries were not aligned with any particular institution, and there were often only ambiguous distinctions made between barbers, surgeons, and bleeders. For official physicians and surgeons, the standards were raised in Spain and its colonies in the eighteenth century, partly through the creation of surgical colleges like the Real Escuela de Cirugía in New Spain. In practice, however, some parts of New Spain, like Guatemala, established medical schools but graduated few students until the end of the eighteenth century. Accordingly, up until 1793 it was the Audiencia of Guatemala—the high court—that licensed Guatemalan medical practitioners rather than the Protomedicato.[23] In both Spain and its colonies, however, distinctions were made between the *cirujano latino* (a university-trained surgeon who had studied Latin) and the *cirujano romancista* (one who did not speak Latin and was trained through apprenticeship).[24] In the Spanish colonial world, the official distinctions between physicians, university-trained surgeons, and apprenticed surgeons largely reflected racial hierarchies and their tensions. As Adam Warren has recently demonstrated in a study of medicine in colonial Peru, Creole and Spanish physicians regularly persecuted black and mulatto surgeons as a means of enhancing their own social status.[25] This was done in the name of establishing clearer professional boundaries, which essentially asserted that university-derived knowledge about the human body was superior to knowledge gained largely from experience. Race and "purity of blood" were thus formative matters that were central both to the medical profession and to informal healing practices in colonial Latin America.

On a practical level, especially in terms of the root causes of infection, the body of knowledge attained at the university was not necessarily better than that derived from experience. As Antonio Barrera-Osorio shows in his work, medicine as practiced (and hybridized) in the New World strengthened medical empiricism in the early modern Iberian world, which in turn "helped break the late medieval and humanist dependence of knowledge upon textual interpretation and exegesis."[26] Spanish colonial au-

[21] Luz María Hernández Sáenz, *Learning to Heal: The Medical Profession in Colonial Mexico, 1767–1831* (New York, 1997), 76.

[22] Fields, *Pestilence* (cit. n. 20), 67.

[23] Pamela Voekel, *Alone Before God: The Religious Origins of Modernity in Mexico* (Durham, N.C., 2002), 172.

[24] Hernández Sáenz, *Learning to Heal* (cit. n. 21), 75.

[25] Adam Warren, *Medicine and Politics in Colonial Peru: Population Growth and the Bourbon Reforms* (Pittsburgh, 2010), 12.

[26] Antonio Barrera-Osorio, *Experiencing Nature: The Spanish American Empire and the Early Scientific Revolution* (Austin, Tex., 2006), 2.

thorities favored empirical medicine not only on account of the shortages of licensed surgeons and physicians but also because of the poor results obtained by "learned" practitioners. For instance, in 1755, licensed physicians were called upon to treat Sor María Anna Agueda de San Ignacio, a nun from Puebla who was suffering from excessive pains, physical weakness, headaches, convulsive movements, and aneurysms on her neck.[27] They diagnosed her with "dropsy" and recommended that she undergo repeated bleedings and avoid liquids because of her condition. Sor María Anna's condition worsened; she lost her ability to speak and developed pus-filled sores all over her body. According to historian Asunción Lavrin, who analyzed this case, the doctor put "applications of heat on her stomach, one of which burned her badly. A cupping glass was applied to one foot, and it caused an open sore that never closed. . . . She died on February 15, 1756, at age sixty, despite 'all medical attention,' but possibly a victim of too much of it."[28] Even as doctors and surgeons became more professionalized and improved their craft during the eighteenth and nineteenth centuries, they did not likely enjoy a reputation among the indigenous, African, and mixed-race population that was commensurate with their professional status and credentials.[29]

In the records of the colonial Mexican criminal cases examined here, the courts in practice made few distinctions between surgeons who did and did not speak Latin when requesting expert advice, though we do find that one of the tactics by defense lawyers of the accused was to attack the credentials of those surgeons who were not accredited by the Royal Protomedicato. Such was the case in an early nineteenth-century criminal trial from Mexico City, in which eighteen-year-old indigenous Mariano Marcos accused the slightly older Lorenzo Aguirre of having violently sodomized him. According to Marcos's testimony, one day Aguirre entered Marcos's home and "threw himself upon him and, thrusting his hand into the fly of his pants, grabbed his member and testicles, and he twisted them and forcefully pulled them [*dandole estirones fuertissimos*] as he took off his underwear."[30] Aguirre then penetrated him and "broke his [posterior] orifice with his member." The official report of the surgeon [*reconocimiento del cirujano*] included the following assessment by Don Ysidro Ruiz, a thirty-eight-year-old phlebotomist: "having registered and inspected the [boy's] body, he [Ysidro Ruiz] confirmed that it was in a very bad state due to the act of the delinquent who had ruptured the [boy's] intestines, which is exceedingly difficult to cure, and [he] affirms that this is of vital necessity, but he has been unable to find any means with which to cure him."[31] Despite desperate ministrations by the surgeon and

[27] Asunción Lavrin, *The Brides of Christ: Conventual Life in Colonial Mexico* (Stanford, Calif., 2008), 186.

[28] Ibid., 168.

[29] Lee Penyak argues that there were relatively few changes in the professionalization of medicine during the colonial period, but that many changes are evident after independence; Penyak, "Obstetrics and the Emergence of Women in Mexico's Medical Establishment," *Americas* 60 (2003): 59–85. For other examples in which licensed physicians in the late colonial period contributed only marginally to the health of the population, see Andrew L. Knaut, "Yellow Fever and the Late Colonial Public Health Response in the Port of Veracruz," *Hispanic Amer. Hist. Rev.* 77 (1997): 619–44.

[30] Archivo General de la Nación (hereafter cited as AGN), Mexico, Criminal 98, exp. 2, fols. 25–62: "el dia que llegó a su xacalito Lorenzo Aguirre echandose sobre el y metiendole mano a la portañuela de los calsones le agarro el miembro, y testiculos, y se los torció dandole estirones fuertissimos aquitandole los calsones y aplicandose a el por la parte posterior . . . que Aguirre le rompió el orificio con su miembro."

[31] Ibid., fol. 25: "haviendolo requitado é inxpeccionado le encontro que era cierto estar maleado de la misma suerte que lo acudan al delinquente quien le rompió los intestinos y dificil de curacion, y si afirma ser de necesidad mortal sin hallar auxilio alguno para la curacion."

the boy's mother, Mariano Marcos died one week later from an infection brought on by the injuries he suffered.

In the course of his trial, Lorenzo Aguirre admitted to sodomizing the boy. However, Aguirre's defense lawyer, Don José Agustín Diosdado, sought to defend his client by discrediting the validity of the "expert" testimony provided by Don Ysidro Ruiz. In building his legal argument against the prosecution, Agustín Diosdado correctly asserted that Don Ysidro Ruiz was not a licensed surgeon: "Don Ysidro Ruiz, who performs the offices of medic and surgeon in this town, is effectively neither, but is only a barber; and to that effect, the justification of Your Majesty [the viceroy] can be best served first by certifying whether or not this is true. If Your Majesty determines that the said Ruiz is needed, it is out of necessity, since in this town there is neither a medical expert [*perito*] nor a titled and formally examined university-trained surgeon who can assist with healing and with the *reconocimientos* of the sick and the wounded."[32] Agustín Diosdado argued that in order for the criminal court to accept Ruiz's official report, he should be required to present proof that he had been formally trained in the faculties of medicine and surgery, evidence of his attendance at a university, and his certificate of approval from the Royal Protomedicato. This approach proved ultimately unsuccessful in this case and in others, in part because, as historian Lee Penyak concludes, "judges were forced to rely on the testimonies of noncertified practitioners" because of the overall scarcity of licensed doctors.[33]

In light of what the court took to be the incontrovertible signs of the body and the damning confession of the perpetrator, the court opted not to dismiss the testimony of an unlicensed surgeon. Aguirre, who had also been implicated in the crimes of homicide and bestiality in previous criminal denunciations, was sentenced to ten years of forced labor on public works projects. The judgment in this case hinged on the dynamics that unfolded between different loci of colonial authority—the (unlicensed) medic, the lawyer, and the Spanish or Creole judge. Each side sought to either prove or refute Marcos's accusation by examining the violated body of the victim, the intent of the culprit, or the credentials of the medical expert in the case, who, according to Agustín Diosdado, was "nothing more than a plain barber and bleeder."[34] In this sense, this case illustrates how notions of hegemonic and nonhegemonic masculinities overlapped and intersected with the categories of the natural and unnatural. The violent, predatory, and penetrative masculinity of the aggressor, Aguirre, was "against nature," and therefore punishable, even though the credentials of the unlicensed medic had been called into question.

Thus, we find that the criminal courts recognized that intimate knowledge of the human body could be legitimately based on experience rather than on formal training

[32] Ibid., fol. 40: "Digo: que a el derecho de mi menor conviene acreditar q^e D^n Ysidro Ruis q^e hace oficios de medico y cirujano en este pueblo no lo es efectivam^te, sino solo barbero; y a el efecto la justificacion de VMd ha de servirse lo prim^o de certificar si es esto cierto, y que el valerse VMd de dicho Ruis es por nesesidad, por no haver en este pueblo un perito, ó Facultatibo titulado y examinado q^e asista a las curaciones y reconocimientos de los heridos y enfermos. Y lo otro: de que el referido Ruis bajo de la sagrada religion del juram^to y la protesta que hago de estar solo si lo favorable declare si ha estudiado las facultades de medicina y cirugia: en que colegio, y Universidad ha hecho sus estudios: y si tiene aprobacion o titulo del protomedicato: y en caso de tenerlo lo muestre, y se ponga testimon^o de el en los autos."

[33] Penyak, "Midwives" (cit. n. 7), 253.

[34] AGN, Criminal 98, exp. 2, fol. 46: "El mismo nos confiesa, que no tiene estudios algunos: que no ha cursado medicina, ni cirugia: que no es mas que un puro barbero, y sangrador: y que por lo mismo no tiene titulado, ni aprobacion alguna en aquellas Facultades."

and university credentials. The interactions between hegemonic/nonhegemonic masculinities and the categories of natural/unnatural sexualities could produce, as they
did in the previous cases, multiple and contradictory narratives that reflected their
particular importance in each case. Criminal cases of *estupro* and sodomy—two legally distinct categories that were unsystematically and sporadically punished in New
Spain—are particularly useful in demonstrating how sexual violence and medical
expertise intersected with predatory masculinity and other masculinities as well. The
juxtaposition of (potentially) reproductive sex with sodomy also allows us to reconsider colonial science and masculinities in relation to the natural/unnatural dyad.
Transcripts of the criminal proceedings were shaped in part by those women and men
who were called upon to examine the bodies in question and interpret the physical
evidence. Gender influenced such interactions significantly.

MIDWIVES' KNOWLEDGE, WOMEN'S BODIES, AND HONOR

The obstetrical knowledge of midwives in colonial Mexico, as in other parts of the
early modern world, stemmed largely from their primary role in officiating over
birth—based upon experience rather than formal training. As happened with surgeons, there was an attempt in the seventeenth century to professionalize midwives
and increase oversight of their work—an endeavor that was only minimally successful. Although midwives in colonial Mexico technically fell under the jurisdiction of
the Royal Protomedicato, Sherry Fields notes that throughout much of the colonial
period, the midwife (*partera* or *matrona*) was free to practice her trade, and her role
in assisting women with childbirth was rarely questioned by the Protomedicato: "only
in cases of difficult deliveries, which most often resulted in the death of the mother,
child, or both, was the colonial midwife bound by any sort of legislation; in such a
case it obliged her to seek a surgeon's aid for the suffering parturient."[35] We do know
from some criminal and Inquisition cases that midwives occasionally offered illegal
forms of contraception and herbal abortifacients to women and also provided a variety
of menstrual "regulators"—information that could be found in some medical treatises
or, more commonly, in oral traditions that combined folk Spanish remedies with indigenous, African, and mixed-race knowledge of plants and herbal remedies. As historian Nora Jaffary notes, "'Regulating the menses' was a legally and socially accepted
practice; abortion and infanticide were, of course, illegal."[36] The knowledge of midwives often overlapped with that put forth in medical texts, which were published by
physicians and circulated in the print media in New Spain, including medical works
containing remedies for inducing the resumption of the menses when these had been
"suppressed."[37] Though most midwives and informal healers were illiterate, this did
not necessarily weigh against them in the court's judgments about who did and did not
have expertise, especially in rural areas.[38] The circulation of informal medical knowl-

---

[35] Fields, *Pestilence* (cit. n. 20), 64.
[36] Nora E. Jaffary, "Reconceiving Motherhood: Infanticide and Abortion in Colonial Mexico,"
J. Family Hist. 37 (2012): 3–22, on 9.
[37] Ibid.
[38] Lee Penyak, in a study on midwives in Mexico, notes that between 1754 and 1845, not one of
the fifteen midwives who examined victims of sexual assault knew how to sign her name; Penyak,
"Midwives" (cit. n. 7), 261.

edge in popular culture, especially in the seventeenth and early eighteenth centuries, was structured in ways that often made the court's decision necessarily improvisatory.

It was this intimate experiential knowledge of the female body, obstetrics, and child-birth that enabled midwives—usually older women who were often widowed—to participate as medical examiners in Mexican criminal and ecclesiastical courts. According to male medical authorities in New Spain, midwives had to be "honorable" and knowledgeable so as to minimize the chances that they would either harm the mother or child during childbirth or engage in illegal practices, such as proffering abortifacient herbs or drinks. One Spanish *protomédico*, Doctor Antonio Medina, author of the 1750 *Cartilla nueva, útil, y necesaria para instruirse las Matronas*, suggested a number of "honorable" requisites for the ideal midwife. According to Medina, she must be a good Christian, temperate, healthy, and robust, neither too young nor a virgin (but certainly a mother), and she should be able to read and write.[39] Affirming the gendered hierarchy of medical knowledge, Medina emphasized that the ideal midwife must also be "of a docile disposition and sufficiently inclined to admit the opinion of her superiors, asking for help in opportune times, and seeking consultation from the Medic and the Surgeon in the cases [of childbirth] in which any difficulty occurs."[40] At least in theory, by the mid-eighteenth century, significant shifts had occurred in elite notions of what kinds of knowledge, power, and expertise medical practitioners were supposed to possess. Given that there were no licensing standards for midwives in the sixteenth and seventeenth centuries, reform-minded Bourbon policy makers in the mid-eighteenth century advocated for the licensing of midwives in order to promote the power of the state. To this end, they created a set of qualifications for licensing and provided midwives with professional advice.[41]

At a minimum, however, midwives were to have sufficient training in and knowledge of female anatomy—specifically, the vagina, the uterus, the vulva, the clitoris, the urethra, and the labia majora and minora—to assist in childbirth but also to "establish the [midwife's] knowledge sufficient to make declarations before judges in lawsuits dealing with suspicious virginity, *estupro*, and impotencies; by their lack of it, each day we see many errors committed."[42] According to Medina, the best remedy for midwives' ignorance was to impart to them the knowledge of the university-trained male medics in the colony. He urges further that midwives in doubt about anything "should consult with a trained Medic [*deben las Matronas consultar con Medico sábio*]," thus essentially subjecting female obstetrical knowledge to the control and masculine knowledge of the male medic, which underscores the gendered hierarchy of medical knowledge in New Spain. Regulated or unregulated, midwives were regarded as a subordinate extension of male power. Their personal honor, and that inherent in their office, was thus embedded in the patriarchal honor system.

Models of hegemonic masculinity operated within medico-legal cultures and among colonized subjects in a number of ways: in consideration of the criminal act itself,

[39] Antonio Medina, *Cartilla nueva, útil, y necesaria para instruirse las Matronas, que vulgarmente se llaman Comadres, en el oficio de Partear* (Madrid, 1750), 1–9.

[40] Ibid., 6.

[41] John Jay TePaske, "Regulation of Medical Practitioners in the Age of Francisco Hernández," in *Searching for the Secrets of Nature: The Life and Works of Dr. Francisco Hernández*, ed. Simon Varey, Rafael Chabrán, and Dora Weiner (Stanford, Calif., 2002), 55–64, on 61.

[42] Medina, *Cartilla nueva* (cit. n. 39), 22.

at the level of oral transmission (the recounting of that act by witnesses and victim), and at the level of legal/medical interpretations of the body. Some men felt a sense of entitlement—stemming from their assumptions about patriarchal privilege—over the bodies of women and adolescent males and females that they desired, whether as prospective marriage partners or as individuals with whom they might satisfy carnal impulses. Sonya Lipsett-Rivera has ably shown that "a pattern of associating rape with marriage is frequent among complaints of rape" in central Mexico during the late colonial period, and such patterns stemmed from medieval legal codes such as the thirteenth-century *Siete Partidas* promulgated by Alfonso X.[43] In a significant number of instances, marriage between perpetrator and victim was the ultimate legal outcome of cases in which men forcibly engaged in sexual relations with a younger woman.

It was the forceful act of *estupro*, or "deflowering," that, when brought to light through a formal criminal complaint, thrust violated females into the legal and medical spheres, where official judgments centered less on the act of sexual coercion itself and more on the "honor" and the prior virginity of the plaintiff. The accusation of *estupro*, however, also necessitated a court-mandated medical examination from an older midwife who possessed intimate knowledge of female bodies and pregnancy. And while judges may have "preferred to receive expert opinion from doctors or other males recognized as knowledgeable in medicine, even in cases dealing with sex crimes against women," midwives were regularly called to court to examine women and testify, in part to preserve female "honor" by not subjecting them to the male medical gaze, as we saw in the 1684 case of Angelina García's *estupro*.[44] Thus, paradoxically, while midwives were called upon by courts, and their testimonies shaped the legal narratives of predatory masculinity and sexual crimes, licensed medical practitioners were working to devalue and delegitimize unlicensed midwives and experiential female medical knowledge in a variety of ways.[45] As Penyak observed in his study of obstetrics in colonial and early national Mexico, "the Protomedicato safeguarded its monopoly on scientific learning at the expense of the general population and purposely stifled female participation in the medical establishment."[46]

Over the course of the eighteenth century, the long attempt to control midwives and wrest the monopoly of obstetrics from them began to produce certain results in the juridical sphere. Though, as in the 1684 case we started with, it was not uncommon for a midwife to be consulted in criminal cases of *estupro* without the support or interference of other doctors, a 1795 criminal case from a small town outside Tlaxcala demonstrates that the testimony of the midwife was supplemented by other, male voices, including that of the surgeon. This case featured the rape of a preadolescent female by a nineteen-year-old Indian male. On 23 September 1795, a number of individuals from the pueblo of Santa Ysabel Xiloxostla, including the town notary, presented the four-year-old Victoriana Severina, a "young Indian girl . . . injured in her immodest parts [*partes pudendas*]," to colonial officials in Tlaxcala, denouncing an indigenous man, Juan Ramos, for the offense. The community governor of Tlaxcala, Francisco de Lissa, mandated that the girl be examined "by an expert surgeon and by a matron or midwife [*matrona o comadre*] . . . who are to declare what is of note regarding the

---

[43] Lipsett-Rivera, "Intersection" (cit. n. 2), 560.
[44] Penyak, "Midwives" (cit. n. 7), 253.
[45] Ibid.
[46] Penyak, "Obstetrics" (cit. n. 29), 62.

girl's virginity, and whether or not it [the hymen] has been lacerated."[47] As numerous historians have noted, in rape cases masculine violence and coercive acts per se played a less important role in judicial proceedings than did the act of lacerating the hymen. Hence, the court's gaze fell upon the girl's virginity and her "honor" rather than on the violent acts perpetrated on her body. In so doing, the court indicated that predatory male sexuality was a secondary and perhaps even a natural behavior, even when directed toward a young child.

Upon examining Victoriana Severina, surgeon Miguel Sandoval attested that her insides were "lacerated and effusing blood from the uterus, demonstrating that she had cohabitated with a man whose member had torn numerous fibers and veins, causing the said Victoriana to suffer a fever and other symptoms," all of which placed her in evident danger of losing her life.[48] Rosalia Gómez, a midwife who testified that the four-year-old girl was "torn, injured, and [now] totally lacking in her virginity," substantiated these observations, indicating in part that while the medical profession throughout Spain's colonies devalued female medical expertise, the criminal court system nonetheless continued to rely heavily on the testimonies of midwives into the late colonial period.[49] In his defense, the suspect, Juan Ramos, told authorities that he had merely kicked the young girl upon encountering her in a field while cutting sugarcane. The case came to an abrupt end on 20 October, when the girl's parents alerted authorities that they had informally resolved the situation with the culprit and his family. According to Victoriana Severina's parents, because it was "the Christian thing to do," they pardoned Juan Ramos after his parents begged for their son's forgiveness and had offered twenty pesos—a relatively meager sum that was nonetheless comparable to a dowry—as compensation for their daughter's lost virginity.[50] As Sonya Lipsett-Rivera tells us, "The financial settlement compensated the young woman for her loss of virginity and therefore loss of 'market value' in order to make it possible for her to find a husband."[51]

The criminal court referred to the *estupro* of a prepubescent girl as "one of the most atrocious [crimes], hated by the law in its entirety." Yet the court refused to exercise its potential power to prosecute the girl's rapist on its own behalf and accepted the extrajudicial arrangement of the parents. In ways that were at odds with the rhetoric of the offense, as so often happens in cases of violence against women and children, the colonial court accommodated male violence as a normal part of the patriarchal order. This case is a reminder that the convocation of expert witnesses was only part of the mechanism by which the court considered cases involving rape and sexual violence. Historian Catherine Komisaruk, speaking of rape cases in colonial Guatemala, rightly notes that "the judicial system's treatment of sexual offenses (and its failure to recognize sexual offenses) has partly occluded our study of sexuality and of violence in colonial society."[52] Violent masculinity was accepted as a natural part of colonial life. It was a product of the hierarchy of hegemonic masculinities that operated within the colonial patriarchy with its underlying logic of honor and purity of blood [*limpieza de sangre*]

---

[47] Archivo Histórico del Estado de Tlaxcala, caja 44, exp. 39, fol. 1v.

[48] Ibid., fol. 2.

[49] Ibid., fol. 3.

[50] The silver peso, a hard currency in the Iberian Atlantic world, was roughly equal to the American dollar. Given that some families whose daughters had been raped received financial recompense of one hundred pesos or more, the twenty-peso compensation is relatively small.

[51] Lipsett-Rivera, "Intersection" (cit. n. 2), 571.

[52] Catherine Komisaruk, "Rape Narratives, Rape Silences: Sexual Violence and Judicial Testimony in Colonial Guatemala," *Biography* 31 (2008): 369–96, on 371.

to distinguish between men, women, and "impure" races. María Elena Martínez has pointedly asserted that "Spanish notions of purity and impurity of blood were fictions, ideological constructs based on religious and genealogical understandings of difference that despite their invented nature were no less effective at shaping social practices, categories of identity, and self-perceptions."[53] If we push this analysis a little further, we can see how the hegemonic nature of predatory masculinity expressed through male privilege and sexual violence is also a fiction, preserved in the archives by individuals who played the roles ascribed to them in the social hierarchy. Yet these effects had very real consequences in terms of shaping social practices and gendered behavior.

Predatory male sexuality thus found its place within a supposedly "natural" hierarchy of gender, race, class, and age in colonial society. Violent masculinity was simultaneously criminalized and implicitly tolerated at the judicial level. The role of the medical expert, then, was far from definitive, given the fact that judges, colonial authorities, and families often opted to selectively turn a blind eye toward—and thereby legitimize—instances of male sexual violence even when they violated countervailing codes of female honor and virginity. The courts, as we have seen here, did not often consider the injuries of the victim. In view of this prevailing paternalism (which claimed to safeguard women's modesty while exposing them to extraordinary instances of violence), we need to look at cases that criminal courts found extraordinary, and those it did not. From the theoretical standpoint as expressed in official law, certain instantiations of predatory masculinity that upset colonial order and hierarchies—public drunkenness, *estupro*, homicide, sodomy, and rebellion—were clearly criminal. But in practice, local criminal courts had room to selectively apply and interpret the law. As in Victoriana Severina's case, we cannot determine when the court would or would not act to punish the offender. What we do see, however, is that despite the legal and moral rhetoric, the court rarely chose to exercise its right to pronounce the death penalty for the perpetrators of rape. Furthermore, as the colonial period progressed, punishments for rape and many other sexual crimes became less severe. For example, there is no evidence after the late seventeenth century that any individual was put to death for the crime of sodomy, whereas in the fifteenth and sixteenth centuries, the death penalty was not uncommon.[54] Because I am interested here in how instances of violent masculinity intersected with colonial order and medical expertise, I move from the sphere of the female medical practitioner/victim to that of the male surgeon/victim in cases of violent and coercive sodomy. This shift in focus allows us to highlight the continuities and connections between the sexual subjugation of women and girls and the bodies of men and boys.

CORPOREAL SIGNS AND THE "NEFARIOUS SIN"

The narratives and grounds for assessing predatory masculinity shifted when the central corporeal act revolved around what colonial secular and ecclesiastical officials

[53] María Elena Martínez, *Genealogical Fictions: Limpieza de Sangre, Religion, and Gender in Colonial Mexico* (Stanford, Calif., 2008), 61.

[54] Well into the nineteenth century, local secular courts frequently pronounced the death penalty for sodomy, but upon appeal these sentences were typically overturned by the highest criminal courts in the viceroyalty of New Spain—the *audiencias* of Santo Domingo, Mexico, Guatemala, and Guadalajara. See Zeb Tortorici, "Contra Natura: Sin, Crime, and 'Unnatural' Sexuality in Colonial Mexico, 1530–1821" (PhD diss., Univ. of California, Los Angeles, 2010) and my current book manuscript, "Sins against Nature: Sex, Colonialism, and Historical Archives in New Spain, 1530–1821."

deemed sins "against nature" [*contra natura*]. It might come as a surprise to learn that in New Spain, the Holy Office of the Mexican Inquisition by and large did not have jurisdiction over the "nefarious sin" [*pecado nefando*] of sodomy, unless a priest was involved or an act of heresy had been committed.[55] The "sins against nature"—masturbation, bestiality, male-female intercourse in an inappropriate position, and copulation with a member of one's own sex—all largely fell outside of the jurisdictional boundaries of the ecclesiastical courts in New Spain, although some priests and branches of the Inquisition incorrectly prosecuted sodomy anyway.[56] This limited jurisdiction stands out in contrast to much of the early modern Iberian world, including the Spanish cities and municipalities of Valencia, Barcelona, Zaragoza, and Palma de Mallorca, as well as Portugal and its overseas colonies of Brazil and Goa, where sodomy and bestiality fell under the jurisdiction of the Inquisition. Between 1540 and 1700, for example, the tribunals of the Inquisition in Spain prosecuted 380 cases of sodomy in Valencia, 791 cases in Zaragoza, and 433 cases in Barcelona.[57] Similarly, between 1587 and 1794, the Portuguese Inquisition tried some 400 individuals for sodomy, about thirty of whom were executed for their crimes.[58]

In terms of mapping out a history of these geographically indefinite jurisdictions over sodomy, it is relevant that in 1505, Ferdinand II of Aragon initially placed sodomy under the jurisdiction of the Inquisition. The famed 1469 marriage of Ferdinand II of Aragon and Isabella of Castile brought stability to these Iberian kingdoms and eventually laid the foundations for the political unification of Spain. In 1509, Ferdinand signaled a major shift in his approach to sexual crimes when he decided to place sodomy under the jurisdiction of secular authorities throughout much of the Iberian Peninsula.[59] This change in policy was enacted so that the Inquisition was not diverted from its primary task of hunting down and prosecuting heretics and Judaizers in the Iberian Peninsula.[60]

As a consequence of this shift in the legal approach to sodomy, early sixteenth-century secular courts gained jurisdiction over sodomy cases throughout the Spanish kingdoms of Castile, Granada, and Seville. Because the recently discovered territories of the New World—the Indies—had been incorporated into the Crown of Castile in the sixteenth century, the Castilian legal system, along with its administrative and judicial bureaucracies, was established throughout those vast territories.[61] Despite the assertions by one historian that "over the course of the colonial period, both secular and ecclesiastical authorities held jurisdiction over sodomy cases in the tribunals of

[55] The same was true for other regions of the Spanish Americas, including Peru and Nueva Granada, where sodomy also fell largely under the jurisdiction of secular courts as opposed to the Inquisition.

[56] Tortorici, "Against Nature" (cit. n. 4), 167.

[57] Federico Garza Carvajal, *Butterflies Will Burn: Prosecuting Sodomites in Early Modern Spain and Mexico* (Austin, Tex., 2003), 71.

[58] Merry Wiesner-Hanks, *Christianity and Sexuality in the Early Modern World: Regulating Desire, Reforming Practice* (New York, 1999), 126.

[59] Rafael Carrasco, *Inquisición y represión sexual en Valencia: historia de los sodomitas (1565–1785)* (Barcelona, 1985), 11.

[60] Louis Crompton, *Homosexuality and Civilization* (Cambridge, Mass., 2003), 295.

[61] John Elliott, *Empires of the Atlantic World: Britain and Spain in America, 1492–1830* (New Haven, Conn., 2006), 127. Initially, the Council of Castile directly administered the Indies, but in a 1524 cédula, Charles V established the Council of the Indies as the primary means of dealing with and administrating the Indies. See Cutter, *Legal Culture* (cit. n. 11), 49. The 1680 Recopilación de las leyes de Indias was also based on, but not identical to, the 1567 Recopilación of the laws of Castile that was published by order of Phillip II.

New Spain," in reality, the law was clear: the Mexican Inquisition was allowed to prosecute cases of sodomy only when they involved some overt heresy (like solicitation in the confessional), blasphemous, heretical propositions (such as asserting that "sodomy is not a sin"), or the participation of a priest.[62] The tribunal of the Inquisition in Mexico was informed in 1580 that Rome, despite the Aragonese precedent, would no longer allow sodomy to be tried in Mexican ecclesiastical courts.[63]

Because both sodomy and rape cases fell largely under the jurisdiction of the secular courts in New Spain, they make for an interesting comparison. Since neither crime was technically defined as heretical, the ecclesiastical courts in New Spain rarely had jurisdiction over such crimes. Similar to cases of *estupro*, after the initial fact-finding phase of a judicial inquiry into the crime of sodomy had been completed and the suspect imprisoned, the primary mode of determining whether or not an act of anal penetration had been consummated was, as with rape cases, the medical inspection [*reconocimiento del cirujano*] by a trained medic or, more commonly, a surgeon. Male medical practitioners in instances of sodomy sought, through the medical examination, any physiological irregularities in the colon, rectum, or anal canal, which they read as proof of anal penetration. Given that midwives were almost never called to inspect the bodies of the boys who were suspected victims of sodomy, we already find one significant point of departure in terms of who had the authority to read and interpret the signs of the bodies in question. While male medics and surgeons regularly gazed upon and assessed the bodies of both men and women, midwives typically possessed informal authority to examine only female bodies. In this sense, when the body in question had been forcibly penetrated, medics and courts rendered male and female bodies (il)legible in distinct ways. Midwives and medics were obliged to testify about whether a female's hymen—the physical proof—remained intact after an act of *estupro*. No such corporeal "proof" existed for the male body, and the assessments of medics showed that trying to determine whether a male body had been penetrated was exceedingly difficult except in cases of extreme violence.

While the corporeal signs pointing to Lorenzo Aguirre's violent penetration of Mariano Marcos were irrefutable, the physiological signs of sodomy, more often than not, proved to be ambiguous, making it difficult for the court to determine whether or not a "nefarious" act had in fact taken place. One example comes to us from the notarial archive of Puebla from 1714 in which an indigenous man, Cristóbal de Contreras, alerted authorities that one night another indigenous man, Antonio Pérez, had invited him to drink the fermented alcoholic *pulque* and subsequently tried to kiss, hug, and touch him in an effort to incite him to have sex. According to Contreras's accusation, Pérez pleaded to him, "fuck me, man" [*hódame hombre*], but Contreras defended himself physically and verbally and resisted these advances, telling him, "I am not of that type, and I do not want to do this."[64] At that point, the records show that Contreras took matters into his own hands, informing secular authorities of Pérez's behavior so that he could be tried and punished. However, when surgeons examined Antonio Pérez, presumably in an attempt to see if he had previously committed the

[62] Garza Carvajal, *Butterflies* (cit. n. 57), 71.

[63] William Monter, *Frontiers of Heresy: The Spanish Inquisition from the Basque Lands to Sicily* (Cambridge, 1990), 287.

[64] Archivo General de Notarías de Puebla, caja 39A, 1700–1800 Civil, Criminal, Testamentos, unnumbered exp., fols. 1–8.

pecado nefando, they filed a report stating that although he was lacking some folds of the anal skin and he had barely discernible fissures in his lower intestines, his anal region was largely in its "natural state." Doctors also noted that such physiological signs may have occurred naturally and therefore could not be taken as signifiers of sodomy. Another doctor who examined him asserted, "as long as the witnesses have not seen the blood of the nefarious act upon bed sheets or clothing, the crime cannot be proven."[65] Here, given that violence in a literal sense had not occurred in this case, the alleged "predator"—rather than the potential victim—was subjected to medical examination. Despite the fact that medical examinations proved inconclusive and the only evidence against the suspect was Contreras's initial accusation, Pérez spent four years in prison (due in part to the slow wheels of colonial justice) and was forced to pay the costs of the trial upon his release. In this case, there was no tangible evidence of sodomy, but the court was equally concerned with the signs of the intent to commit sodomy. Here, the criminal court rendered predatory masculinity—when framed as "against nature" and directed toward the body of another male—a punishable offence. This stands in contrast to cases of male-female sexual aggression, in which allegations of the mere intent to commit an act of *estupro* were inadmissible and not punishable in a court of law. Masculinity thus intersected with medicalized readings of the body in complex ways. This case also contrasts with cases of coercive sodomy in which there was ample physiological evidence of the crime.

In the absence of legible signs and evidence of sodomy on the body in question, medical practitioners and legal thinkers devised other (problematic) ways of ascertaining whether or not an individual was accustomed to committing the "sin against nature" with other men. As Garza Carvajal states, "The Spanish courts, in their attempt to prove the abominable nature of sodomy, sometimes resorted to the use of science to quantify their discursive descriptions. Some courts subjected accused sodomites to humiliating physical examinations."[66] The 1764 Mexican legal treatise, the *Libro de los principales rudimentos tocante a todos juicios, criminal, civil y executivo* (Book of the basic principles regarding all criminal, civil, and executive trials), for example, outlined that one mode of determining guilt in sodomy cases was to "examine [the suspect's posterior] with an instrument that surgeons carry with them, and in its absence, with a hen's egg [*huebo de Gallina*] that is large and will be inserted in the posterior eye [*ojo de atras*, anus] to see if it disappears, and this examination will be related to the judge."[67] While the use of this recommended technique has yet to be documented in sodomy cases, medical practitioners certainly did use special instruments to prod and (re)penetrate the bodies—thereby reenacting invasion of the

---

[65] Ibid., fol. 24: "que en quanto no ben los testigos la sangre de tan nefando acto en sabanas [y] camisas, no se puede comprobar semejante delicto."

[66] Garza Carvajal, *Butterflies* (cit. n. 57), 105.

[67] Charles Cutter, ed., *Libro de los principales rudimentos tocante a todos juicios, criminal, civil y executivo: Año de 1764* (Mexico City, 1994), 38: "esta causa continuadamente se hazen por denuncia que se da al juez de haberlos cogido en el Pecado y assi se examina al Denunciate, y demas Testig.s que se hallaren para la preuba del Delicto, se les toman sus Declaraciones a los Reos Separados si uno niega se carea con otro si este confiesa sino se carea con el Denunciante y testig.s se reoconoserlo es co vn instrum.to que traheen los ciruxanos, y no habiendolo con vn heubo de Gallina que sea largo el qual se le pone en el ojo de atraz, y se sume, y a este reconosim.to se a de ayar el juez, y SS.no no para dar fee, lo qual no sucede en las Estrupadas, q. se reconocen sin q. esten precentes." I wish to thank Chad Thomas Black for this reference.

examined body—of those who were purportedly forcibly penetrated by older males. In this sense, medical invasiveness on the bodies of male victims of coercive sodomy (or female victims of *estupro*) could supplement the initial sexual violence in many cases.

Equally important in the judicial realm was the absence of such signs, as verified by surgeons, which could be used in court to disprove allegations of sodomy—though these criteria were often equally unreliable. In cases of sexual coercion and rape, medical examinations of the victims, as we have seen, were indispensable in proving or contradicting the claims of the aggrieved. In 1794 in Guatemala, for example, Paulo Jiménez, a fifty-year-old mestizo, was accused of having committed sodomy with José María Clemente, a boy of seven or eight years of age.[68] The boy's grandfather, who had been told that Jiménez hugged and kissed his grandson at night, filed the charges. The conflicting testimonies of the boy, who constantly changed his story, were deemed unreliable, and the surgeon who examined the boy found nothing abnormal and declared that the boy had no pain or discomfort in his backside, further calling into question the alleged rape. The boy eventually admitted that though the defendant had only fondled him, his mother persuaded him to tell authorities that Jiménez had penetrated him.[69] There is clearly a back story to this case—one that is not included in the transcripts of the trial—which points to how false or exaggerated accusations of sodomy were occasionally used as political tools to dishonor an individual and stain his reputation.[70] Given the boy's wavering testimony, the lack of medical proof, and the role of the boy's mother in fabricating the charge of sodomy, Jiménez was set free and absolved of the charges against him. Although he may have fondled the boy, medical evidence cast doubt on the initial allegations of sodomy.

There were, to be sure, a number of methodological problems associated with the reliance on medical examinations to confirm or disprove sodomy. Common physical ailments, such as hemorrhoids or ulcers, that cause inflammation and bleeding in the anal area may have been incorrectly interpreted as signs of anal intercourse. This, however, did not stop many medical practitioners from interpreting symptoms such as blisters, scars, lacerations, inflammation, intestinal irritation, or a "loose" anus as indications of sodomy. Speaking to the ambiguity of such evidence, in an incomplete 1780 case from Mexico City, in which a Spaniard, Leandro Hurtado de Mendoza, and an indigenous man, Pedro Joseph Pinedo, were accused of sodomy, medical doctors found neither lesions nor "irregularities" in Pinedo's anus or intestines, thereby demonstrating to their satisfaction that the crime had not taken place. The two surgeons in charge of the medical examination, Don Josef Medina and Don Manuel Josef Rivillas, both approved by the Protomedicato, criticized the jurisprudential reliance on "very

[68] Archivo General de Centro América [Guatemala], A2.2, leg. 175, exp. 3471.

[69] Ibid. The boy repeatedly changed his mind about whether or not he had been hugged, kissed, or penetrated by Jiménez. In the end, he asserted that "when they lived together he slept with him, and at night the Old man [Jiménez] would rub his 'little bird' against his own [quando vivian juntos dormia con él, y por las noches el Viejo [Jiménez] le ponia su pajarito arrimando al suio]." The boy also said that on three other occasions on a nearby hillside, Jiménez took off his clothes and had the two put their penises together. While same-sex touching or fondling was technically illegal, acts that did not culminate in anal penetration were prosecuted less frequently because they might not be defined technically as sodomy.

[70] For more on sodomy as a political tool, see Geoffrey Spurling, "Honor, Sexuality, and the Colonial Church: The Sins of Dr. González, Cathedral Canon," in *The Faces of Honor: Sex, Shame, and Violence in Colonial Latin America*, ed. Lyman L. Johnson and Sonya Lipsett-Rivera (Albuquerque, N.Mex., 1998), 45–67.

suspicious" tangible evidence of sodomy such as "bloodied sheets or a shirt stained with blood" because, according to the doctors, "in practice, innumerable ailments could be responsible for the corruption of the bodily humors."[71] This critique, articulated by Medina and Rivillas, of court reliance on tangential bodily signs to prove or disprove allegations of sodomy demonstrates clearly that the observation of the body and its emissions could be indeterminate. Thus, for courts trying to establish proof of sodomy through physiological irregularities, sodomy did not always lead to the physical irritations and swelling that medics looked for in the body of the penetrated partner. As we have already seen, however, in cases of overt sodomitical violence, medical examinations were indispensable in constructing an archival narrative.

COLONIAL VIOLENCE, CORPOREAL AMBIGUITY

Cathy McClive rightly notes that "'women's secrets' have long been the focus of gender and medical history for the late medieval and early modern periods at the expense of investigation into male corporeality."[72] Criminal cases that concern predatory masculinity through the violent acts of *estupro* (with female victims) and coercive sodomy (with male victims) provide a partial corrective to this historical bias on female corporeality in the history of medicine. Readings of the body by medics, surgeons, and midwives ultimately served to "legitimate medicine's authority," and to legitimate their own authority as medical practitioners in the colonies.[73] More importantly, licensed medics, midwives, and informal healers with sufficient knowledge of bodies in sickness and health functioned as intermediaries within the context of colonial jurisprudence and judicial praxis. In relying on female and male medical experts to read the signs of the violated body, criminal and ecclesiastical authorities in the Iberian world accepted to a large extent the narratives of gender, criminality, and victimhood that midwives and surgeons constructed in their interpretations of bodies and events.

In conclusion, I want to assert that colonized bodies were riddled with ambiguities but suggest that this speaks to larger (and no less ambiguous) intersections of bodies, desires, medical expertise, and criminal courts in the early modern and nineteenth-century Iberian world. A masculine prerogative over the bodies of others was enacted at multiple levels of the judicial process. We can assume that when a crime was actually committed, there was first an initial (sexual) violation by an older male, next by the (medical) subjection of that individual to the prying hands of surgeons and/or midwives, and finally a punishment handed down by the court, which was often more interested in upholding notions of family honor and colonial order than in the

[71] AGN, Indiferente Virreinal 1182, exp. 31: "Dixeron que inspeccionado y reconocido el podex de Pedro Josef Oineda, o vaso exterior del intestino recto, y tomadole su indicacion como el caso pide pasaron al reconocimiento del ano, el que hallaron ilexo pues no tenia señal de esquimosis o volgarmente cardenal ni plaga, ni señal ninguna pues segun Josef Mercado jurisconsulto que la sodomia se prueba cuando los testigos depusieren haver visto las sabanas sangrientas, o la camisa teñida de sangre, y segun la opinion de Bernardo Botallo dize que por manera alguna no puede el arte de la cirujia declarar en este asunto pues estos son indicios mui sospechosos para comprobar lo que el derecho [pages are disintegrated] . . . Pues por causa interna se experimenta en la practica innumerables enfermedades por la corrupcion de los humores que lo que llevan dicho es lo que hayan a todo su leal saver y la verdad por el juramento que fecho tienen en que se afirmaron ratificaron y lo firmaron."

[72] Cathy McClive, "Masculinity on Trial: Penises, Hermaphrodites, and the Uncertain Male Body in Early Modern France," *Hist. Workshop J.* 68 (2009): 45–68, on 45.

[73] Martha Few, "'That Monster of Nature': Gender, Sexuality, and the Medicalization of a 'Hermaphrodite' in Late Colonial Guatemala," *Ethnohistory* 54 (2007): 159–76, on 161.

well-being of the victim. The fact that rape was rarely punished severely by criminal courts in New Spain—the 1684 death sentence meted out to Domingo de Silva being an extreme exception to this rule—suggests that violent masculinity buttressed gendered colonial hierarchies while simultaneously threatening to disrupt them.

Colonial authorities and ecclesiastics certainly viewed *sodomía* and *estupro* as inherently dangerous and disorderly practices that had the capability of disrupting colonial hierarchies of race, honor, and family. But as we have seen, medical practitioners in conjunction with the criminal courts constructed narratives about perpetrators and victims of violent sex acts that normalized predatory masculinity while nonetheless criminalizing it. It is clear that medical observations and the ways that they were employed in colonial Mexico's judicial process "worked to establish what was 'natural' in terms of sexual activity."[74] There is a way then in which medical testimony reinforced both legal discourse and social practice to normalize predatory masculinity, even when this involved the "unnatural" crime of sodomy.

Colonial violence, we can conclude, was enacted at multiple levels on the bodies of the female and male children and adolescents who were thrust into the bureaucratic legal sphere—and thus into the archive—as a result of instances of predatory masculinity and coercive sex acts. Invasive internal medical examinations could supplement the initial sexual violence in the cases and may have been awkward, embarrassing, and uncomfortable for the individual being examined. But, we should note, such examinations did try to corroborate the facts and get to the bottom of the allegations, and they did not ignore the plight of the victim. Given that one of the goals of the colonial state was to protect the honor of "honest" women, if we think back to the case with which I opened this essay, it is perhaps paradoxical that the midwife who was called upon to preserve the "modesty" of Angelina García also delivered explicit information to a group of men within the court, which contravened rules of modesty in another sense. Midwives, however, often appeared to sympathize with the violated girls who were brought before them for examination.

Where then does this leave us in terms of the intersections of science and masculinity in the context of colonial New Spain? Acts of sexual violence perpetrated on the bodies of girls in cases of *estupro* and on the bodies of boys in cases of coercive sodomy were corroborated through medical examinations of the bodies of the victims (and occasionally the perpetrators). These medical examinations of the violated bodies in question functioned within the larger framework of long-standing tensions between male and female knowledge in the fields of obstetrics and colonial medicine. Leah DeVun in her contribution to this volume discusses the authority of male medics and surgeons to regulate sex and sexual difference in medieval Europe, thus calling into question orthodox narratives in the history of sex and sexuality. DeVun notes that courts in medieval Europe prosecuted female and other illicit healers who practiced medicine without a license, making health care a contested site that pitted men against women and men against other men within the spheres of scientific knowledge, credentials, and medical practice. In contrast to the medieval context analyzed by DeVun, the criminal and ecclesiastical courts of New Spain relied on the obstetrical knowledge of a variety of medical experts. Yet, in line with DeVun's analysis, in the context of New Spain, a science of sex (and of policing sexual violence) was emerging to engage with

[74] Ibid., 171.

predatory masculinity "at precisely the moment that male medical practitioners were so intent on policing the masculinity of their own profession."[75]

This deliberate exclusion of women from the formal profession of medicine is a masculinized enactment of policing not only the boundaries of colonial medicine but of gender itself. The processes through which the colonial state prosecuted certain predatory sexual crimes by men simultaneously criminalized and validated this form of masculinity. Such forms of predatory sexual violence were decidedly colonial in that they involved social actors whose gendered and racial configurations were located within complex hierarchies that were partly determined by the patriarchal and racially mixed context in which they lived. In the colonial legal and medical spheres, while masculinity typically trumped femininity, masculine hegemonies were organized hierarchically according to degrees of Spanish blood and male prerogative (within the formal field of medicine) over female knowledge of science, medicine, and the body, despite the fact that, in practice, criminal courts regarded the testimonies of female experts as on par with those of male physicians and surgeons. Historian Antonio Barrera-Osorio has shown that in its overseas colonies the Spanish crown institutionalized modes of scientific observation, empirical practices, methods, and procedures—"practices and institutions [that] were linked to the development of Spain's long-distance strategies for controlling the New World."[76] If we look at the broader picture of the production of knowledge about bodies, lands, and peoples, then we can see how the scientific and medical observations of bodies in these and other colonial criminal cases were linked to empirical strategies of surveillance of colonial subjects and lands. The issue of control was never far behind that of medical authority and the right to read and interpret the signs of the body.

Gendered violence in colonial society was a part of everyday life, marriage, and courtship, yet there exists a danger of reifying a particular type of violent masculinity when relying upon judicial texts dealing specifically with *estupro* and coercive sodomy to re-create colonial histories of gender relations through notions of hegemonic masculinity. Predatory masculinity in New Spain was normative to a degree, but it was also criminalized and contested. The hegemony of predatory masculinity could subtly and paradoxically be upheld by the rulings of judges and criminal courts, which were more intent on upholding patriarchy with its underlying and misogynistic notions of "honor" and "purity of blood." On the other hand, witnesses, fathers and mothers of the victims, the victims themselves, surgeons and midwives, and even friends and family members of the perpetrators often did directly challenge the hegemonic nature of predatory masculinity. Violent and predatory masculinity could thus be both hegemonic and nonhegemonic, and it could be enacted at the level of the sexual crime itself, the medicalized readings of the body, and the judgment and rulings of the court.

We should, therefore, heed Sonya Lipsett-Rivera's observation that one especially rare image in colonial Mexican archives is that of "the loving father."[77] Violent and predatory masculinity was simply one of multiple overlapping masculinities that, for reasons tied to the nature of the sources and the methodologies of cultural history

[75] Leah DeVun, "Erecting Sex: Hermaphrodites and the Medieval Science of Surgery," in this volume, 21.

[76] Barrera-Osorio, *Experiencing Nature* (cit. n. 26), 7.

[77] Sonya Lipsett-Rivera, "Introduction: Children in the History of Latin America," *J. Family Hist.* 23 (1998): 221–4, on 221.

based on criminal and ecclesiastical archives, have been highlighted by historians more often than other, less accessible masculinities. A multiplicity of masculinities existed in colonial Latin America, and while some scholars have emphasized corporeal ambiguity, asserting that the embodiment of masculinity could be "fragile and uncertain," my focus on how bodies were read and interpreted in the context of criminal cases of *estupro* and coercive sodomy highlights a particular predatory masculinity that, at least on the surface, was anything but fragile.[78] Masculinity was, nonetheless, highly contested and perhaps even rendered "fragile and uncertain" when it entered the realms of colonial medicine, science, and obstetrics.

[78] McClive, "Masculinity on Trial" (cit. n. 72), 48. Susan Migden Socolow, among other scholars, echoes similar beliefs that in colonial Latin America, the "standards of masculinity combined with a subculture of violence to make women the victims of men and other women, usually kinfolk. Although violence against women varied from group to group, it was universally accepted." Socolow, *The Women of Colonial Latin America* (New York, 2000), 163.

Notes on Contributors

Frances Bernstein is Associate Professor of History at Drew University. Her research and teaching focus on the history of Russia and the USSR, sexuality and the body, medicine, and disability. She is the author of *The Dictatorship of Sex: Lifestyle Advice for the Soviet Masses* (DeKalb, Ill., 2007), as well as coeditor and contributor to *Soviet Medicine: Culture, Practice, and Science* (with Christopher Burton and Dan Healey; DeKalb, Ill., 2010). She is currently working on a book on the medical treatment and cultural representation of Soviet soldiers disabled during World War II.

Leah DeVun is Associate Professor of History at Rutgers University. She is the author of *Prophecy, Alchemy, and the End of Time* (New York, 2009), as well as articles in *Journal of the History of Ideas*, *Radical History Review*, and *GLQ*. She is currently at work on a history of "hermaphrodites" from the ancient world to the Renaissance.

Nathan Ensmenger is an Associate Professor in the School of Informatics and Computing at Indiana University. His research focuses on the social and cultural history of software and software workers, and questions of gender and identity in computer programming. His 2010 book, *The Computer Boys Take Over: Computers, Programmers, and the Politics of Technical Expertise*, explored the rise to power of the "computer expert" in American corporate, economic, and political life. He is one of the coauthors of the most recent edition of the popular *Computer: A History of the Information Machine*. He is currently working on a book exploring the global environmental history of the electronic digital computer.

Nathan Ha received his PhD in 2011 in History of Science at Princeton University and was a Postdoctoral Fellow at the University of California, Los Angeles, Institute for Society and Genetics from 2011 to 2014. His research focuses on the history of the biomedical sciences, especially twentieth-century genetics, American cultural history, and the history of gender and sexuality.

Whitney Laemmli is a doctoral candidate in History and Sociology of Science at the University of Pennsylvania and has been a fellow at the Max Planck Institute for the History of Science. Her dissertation, "The Choreography of Everyday Life: Rudolf Laban and the Analysis of Modern Movement," follows a system for recording bodily movement on paper from its birth in Weimar dance to its use in projects of corporate capitalism, psychiatric management, and anthropological investigation in the United States and the United Kingdom. She has also written a history of the modern ballet pointe shoe for *Technology and Culture*.

Eugenia Lean is Associate Professor of History at Columbia University. Her book *Public Passions: The Trial of Shi Jianqiao and the Rise of Popular Sympathy in Republican China* (Berkeley and Los Angeles, 2007) shows how a high-profile crime of female passion gave rise to the moral, judicial, and political authority of "public sympathy" in Republican-era China. Her current project, "Manufacturing Modernity in Early Twentieth-Century China: Chen Diexian, a Man-of-Letters in an Age of Industrial Capitalism," examines polymath Chen Diexian, a professional writer/editor, science enthusiast, and pharmaceutical industrialist, to explore the cultural, social, and intellectual dimensions of industrialization in China.

Beth Linker is Associate Professor in the Department of the History and Sociology of Science at the University of Pennsylvania. She is the author of *War's Waste: Rehabilitation in World War I America* (Chicago, 2011) and coeditor of *Civil Disabilities: Citizenship, Membership, Belonging* (with Nancy J. Hirschmann; Philadelphia, 2015). She has published over a dozen articles and reviews that pertain to the history of disability, gender, health policy, the body, and medicine. She is currently researching two book projects—"Slouch: The Rise and Fall of American Posture" and "Making the Cut: Excess Surgery in the United States."

Erika Lorraine Milam is Associate Professor of History and History of Science at Princeton University. Her book, *Looking for a Few Good Males: Female Choice in Evolutionary Biology* (Baltimore, 2010), explored the history of courtship in evolutionary theory, from Charles Darwin in the mid-nineteenth century to sociobiology in the 1970s. Her current research turns to scientific and popular debates over instinctual aggression as a key facet of human nature in the later Cold War.

Robert A. Nye is the Emeritus Horning Professor of the Humanities and Professor of History at Oregon State University. His books include *Masculinity and Male Codes of Honor in Modern France* (Oxford, 1993), *Crime, Madness and Politics in Modern France* (Princeton, N.J.,

1984), and *Sexuality* (Oxford, 1999). He has published sixty scholarly articles and chapters. He is presently working on dueling protocols, violence, international law and arbitration, and the rules of war.

Michael S. Reidy is Professor of History in the Department of History and Philosophy at Montana State University. He is the author of *Tides of History: Ocean Science and Her Majesty's Navy* (Chicago, 2008) and coauthor of *Exploration and Science* (with Gary Kroll and Erik M. Conway; Santa Barbara, Calif., 2006) and *Communicating Science* (with Alan G. Gross and Joseph E. Harmon; Oxford, 2000). He is also the co–general editor of the ongoing sixteen-volume *Correspondence of John Tyndall* and recently coedited a volume on Tyndall entitled *The Age of Scientific Naturalism: Tyndall and His Contemporaries* (with Bernard Lightman; London, 2014). His current research focuses on how the advance of mountaineering changed the practice of science in the nineteenth century.

Michael Robinson is Associate Professor of History at Hillyer College, University of Hartford. His work focuses on exploration and its place in the cultural imagination. He is the author of *The Coldest Crucible: Arctic Exploration and American Culture* (Chicago, 2006) and *The Lost White Tribe: Explorers, Scientists, and the Theory That Changed a Continent* (New York, 2016).

Alexandra Rutherford is Professor of Psychology at York University in Toronto. She is the author of *Beyond the Box: B. F. Skinner's Technology of Behavior from Laboratory to Life, 1950s–1970s* and is currently working on a book that examines the relationships among feminism, psychological science, and policy in late twentieth-century America. She is the coeditor of the *Handbook of International Feminisms: Perspectives on Psychology, Women, Culture, and Rights* and directs the Psychology's Feminist Voices digital archive and oral history project (www.feministvoices.com). She is a fellow of the American Psychological Association and a past president of the Society for the History of Psychology.

Mary Terrall is Professor of History at the University of California, Los Angeles. She has published widely on the history of eighteenth-century science and has a long-standing interest in gender and science. She is the author of *The Man Who Flattened the Earth: Maupertuis and the Sciences in the Enlightenment* (Chicago, 2002) and, most recently, of *Catching Nature in the Act: Réaumur and the Practice of Natural History in the Eighteenth Century* (Chicago, 2014).

Zeb Tortorici is Assistant Professor of Spanish and Portuguese Languages and Literatures at New York University. His work focuses on the connections between sexuality, colonialism, and historical archives in Latin America. With Martha Few he recently coedited *Centering Animals in Latin American History*, and with Daniel Marshall and Kevin Murphy he coedited two special issues of *Radical History Review* on the topic of "Queering Archives." His forthcoming anthology, *Sexuality and the Unnatural in Colonial Latin America*, will be published by the University of California Press in 2016.

Index

SUGGESTIONS FOR CONTRIBUTORS TO OSIRIS

OSIRIS is devoted to thematic issues, conceived and compiled by guest editors who submit volume proposals for review by the OSIRIS Editorial Board in advance of the annual meeting of the History of Science Society in November. For information on proposal submission, please write to the Editor at osiris@etal.uri.edu.

1. Manuscripts should be submitted electronically in Rich Text Format using Times New Roman font, 12 point, and double-spaced throughout, including quotations and notes. Notes should be in the form of footnotes, also in 12 point and double-spaced. The manuscript style should follow *The Chicago Manual of Style*, 16th ed.

2. Bibliographic information should be given in the footnotes (not parenthetically in the text), numbered using Arabic numerals. The footnote number should appear as superscript. "Pp." and "p." are not used for page references.

   a. References to books should include the author's full name; complete title of book in *italics*; place of publication; date of publication, including the original date when a reprint is being cited; and, if required, number of the particular page cited (if a direct quote is used, the word "on" should precede the page number). *Example*:

   [1] Mary Lindemann, *Medicine and Society in Early Modern Europe* (Cambridge, 1999), 119.

   b. References to articles in periodicals or edited volumes should include the author's name; title of article in quotes; title of periodical or volume in *italics*; volume number in Arabic numerals; year in parentheses; page numbers of article; and, if required, number of the particular page cited. Journal titles are spelled out in full on the first citation and abbreviated subsequently according to the journal abbreviations listed in *Isis Current Bibliography*. *Example*:

   [2] Lynn K. Nyhart, "Civic and Economic Zoology in Nineteenth-Century Germany: The 'Living Communities' of Karl Möbius," *Isis* 89 (1999): 605–30, on 611.

   c. All citations are given in full in the first reference. For succeeding citations, use an abbreviated version of the title with the author's last name. *Example*:

   [3] Nyhart, "Civic and Economic Zoology" (cit. n. 2), 612.

3. Special characters and mathematical and scientific symbols should be entered electronically.

4. A small number of illustrations, including graphs and tables, may be used in each volume. Hard copies should accompany electronic images. Images must meet the specifications of The University of Chicago Press "Artwork General Guidelines" available from the Editor.

5. Manuscripts are submitted to OSIRIS with the understanding that upon publication copyright will be transferred to the History of Science Society. That understanding precludes consideration of material that has been previously published or submitted or accepted for publication elsewhere, in whole or in part. OSIRIS is a journal of first publication.

OSIRIS (ISSN 0369-7827) is published once a year.

Single copies are $33.00.

Address subscriptions, single issue orders, claims for missing issues, and advertising inquiries to *Osiris*, The University of Chicago Press, Journals Division, PO Box 37005, Chicago, IL 60637.

Postmaster: Send address changes to *Osiris*, The University of Chicago Press, Journals Division, PO Box 37005, Chicago, IL 60637.

OSIRIS is indexed in major scientific and historical indexing services, including *Biological Abstracts, Current Contexts, Historical Abstracts*, and *America: History and Life*.

Paperback edition, ISBN 978-0-226-26761-6

Osiris

A RESEARCH JOURNAL DEVOTED TO THE HISTORY OF SCIENCE AND ITS CULTURAL INFLUENCES

A PUBLICATION OF THE HISTORY OF SCIENCE SOCIETY

EDITOR
ANDREA RUSNOCK
University of Rhode Island

COPY EDITOR
BARBARA CONDON
Wellesley, MA

PAST EDITOR
KATHRYN OLESKO
Georgetown University

PROOFREADER
JENNIFER PAXTON
The Catholic University of America

OSIRIS EDITORIAL BOARD

FA-TI FAN
State University of New York, Binghamton

PAMELA LONG
Independent Scholar

W. PATRICK MCCRAY
University of California, Santa Barbara

H. FLORIS COHEN
Utrecht University
EX OFFICIO

SUMAN SETH
Cornell University

EDNA SUÁREZ-DIAZ
Universidad Nacional Autónoma de México

HSS COMMITTEE ON PUBLICATIONS

KATHARINE ANDERSON
York University

SORAYA DE CHADAREVIAN
University of California, Los Angeles
CHAIR

FLORENCE HSIA
University of Wisconsin, Madison
SECRETARY

MICHAEL GORDIN
Princeton University

MATTHEW JONES
Columbia University

JANET BROWNE
Harvard University
EX OFFICIO

EDITORIAL OFFICE
DEPARTMENT OF HISTORY
80 UPPER COLLEGE ROAD, SUITE 3
UNIVERSITY OF RHODE ISLAND
KINGSTON, RI 02881 USA
osiris@etal.uri.edu